Design of
Steel Structures

About the Authors

Jay Shen, Ph.D., P.E., S.E., has taught design of steel structures at the Illinois Institute of Technology and at Iowa State University, where he serves as the Director of Graduate Study in the Department of Civil, Construction and Environmental Engineering. He has been involved in teaching, researching, and consulting in steel structures for the last 35 years, and has published extensively in performance-based earthquake engineering and innovations in steel structures. Dr. Shen was the 2017 recipient of the Joseph C. and Elizabeth Anderlik Faculty Award for Excellence in Undergraduate Teaching at Iowa State University, and a 1997 University Excellence in Teaching Award. He received his Ph.D. from the University of California at Berkeley in 1992.

Bulent Akbas, Ph.D., has been actively engaged in research, teaching, and consulting in the field of structural engineering for over 30 years. He has been a member of the faculty since 1997 at Gebze Technical University in Turkey, where he serves as the Director of the Earthquake and Structural Engineering Graduate Program and the Chair of the Department of Civil Engineering. Dr. Akbas has performed research on nonlinear analysis and behavior of steel structures as well as on performance-based design of structures, structural health monitoring, soil-structure interaction, silos, and tanks. He received his Ph.D. from the Illinois Institute of Technology in 1997.

Onur Seker, Ph.D., is currently an Assistant Professor at Gebze Technical University in Turkey. He earned his M.Sc. degree in 2011 from Yildiz Technical University, where he completed his undergraduate studies in 2009. He studied structural engineering at Iowa State University and received his Ph.D. in 2016. Dr. Seker's research includes various projects related to seismic behavior of steel structures and structural components.

Mahmoud Faytarouni, Ph.D., is a Research Structural Engineer in the Steel Research Group at Genex Systems, supporting research work for the Federal Highway Administration (FHWA) at Turner-Fairbank Highway Research Center (TFHRC) in Virginia. He earned a Ph.D. in structural engineering from Iowa State University, an M.Engr. in engineering mechanics from Iowa State University, an M.Sc. in structural engineering from Milwaukee School of Engineering, and a B.S. in civil engineering from Beirut Arab University. Dr. Faytarouni's research interests include performance-based earthquake engineering, seismic evaluation and retrofit of steel buildings, bridge engineering, and fatigue and fracture of structures.

Design of Steel Structures

Jay Shen

Bulent Akbas

Onur Seker

Mahmoud Faytarouni

New York   Chicago San Francisco
Athens London Madrid
Mexico City Milan New Delhi
Singapore Sydney Toronto

Library of Congress Control Number: 2021931346

Design of Steel Structures

1 2 3 4 5 6 7 8 9   CCD   26 25 24 23 22 21

ISBN 978-1-260-45233-4
MHID 1-260-45233-6

This book is printed on acid-free paper.

Sponsoring Editor
Ania Levinson

Editing Supervisor
Stephen M. Smith

Production Supervisor
Lynn M. Messina

Acquisitions Coordinator
Elizabeth M. Houde

Project Managers
Touseen Qadri and
Rishabh Gupta, MPS Limited

Copy Editor
Bindu Singh, MPS Limited

Proofreader
Heather Mann

Indexer
Jerry Ralya

Art Director, Cover
Jeff Weeks

Composition
MPS Limited

Contents

Preface

The purpose of this book is to provide a primary textbook for an undergraduate course in the design of steel structures. To facilitate effective teaching and learning, the book offers a rational balance between basic theories and their engineering applications, concise and easy-to-understand explanations of the design and behavior of basic steel members and systems, ample practical examples, and companion simulation movies illustrating structural limit states and failure.

Having a balanced knowledge of theory and application is particularly important for senior undergraduates because it can empower them to grow professionally and obtain their professional licenses earlier. However, constrained by contact hours in a typical 16-week semester, most of our fellow instructors find it is very challenging to select a suitable textbook, even though a large number of textbooks on structural steel design are available. Generally, existing textbooks often fall into one of two distinctive groups: Group 1 has coverage of all subjects related to steel structures with a focus on the theoretical background of design formulas. These books include extensive coverage of how the design specifications were developed, with limited application to engineering practice. They are intended for both undergraduate programs with solid prerequisite preparation and graduate students in a multiple-level course series, and also serve as a good reference for graduate students involved in research. However, they are not suitable for the first steel design class in terms of balance of theory and application. A fact is that the majority of the topics in these books are unable to be covered in a semester-long course. Group 2 has full coverage of all subjects included in the design specification, with a focus on practice. The textbooks in this group tend to discuss the design procedures in the current design specification with the proper number of examples, but little or no theoretical background is provided. This type of textbook may help students get familiar with current provisions well. Lack of basic understanding of fundamental concepts behind design formulas limits students' continuous growth in practice and imposes risk of malpractice.

Differently, this textbook introduces a contemporary methodology in the deliberation of theory too much space here. Our teaching experience has shown that theoretical coverage only using mathematical equations has hindered our engagement with students and discouraged them from studying these subjects actively. To address this issue, in this textbook a combination of concise deliberation of fundamental concepts and visual presentation of structural behavior is introduced. To facilitate sound understanding and conceptualization of failure modes and limit states observed in laboratories and real-world structures, in this textbook major failure modes are presented visually with simulations generated by the nonlinear finite-element method (FEM), together with explanations connecting theory and application. Practical examples have been selected carefully, based on the authors' experiences in teaching, practice, and conducting research in steel buildings. These examples have been proven effective in explaining design requirements and process, and in furthering understanding of theory. They are derived from typical steel building structures that students can see in real life, thus motivating students to learn the subject.

Acknowledgments

The senior author would like to thank all students who took his steel design classes in Chicago, Berkeley, and Ames over the last 30 years. All of us would like to express our gratitude to Charlie Carter and the American Institute of Steel Construction for leadership in shaping national standards in steel design and construction and constant support in educating a new generation of structural steel design professionals. Grateful acknowledgment is also due to Ania Levinson of McGraw Hill, and to Touseen Qadri, Rishabh Gupta, and the editorial staff of MPS Limited for their dedicated editing and production of this publication. Last but not least, we want to thank our families for their support throughout the journey of making this book project a reality.

Jay Shen
Bulent Akbas
Onur Seker
Mahmoud Faytarouni

CHAPTER 1

Introduction

1.1 Introduction to Structural Steel Design

Structural design starts with planning, analysis, design, and ends with construction. It is the art and science of using knowledge of mechanics of materials, statics, dynamics, and understanding of the behavior of structures to select material and geometry of structures to carry intended forces in a safe and economical way. Safety and serviceability are provided through *strength and stiffness*, whereas economical solution might be achieved through complying with the *minimum local code requirements, using standard sections and locally available materials, similar detailing*, and *sound construction methods*. A structural engineer should always be aware of the fabrication equipment and transportation, and should check compatibility with other trades such as mechanical and architectural aspects. The basic steps involved in structural design are summarized in Fig. 1.1. Preparing architectural plans, and deciding on the function/functions of the structure, is the architect's responsibility. However, a team study by architects, structural engineers, architectural engineers, and geotechnical engineers is needed to determine framing layout and materials, and select the gravity and lateral force–resisting systems (LFRSs). Then, the engineer takes the lead and starts working on calculating the loads, both gravity and lateral, that the structure might be subjected to, preliminary design of structural members, and constructing a numerical model of the structural framing system. The last step is to ensure that the demand for structural members due to loading does not exceed the capacity of the members, considering the strength and serviceability conditions. If some members do not have sufficient capacity, then redesign becomes necessary. Thus, structural design is an iterative process (Fig. 1.1).

1.2 Structural Steel as a Building Material

Steel, as we know today, has been used as a structural material since the late 18th century. Before that, iron was the primary structural material that was used to make connections in masonry or wooden structures. The Industrial Revolution in the second half of the 18th century made production of iron cheaper. A special form of iron, cast iron (Fig. 1.2*a*), which was stronger than bricks, stone, wood, etc., at the time, was used in constructing light structures, long-span bridges, greenhouses, etc. Cast iron is hard and brittle, and cannot be welded. The carbon content in cast iron is less than 3–4 percent. Iron Church is a good example of cast iron construction, which was completed in 1898 (Fig. 1.3). A relatively ductile wrought iron became popular in the second half of the 18th century, and it totally replaced cast iron (Fig. 1.2*b*). The manufacturing

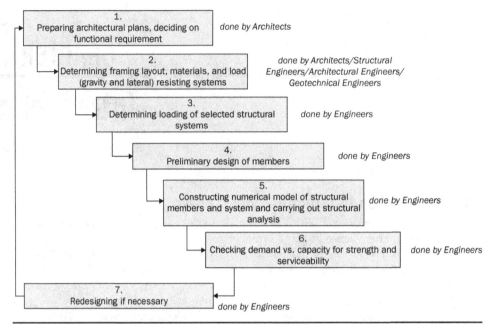

FIGURE 1.1   Steps involved in structural design

FIGURE 1.2   (a) A cast iron wagon wheel, and (b) Eiffel Tower constructed from wrought iron

technology enabled production of rolling shapes, especially I-shapes, using wrought iron. Compared with cast iron, wrought iron is softer and malleable, having a carbon content of less than 0.2 percent. Following the invention of the Bessemer converter in 1870, iron products had become one of the primary structural materials. During 1890s, steel with carbon content between 0.2 and 2 percent and small amounts of phosphorous, silicon, sulfur, and oxygen replaced wrought iron as the principal building material (AISC, 2007). All these chemical elements provide steel with consistent quality and grade. Structural steel used in structures, designated as structural carbon steel or alloy steel, is known as mild steel and has carbon content between 0.2 and 0.25 percent. Increasing carbon content in steel provides it more strength, hardness, and corrosion-resistance, but decreases its ductility and weldability as well as malleability. Adding

Figure 1.3 Iron Church (Bulgarian St. Stephen Church), Istanbul

controlled amount of chromium, cobalt, copper, molybdenum, nickel, tungsten, and vanadium to the carbon steel creates alloy steel. Stainless steel and weathering steel are two types of alloy steel.

To determine weldability of structural steel, equivalent carbon content or carbon equivalent (CE) is used, as suggested by the International Institute of Welding (IIW). The formula is as follows:

$$CE = \%C + \frac{\%Mn + \%Si}{6} + \frac{\%Cr + \%Mo + \%V}{5} + \frac{\%Cu + \%Ni}{15} \leq 0.5 \quad (1.1)$$

where C, Mn, Si, Cr, Mo, V, Cu, and Ni are carbon, manganese, silicon, chromium, molybdenum, vanadium, copper, and nickel contents, respectively, by weight. For cast iron, CE is used to distinguish the grey and white irons. Grey iron contains graphite due to high cooling rate and is hard and brittle that cannot be machined, whereas white iron holds carbon as cementite due to low cooling rate and is machinable (Yescas-Gonzalez and Bhadeshia, 2020). CE for cast iron may be defined as

$$CE = \%C + \frac{\%Si + \%P}{3} \quad (1.2)$$

where P is phosphorous by weight.

Even though structural steel has been used as a structural material for over a century, structural engineering design applications could not keep pace with the improved manufacturing and construction practice. Since the end of the 19th century to the 1970s, a number of steel bridges collapsed due to human errors, new technologies, materials, loads, etc. (Fig. 1.4). An interstate highway bridge (I-35W) in downtown Minneapolis loaded with rush-hour traffic at 6:05 pm dropped more than 60 ft into the Mississippi River, sending at least 50 vehicles and passengers into the water. Since 1993, the bridge had been inspected annually by Mn/DOT. In the years prior to the collapse, several

A   B

FIGURE 1.4   (a) Minneapolis bridge, and (b) parts of the collapsed bridge (August 1, 2007)

reports cited problems with the bridge structure. In 1990, the federal government gave I-35W bridge a rating of "structurally deficient," citing significant corrosion in its bearings. Approximately 75,000 other US bridges had this classification in 2007.

Example 1.1

A cast iron church from the late 1800s (Fig. 1.5) which had substantial corrosion in its structural elements was closed to visitors in early 2010. It was restored after 7 years of restoration work and was reopened to visitors in 2018. The chemical analyses during the restoration work revealed the following chemical elements by weight in percentage:

Element	%
Carbon	0.0700
Manganese	0.3550
Silicon	0.00842
Chromium	0.0010
Molybdenum	0.0137
Vanadium	0.0010
Copper	0.0128
Nickel	0.0239
Phosphorous	0.03240

FIGURE 1.5

Compute the equivalent carbon content (CE) of the material.

Solution:
Using Eq. (1.2),

$$CE = \%C + \frac{\%Si + \%P}{3}$$

$$CE = 0.0700 + \frac{0.00842 + 0.03240}{3} = 0.0836$$

1.3. Material Properties of Steel and Cross-Sectional Shapes

Three main structural steel materials are available for building and bridge construction, namely, carbon steels, high-strength low-alloy steels, and low-alloy steels (Table 1.1). The density of steel is 490 lb/ft³ (for comparison, density of concrete = 150 lb/ft³). Carbon steels have a definite yield point. Typical stress–strain curve for structural carbon steels for tension is shown in Fig. 1.6. Below the yield stress, steel is linearly elastic and has the same response in both tension and compression. Material properties needed in structural design are as follows:

1. Modulus of elasticity (Young's modulus), $E = 29,000$ ksi $(=F_y/\varepsilon_y)$
2. Specified minimum yield stress, F_y (ksi)
3. Specified minimum tensile strength, F_u (ksi)

Type	ASTM designation	F_y Minimum yield stress (ksi)	F_u Minimum Tensile strength (ksi)	Notes
Carbon steels	A36	36	58–80	Primarily used in buildings
	A53 Grade B	35	60	For pipes
	A500 (Grade A, B, or C)	33, 42, 46	45, 58, 62	For structural tubing
High-strength low-alloy steels	A242	42–50	63–70	In bridge construction where corrosion resistance is desired
	A572 Gr. 50	50	65	Primarily used in buildings
	A588	50	70	Corrosion resistance is 4 times that of A36
	A992	50	65	Having a maximum yield-to-tensile strength ratio of 0.85
	A913	50	60	For all structural shapes
Alloy steels	A514 Gr. 100	100	110–130	Not available in hot-rolled shapes
	A709 Gr. 100	90	100–130	

TABLE 1.1 Structural Steel Types and Their Designation

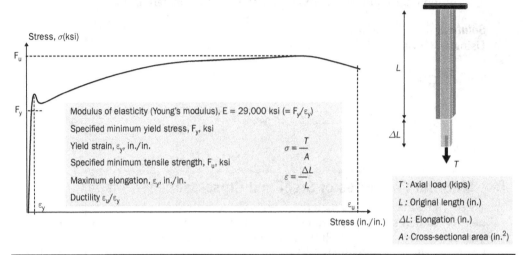

FIGURE 1.6 Typical stress–strain curve for structural steel

Steel grade	R_y	R_t
ASTM A36/A36M	1.5	1.2
ASTM A572 Gr. 50	1.1	1.1
ASTM A992	1.1	1.1
ASTM A500/A500M Gr. B	1.4	1.3
ASTM A500/A500M Gr. C	1.3	1.2
ASTM A572/A572M Gr. 42	1.3	1.0

TABLE 1.2 R_y and R_t Values for the Most Commonly Used Steel Grades

The yield stress for any steel type will likely be higher than the specified minimum value (F_y). When required, expected yield stress, F_{ye}, or expected tensile strength, F_{ue}, can be obtained by using R_y and R_t values, respectively (Table 1.2).

$$F_{ye} = R_y F_y \tag{1.3a}$$
$$F_{yu} = R_t F_u \tag{1.3b}$$

Since the 1994 Northridge earthquake, A36 carbon steel is not the first choice for structural engineers, but A992. The mechanical properties of this new grade of steel are quite similar to that of A572 Grade 50, but it has a maximum tensile-to-yield strength ratio of 0.85 (Table 1.1).

Structural steel shapes are either hot-rolled or built-up shapes. Hot-rolled shapes in desired cross-sectional shapes (Fig. 1.7) are obtained by squeezing the hot steel through rollers. Standard hot-rolled shapes are wide-flange W-shapes, M-shapes, S-shapes, C-shapes (or channels), angles, WT-shapes (or structural tees), hollow sections, and pipes. Each shape has a special designation. For W-, M-, S-, C-, WT-shapes, the first number refers to the nominal depth, and the second number is the weight of the element (Fig. 1.7). Similarly, for angles and hollow sections, the first two numbers are the leg sizes and side widths, whereas the third number is the thickness (Fig. 1.7).

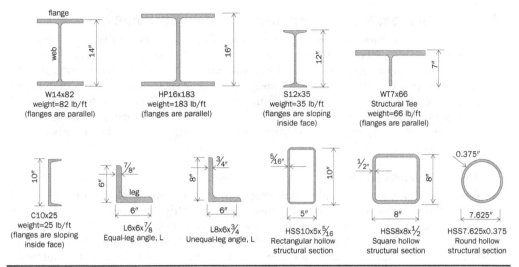

FIGURE 1.7 Cross-sectional shapes

Example 1.2

Determine the expected yield stress and expected tensile strength for A36, A500 Gr. C, and A992 steel grades to be used in an office building.

Solution:

Yield stresses: $F_y = \begin{cases} 36 \text{ ksi} & \text{for A36} \\ 46 \text{ ksi} & \text{for A500 Gr. C} \\ 50 \text{ ksi} & \text{for A992} \end{cases}$

Tension strength: $F_y = \begin{cases} 58 \text{ ksi} & \text{for A36} \\ 62 \text{ ksi} & \text{for A500 Gr. C} \\ 65 \text{ ksi} & \text{for A992} \end{cases}$

$R_y = \begin{cases} 1.5 & \text{for A36} \\ 1.3 & \text{for A500 Gr. C} \\ 1.1 & \text{for A992} \end{cases}$

$R_t = \begin{cases} 1.2 & \text{for A36} \\ 1.2 & \text{for A500 Gr. C} \\ 1.1 & \text{for A992} \end{cases}$

Expected yield stresses: $F_{ye} = R_y F_y = \begin{cases} 1.5 \times 36^{\text{ksi}} = 54 \text{ ksi} & \text{for A36} \\ 1.3 \times 46^{\text{ksi}} = 59.8 \text{ ksi} & \text{for A500 Gr. C} \\ 1.1 \times 50^{\text{ksi}} = 55 \text{ ksi} & \text{for A992} \end{cases}$

Expected tension strengths: $F_{yu} = R_t F_u = \begin{cases} 1.2 \times 58^{\text{ksi}} = 69.6 \text{ ksi} & \text{for A36} \\ 1.2 \times 62^{\text{ksi}} = 74.4 \text{ ksi} & \text{for A500 Gr. C} \\ 1.1 \times 65^{\text{ksi}} = 71.5 \text{ ksi} & \text{for A992} \end{cases}$

1.4   Structural Systems and Elements in Steel Structures

Three main structural framing systems are available in steel structures, namely, (a) moment frames, (b) concentrically braced frames, and (c) eccentrically braced frames. These structural systems can be used in combination with shear wall systems such as reinforced concrete (R/C) wall with steel columns, steel wall with steel columns, or composite walls with steel and R/C framing. Structural members in these structural framing systems will be subject to internal forces. These three structural framing systems, internal forces, and the most important cross-sectional properties are briefly discussed in the following text.

1.4.1   Moment Frames

Beams and columns are rigidly connected in moment frames (MF) to transfer lateral forces through flexure and shear in these elements. MFs are seismically designed in such a way that yielding occurs at the beam ends through the formation of plastic hinges (Figs. 1.8 and 1.9). AISC 341 requires that MFs sustain large inelastic deformations without failure of beam-to-column connections. These large inelastic deformations are maintained through beam flexural yielding.

Lateral stiffness of MFs is the smallest among all-steel structural framing systems. That is why the design of a typical MF is governed by the story drift requirement as specified by ASCE 7. Beams and girders are required to satisfy the maximum load effect obtained from the load combinations specified in ASCE 7 (load combinations 1.7a through 1.7e) (Fig. 1.10). The required strength of a column in an MF should satisfy the greater of the following requirements:

a. The load effect obtained from the load combinations specified in ASCE 7 (load combinations 1.7a through 1.7e).

b. The compressive axial strength and tensile strength as determined using the overstrength seismic load (load combinations 1.7f and 1.7g). Bending moments are allowed to be ignored in applying this requirement.

Columns at a beam-column joint must be stronger than beams. This principle is satisfied by taking the ratio of sum of plastic moment capacities of columns at the joint to the sum of plastic moment capacities of beams at the intersection of the beam and column centerlines.

1.4.2   Concentrically Braced Frames

Concentrically braced frames (CBF) are considered to be truss systems resisting lateral forces by truss action and the most cost-effective seismic force–resisting system (SFRS) (Fig. 1.11). Brace members in a typical CBF are expected to dissipate energy through inelastic action by yielding in tension and buckling in compression. Thus, the seismic response of a CBF is highly dependent on the inelastic cyclic behavior of the braces. However, the braces yielding in tension are the main source of ductility, because buckling is a nonductile failure mode. CBFs are less ductile than moment frames and have lower response modification coefficient, R, values, but they meet the lateral stiffness and strength requirements easily. The design of a CBF can easily be performed by hand, and no structural software is needed to calculate the internal forces in the structural elements (braces, beams, columns).

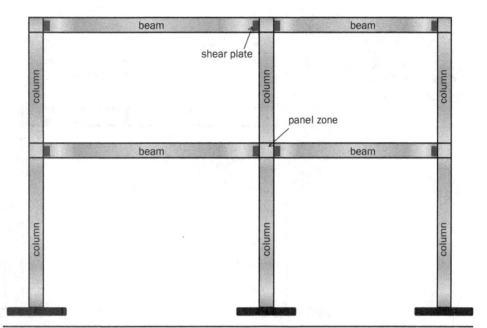

FIGURE 1.8 A moment frame

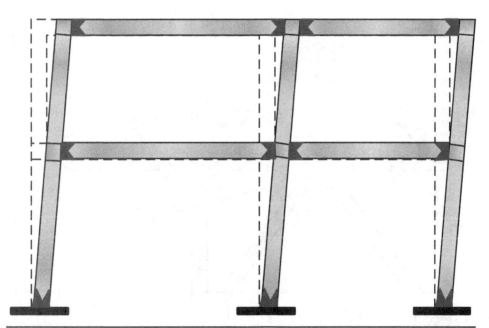

FIGURE 1.9 Yielding at the beam ends of a seismically detailed moment frame and formation of plastic hinges

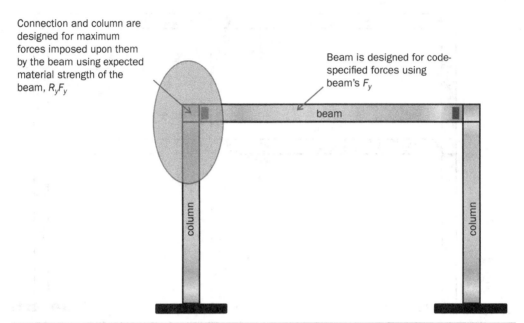

Connection and column are designed for maximum forces imposed upon them by the beam using expected material strength of the beam, R_yF_y

Beam is designed for code-specified forces using beam's F_y

beam

column

column

FIGURE 1.10   Use of expected material strength for a moment frame design

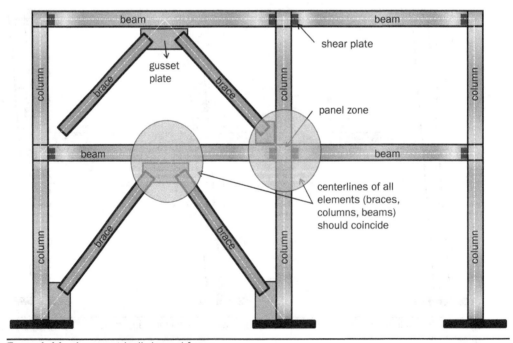

beam

beam

shear plate

gusset plate

brace

brace

column

column

column

panel zone

beam

beam

column

column

brace

brace

centerlines of all elements (braces, columns, beams) should coincide

column

column

FIGURE 1.11   A concentrically braced frame

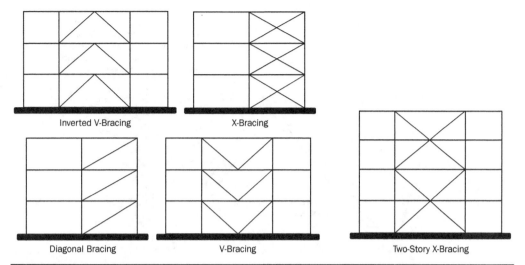

Figure 1.12 Bracing configurations

Different bracing configurations are shown in Fig. 1.12 (V-, inverted V-, X-, diagonal, and two-story X-bracing). Even though the most common types of bracing configurations are V- and inverted V-bracing, two-story X-bracing configuration has become quite popular recently. The main reason for this is that a large unbalanced force due to the difference in yielding and buckling strengths of braces develops in V- and inverted V-bracings, which results in heavy and deep beam sections. On the other hand, in a two-story X-bracing configuration, there is no need to consider unbalanced force in the beam since the unbalanced forces below and above the beam will have the same magnitude, but act in opposite directions.

The first step in sizing the structural elements is to design the braces first in a CBF. Brace design is carried out by the lateral forces specified in ASCE 7. Then, beams and columns are designed for the maximum forces imposed upon them by the braces (Fig. 1.13). In determining the capacity-limited horizontal seismic load effects, E_{cl}, in girders and columns, the following two structural analysis cases are required to be carried out:

1. A structural analysis in which all braces are assumed to resist forces corresponding to their expected strength in compression or tension.
2. A structural analysis in which all braces in tension are assumed to resist forces corresponding to their expected strength and all braces in compression are assumed to resist their expected post-buckling strength.

1.4.3 Eccentrically Braced Frames

Unlike concentrically braced frames (CBFs), eccentrically braced frames (EBFs) resist lateral forces by a combined frame and truss action of beams, columns, and braces (Fig. 1.14), i.e., it is a hybrid system. To form an EBF, a beam segment should be isolated

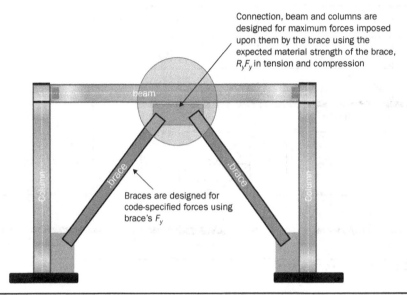

FIGURE 1.13 Use of expected material strength in concentrically braced frame design

either at the midspan or adjacent to the column face (Figs. 1.14 and 1.15). The isolated segment is called a link, and within the link, additional stiffeners need to be provided. The primary source of inelastic behavior is the links that develop ductility. EBFs are both ductile and highly stiff structural framing systems. Since EBFs allow larger frame openings for architectural use, it might be an alternative structural framing option, keeping in mind that it is not the most cost-effective system due to increased detailing and connection costs.

The design of EBFs starts with the design of links for code-specified forces using links' F_y. Connections, beam outside the link, braces, and columns are then designed for maximum forces imposed upon them by the fully yielded and strain hardened links (Fig. 1.16).

1.4.4 Primary Internal Forces

A typical steel structural member might be subjected to one or more of the four primary internal forces due to external loadings: (1) axial tension force, (2) axial compression force, (3) bending moment, (4) torsion. It is very rare that one member is only subject to one type of internal forces. In some cases, we might consider only one of them if it is so much dominant. In a steel structure, all structural members are connected to each other by various types of connection [pin connection (or simple or shear connection), rigid connection, semi-rigid connection]. Thus, the design of steel structures consists of designing individual components and their connections (Fig. 1.17).

For members subject to an axial tension force, cross-sectional area, A_g, is the most important sectional property. If the member is subject to axial compression force, the radius of gyration, r, and cross-sectional area, A_g, become important in design. The cross-section types for tension and compression members are similar, but high r/A_g values are required for compression members. For members subject to bending, the

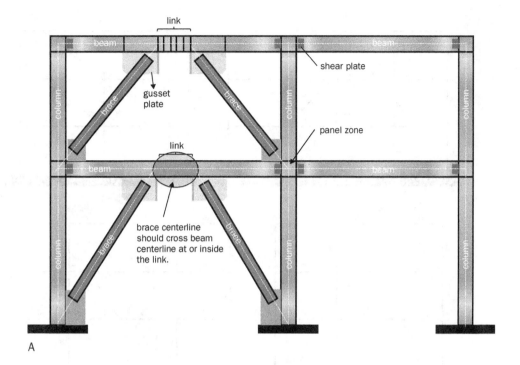

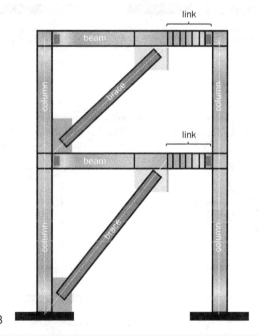

FIGURE 1.14 Eccentrically braced frames: (*a*) link at midspan, (*b*) link adjacent to column face

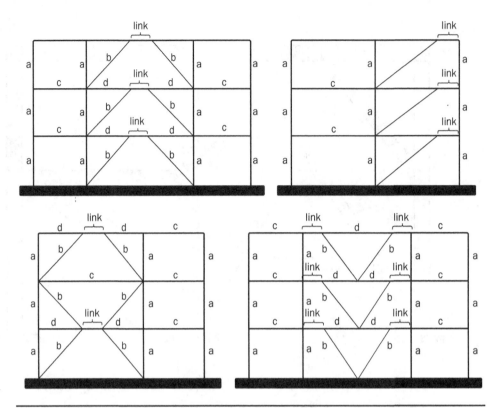

FIGURE 1.15   Eccentrically braced frame configurations (a: column, b: brace, c: beam, d: beam outside link)

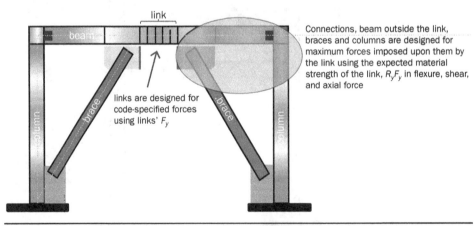

links are designed for code-specified forces using links' F_y

Connections, beam outside the link, braces and columns are designed for maximum forces imposed upon them by the link using the expected material strength of the link, $R_y F_y$ in flexure, shear, and axial force

FIGURE 1.16   Use of expected material strength in eccentrically braced frame design

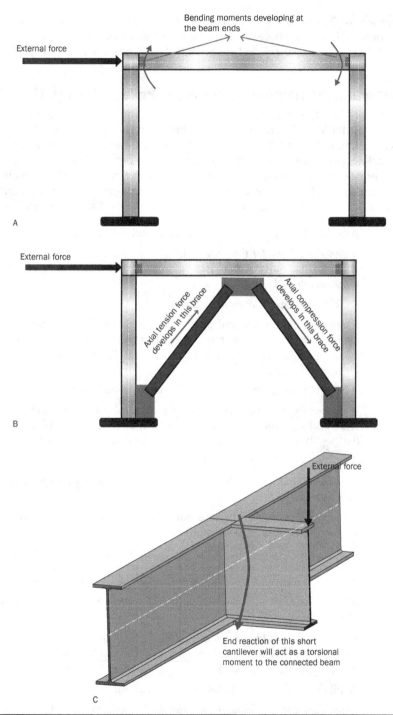

FIGURE 1.17 External loads and corresponding internal forces: (*a*) bending moment developing at beam ends due to external lateral load in a moment frame, (*b*) axial tension and compression forces developing in braces due to external lateral load in a concentrically braced frame, (*c*) torsion developing in a beam due to the concentrated load at the end of a connected cantilever beam

moment of inertia, I, and modulus of elasticity, E, are the most important sectional properties. In cases where the structural member is subject to torsion (not intentionally), torsional constant, J, and shear modulus, G, become important.

1.5   Design Specifications and Design Philosophies (LRFD/ASD)

Codes are legally enforceable documents providing minimum design requirements to determine the loads for structural design for buildings and building-like structures as well as the minimum acceptable level of safety. On the other side, specifications are defined as guidelines providing engineers requirements that help to design and construct safe and economic structural members, systems, and connections. Our job as structural engineers is *to understand* the structural behavior and *apply* the specification and code accordingly. Specifications most commonly used in the design of steel structures are listed below:

- AISC 360 Specification for Structural Steel Buildings

  It provides criteria for the design, fabrication, and erection of structural steel members, buildings, and building-like structures and is published by American Institute of Steel Construction (AISC).

- AISC Steel Construction Manual

  It provides dimensions and properties of structural steel shapes as well as design manuals for structural members such as beams, compression and tension members, beam-columns, and connections. It is one of the primary sources in steel design published by AISC.

- AISC 341 Seismic Provisions for Structural Steel Buildings

  It is used together with AISC 360 for connection detailing and member design for structural steel and composite systems in seismic regions published by AISC. It also outlines the requirements for the qualification of seismic moment connections.

- AISC Seismic Design Manual

  It provides guidance on the use of AISC 341 with many design examples and is also published by AISC.

- AISC 358 Prequalified Connections for Special and Intermediate Steel Moment Frames for Seismic Applications

  It is a companion of AISC 341 providing, in its latest version, nine prequalified beam-to-column moment connections for special and intermediate moment frames published by AISC. It also outlines step-by-step design procedure of each connection.

- ASCE 7 Minimum Design Loads and Associated Criteria for Buildings and Other Structures.

  It is a code for buildings and other structures to determine the loads (dead, live, soil, flood, tsunami, snow, rain, atmospheric ice, earthquake, wind) for structural design. It is published by American Society of Civil Engineers.

- AASHTO LRFD Bridge Design Specification

  It includes the requirements for the design of highway bridges and bridge-related structures by providing design specification not only for steel but also reinforced concrete and timber. It is published by American Association of State Highway and Transportation Officials for Bridges.

- AWS American Welding Society

  It provides specific requirements for fabricating and erecting welded steel structures and outlines groove welds, fillet welds, and plug and slot welds.

- ASTM American Society for Testing and Materials

  It provides technical standards for a wide range of materials, products, systems as well as many industries such as metal, constructions, petroleum, etc.

AISC specifications and ASCE 7 listed above are the most important specifications and codes for structural steel design. These will be constantly used and referred to throughout this textbook. The current AISC 360 specification provides two alternative design methods for the design of structural members:

- Allowable strength design (ASD)
- Load and resistance factor design (LRFD)

Both ASD and LRFD methods intend to provide a certain level of probabilistic safety against failure that might occur in a member, connection, or structural system. Whatever design philosophy is applied during the design of the structure, it should provide sufficient safety. It should be noted that

- *In the current AISC specification, the two design methods are combined, i.e., the design provisions are based on limit or strength states for both LRFD and ASD philosophies.*

- **Limit state** *term replaced the term* **failure** *in the current specification.*

Limit states defined in AISC 360 for strength and serviceability to be checked during the design are:

- *Strength Limit States*

  Plastic strength is used and ductility, buckling, fatigue, fracture, etc., limit states are considered for design. Either LRFD or ASD method can be used for design.

- *Serviceability Limit States*

  The element is not allowed to go beyond F_y. The corresponding limit states are related to the daily use of structures such as displacement, floor vibration, permanent deformation, cracking, etc. Neither LRFD nor ASD is needed for the design against these limit states.

The basic difference between ASD and LRFD is best explained by making an analogy to the general engineering design philosophy which is based on the principle that inequality of demand should be less than or equal to the capacity. This principle is stated as follows:

$$R_r \leq R_c \qquad (1.4)$$

where R_r = required design strength due to loading, which is determined by structural analysis (demand)

R_c = design strength that can be tolerated by the structural member which is specified in the specification (capacity)

The left side of Eq. (1.4) represents the demand side of the design equation, which is obtained from structural analyses, whereas the right side of the equation is the capacity which is defined in the specification. Note that both required design strength and design strength consist of a number of uncertainties, such as variations in loads and material properties. Both LRFD and ASD methods are only different in ways how the uncertainties are considered. The LRFD approach follows the general engineering design philosophy and is stated as follows:

$$\underbrace{R_u = \Sigma \gamma_i Q_i}_{\text{demand}} \leq \underbrace{\phi R_n}_{\text{capacity}} \qquad (1.5)$$

where R_u = required strength using LRFD load combinations
R_n = nominal strength of a given limit state, based on specified material strength and member size (Chapters B through K in AISC 360)
Q_i = load effects (dead, live, wind, earthquake, etc.)
ϕ = resistance factor associated with the limit state to account for uncertainties in the determination of R_n
γ_i = load factor to consider load variation during the service of the structure. Its value varies with the load combination and is defined by the code (ASCE 7). The subscript "i" refers to the type of load (dead, live, snow, wind, etc.)
$R_u = \Sigma \gamma_i Q_i$ = factored load effect on the member internal force (axial tension, axial compression, flexure) due to various load combinations
ϕR_n = design strength for LRFD

ASD approach is similar in general, but slightly differs from the general engineering design philosophy and takes the form of

$$\underbrace{R_a = \Sigma Q_i}_{\text{demand}} \leq \underbrace{\frac{R_n}{\Omega}}_{\text{capacity}} \qquad (1.6)$$

where R_a = required strength using ASD load combinations (γ_i = 1.0 for most cases)
R_n/Ω = allowable strength for ASD
Ω = safety factor to consider uncertainties in capacity and demand

The main difference between LRFD and ASD design philosophies is that ASD treats all loads equally in terms of uncertainties using a single load factor and safety factor (generally 1.67 for yield limit states), where LRFD defines multiple load factors for each load effect probabilistically and introduces resistance factors to consider uncertainties in member capacity, i.e., LRFD provides the safety margin by using different load and resistance factors for each load effect and limit states, respectively. The nominal strength in both LRFD and ASD design methodologies assume that the entire cross-section of the structural member reaches yield stress (Fig. 1.18). It should also be noted that

- LRFD seems to be more rational in treating uncertainties.
- LRFD is more consistent with real structural behavior in terms of ultimate capacity in strength.

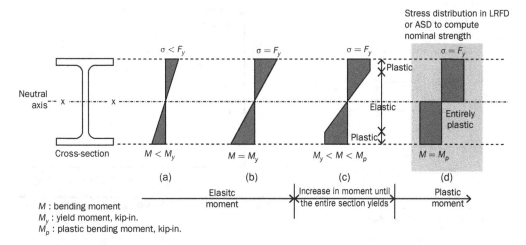

FIGURE 1.18   Normal stress distribution for elastic, partially inelastic, and fully plastic loading of an I-shaped section subject to bending

- LRFD is more flexible for improvement.
- Determination of load and resistance factors in LRFD is a continuous challenge to an improved design specification.
- ASD is still used by many engineers due to experience and education for many decades.
- Both approaches are fundamentally the same in terms of limits states.

Required strength for each structural member has to be obtained from the load combination according to the selected design philosophy at the beginning of the structural design. Throughout this book, LRFD design philosophy will be used. And the LRFD load combinations (ASCE 7) to be used in strength design are as follows:

1. $1.4D$ (1.7a)
2. $1.2D + 1.6L + 0.5L_r$ (1.7b)
3. $1.2D + 1.6L_r + 1.0L$ (1.7c)
4. $(1.2 + 0.2S_{DS})D + \rho Q_E + 0.5L$ (1.7d)
5. $(0.9 - 0.2S_{DS})D + \rho Q_E$ (1.7e)

where D is dead load; L is live load; L_r is roof live load; S_{DS} is design spectral response acceleration parameter at short periods; ρ is redundancy factor; Q_E is effects of horizontal forces resulting from seismic base shear. Seismic loads replace the horizontal seismic loads, ρQ_E, in load combinations (1.7d) and (1.7e) with E_{cl} or E_{mh} $(=\Omega_0 Q_E)$ in special cases such as connection design or element design required by AISC 341, where E_{cl} is the capacity-limited horizontal seismic load effect; E_{mh} is the effect of horizontal seismic forces that includes overstrength; and Ω_0 is overstrength factor (Fig. 1.19 and Table 1.3).

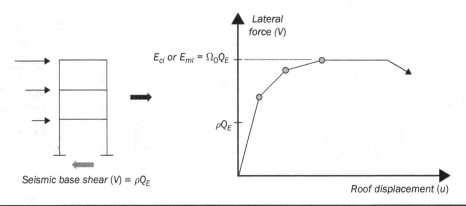

FIGURE 1.19 Capacity-limited horizontal seismic load effect and effect of horizontal seismic force that includes overstrength

Seismic force–resisting systems (SFRS)	Ω_0
Special, intermediate, and ordinary moment frames	3
Special and ordinary concentrically braced frames	2
Eccentrically braced frames	2
Buckling-restrained braced frames	2.5

TABLE 1.3 Overstrength Factors

E_{cl} is equal to the maximum force that can develop in the element obtained through plastic analysis. Thus, the load combinations (1.7d) and (1.7e) become

6. $(1.2 + 0.2S_{DS})D + \Omega_0 Q_E \text{ (or } E_{cl}) + 0.5L$ (1.7f)

7. $(0.9 - 0.2S_{DS})D + \Omega_0 Q_E \text{ (or } E_{cl})$ (1.7g)

E_{mh} in the above load combinations need not exceed E_{cl}. It is clear that the provision allows plastic analysis to determine the maximum horizontal seismic load effect instead of using the effect of horizontal seismic forces that include overstrength, and in some cases it is the only way to do so.

Example 1.3
A hollow structural section (HSS) is subject to a dead load of $D = 175$ kips and a live load of $L = 125$ kips, as shown in Fig. 1.20. Determine the required cross-sectional area (A_g) to design this structural member for both LRFD and ASD ($R_n = F_y A_g$).

Solution:
    a. Use the LRFD design equation to find A_g.

$$R_u = \Sigma \gamma_i Q_i = \phi R_n$$

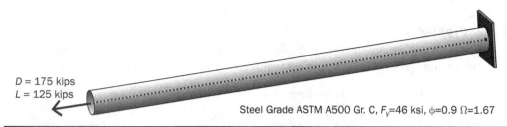

D = 175 kips
L = 125 kips

Steel Grade ASTM A500 Gr. C, F_y=46 ksi, ϕ=0.9 Ω=1.67

FIGURE 1.20   An HSS subject to axial tension

LRFD load combinations that include only D and L are:

$$1.4D$$
$$1.2D + 1.6L$$

Thus,

$$1.4(175^{kips}) = 245 \text{ kips or}$$
$$1.2(175^{kips}) + 1.6(125^{kips}) = 410 \text{ kips} \rightarrow R_u = \Sigma \gamma_i Q_i = 410 \text{ kips}$$
$$\phi R_n = 0.9 F_y A_g = 410 \text{ kips}$$

$A_g = \dfrac{410^{kips}}{0.9(46^{ksi})} = 9.90 \text{ in.}^2$ is the required cross-sectional area for design according to LRFD.

b. Use the ASD design equation to find Ag.

$$R_a = \Sigma Q_i = \frac{R_n}{\Omega}$$

ASD load combinations that include only D and L are:

$$D$$
$$D + L$$

Thus,

$$175 \text{ kips}$$
$$175^{kips} + 125^{kips} = 300 \text{ kips} \rightarrow R_a = \Sigma Q_i = 300 \text{ kips}$$
$$\frac{R_n}{\Omega} = \frac{F_y A_g}{1.67} = 300 \text{ kips}$$

$A_g = \dfrac{(300^{kips})1.67}{46^{ksi}} = 10.89 \text{ in.}^2$ is the required cross-sectional area for the design according to ASD.

Both LRFD and ASD resulted in similar required cross-sectional areas; however, LRFD seems to provide a more economical solution.

1.6 Problems

1.1 A tension-only steel brace, HSS6 × 6 × 3/8, in a single-story frame is subject to the following service and earthquake loading (Steel grade ASTM A500 Grade C):

Dead load: 80 kips (tension)
Roof live load: 30 kips (tension)
Snow load: 25 kips (tension)
Earthquake load: 250 kips (tension)

Please answer the following questions :

1. What is the nominal tensile strength of the brace member?
2. According to LRFD design method:
   a. What is the required brace strength?
   b. What is the design tensile strength of brace ($\phi = 0.90$)?
3. According to ASD design method:
   a. What is the required brace strength?
   b. What is the allowable tensile strength ($\Omega = 1.67$)?

1.2 Define specified minimum yield stress and specified tension strength of steel material. List the minimum yield stress and tensile strength of A36 and A572 steel grades.

1.3 Tensile testing is applied to a steel bar specimen with a diameter of 0.5 in. and length of 12 in. Yield load and failure load are measured as 10 kips and 12 kips, respectively. The length of the specimen just before the failure is measured as 16 in. Please compute the following:

a. Yield stress and yield strain
b. Tension strength
c. Ultimate strain

1.4 Determine the nominal depths, leg size, side widths, and weight of the hot-rolled shapes listed below. Use *AISC Manual* if weight is not given in the shape designation.

W24 × 162	L5 × 3 × $\frac{1}{2}$
S24 × 121	WT120 × 132
HP12 × 84	HSS16 × 4 × $\frac{3}{16}$
C9 × 20	HSS51/2 × 51/2 × $\frac{3}{8}$
L8 × 8 × 1	HSS10 × 0.500

1.5 Determine the expected yield stress and expected tension strength for A572 Gr. 50, A500 Gr. B, and A572 Gr. 42 steel grades.

1.6 Write a python code that computes

  a. Expected yield stress, F_{ye}

  b. Expected tensile strength, F_{ue}, for a given steel grade

1.7 Develop a python code that computes the required design strength and required cross-sectional area of a brace member subject to an axial force according to

  a. LRFD

  b. ASD

Appendix: Project Description

The design of basic steel structural elements (beams and girders, columns, beam-columns, braces, tension members) and connections is intended to be applied on a real six-story building design project. The design steps are given below. First, a general overview and description of the project is given, followed by gravity and lateral load analyses for a moment frame and concentrically braced frame, which are the most common lateral force–resisting systems (LFRSs) in steel construction. Knowing the fact that this book does not introduce earthquake-resisting design of steel structures, some assumptions are made regarding the seismic design requirements throughout the project. The numerical model of the LFRS for the moment frame is needed for structural analysis and design of structural members. The design of the concentrically braced frame is completed by hand calculation. At the end of each chapter (Chapters 2 through 9), the structural elements are to be designed based on the information given below and the topic covered in the corresponding chapter. By the end of the book, the student will have completed the basic design steps of a moment frame and a concentrically braced frame.

1.A1   Overview

The purpose of this project is to carry out structural analysis and design of the structural members of a moment and a braced frame in an office building. The building is located in Los Angeles, California. Particular emphasis will be placed on the development of the framing system, the computation of the loads, the determination of internal forces (axial forces, shear forces, and bending moments), and the design of typical structural members (beams and girders, columns, beam-columns, braces). For this purpose, an office building with a plan dimension of 150′×150′ is proposed. The building has two perimeter lateral force–resisting frames in both directions (see Fig. 1.A1) and consists of six-story frames spaced at 30′. The story height is 13′ except at the first story, where it is 18′.

1.A2   Description of the Six-Story Building

The plan and elevation of the six-story office building are given in Fig. 1.A1. The lateral force–resisting perimeter frames (moment and braced frames) are arranged on the perimeter in both orthogonal directions (Fig. 1.A1a). The columns are assumed to be pinned at the ground level (Fig. 1.A1b). Secondary beams should be placed based on your engineering judgment (a representative secondary beam configuration is shown in Fig. 1.A1a). 3¼″-thick-lightweight concrete with wire mesh on 1½″ metal deck will be used as the composite floor which can be considered to provide a rigid diaphragm to transfer lateral forces in both orthogonal directions. The orientation of the floor system is shown in Fig. 1.A1a.

The lateral force–resisting perimeter frames are to be designed for gravity and lateral loads. Special moment frames (SMFs) and special concentrically braced frames (SCBFs) will be used as lateral force–resisting systems (LFRSs) and designed based on the design requirements for SMFs and SCBFs stipulated in ASCE 7 (2016), AISC 360 (2016), AISC 341 (2016), and AISC 358 (2016). The SMFs and SCBFs on Lines A and F and 1 and 6, respectively, are designed to resist lateral loads while the columns on Lines B through E and 2 through 5 are connected to girders with shear (i.e., simple)

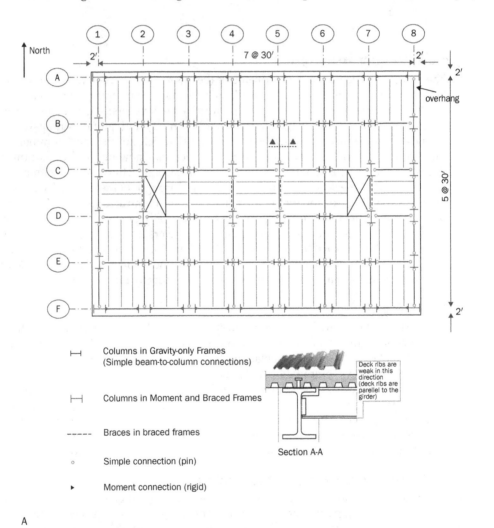

	Columns in Gravity-only Frames (Simple beam-to-column connections)
	Columns in Moment and Braced Frames
-----	Braces in braced frames
○	Simple connection (pin)
▸	Moment connection (rigid)

Section A-A

Deck ribs are weak in this direction (deck ribs are parellel to the girder)

A

Figure 1.A1 Six-story steel building: (a) plan view, (b) elevation of the SMFs on Lines A and F, (c) elevation of the SCBFs on Lines 2, 4, 5, and 7.

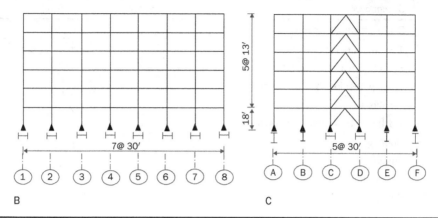

FIGURE 1.A1 (Continued)

connections and form gravity-only frames to support gravity loads (Fig. 1.A1). The contribution of the gravity-only frames for resisting lateral forces is neglected.

Material: $F_y = 50$ ksi steel (A992) is used for wide-flange shape (for beams and columns in SMFs and SCBs)

$F_y = 46$ ksi steel (ASTM A500 Gr. C) is used for round/square HSS

1.A3 Special Moment Frame (SMF) Design

Basic design steps for the design of the moment frame will follow:

1. Calculating gravity loads.
2. Determining lateral loads.
3. Constructing numerical model of the frame.
4. Carrying out structural analysis.
5. Designing for stiffness requirement (drift limitation).
6. Checking Strong Column-Weak Beam (SCWB) requirement in order to prevent possible story mechanisms.
7. Checking the Strength Design requirement (Capacity ≥ Demand).
8. Designing moment connections.
9. Checking whether all the sections are highly ductile to prevent premature local buckling.

1.A3.1 Gravity Loads

Dead Load Dead loads consist of the self-weight of the structural and non-structural members/components. Please fill in Table 1.A1 and calculate the total dead load for a typical floor and roof.

Component	Floor	Roof
3¼″ Lightweight concrete		
1½″ Metal deck		
Steel Beam/Column (self-weight)		
Roofing/Insulation/Elevator equipment		
Partitions		
Mechanical/Electrical components		
Flooring		
Suspended ceiling		
Fireproofing		
Sprinkler		
Total (psf)		

TABLE 1.A1   Dead Loads

Story	Dead load (D) (psf)	Floor area (A) (ft²)	$w_x = D \times A$ (kips)
Roof			
5th			
4th			
3rd			
2nd			
1st			
Effective weight $(W) = \sum w_x$ (kips)			

TABLE 1.A2   Effective Weight of the Building

Live Load   Determine the magnitude of uniformly distributed live loads, L_0, for floors and roof in accordance with Table 4.3-1, ASCE 7.

Floor Live Load = ... psf
Roof Live Load = ... psf

1.A3.2   Lateral Loads

Determine the effective weight of the structure, W, that only consists of dead load given in Table 1.A1 (Table 1.A2).

Story	w_x (kips)	h_x* (m)	$w_x (h_x)^k$	$C_{vx} = \dfrac{w_x (h_x)^k}{\sum w_x (h_x)^k}$	$F_x = C_{vx}V$ (kips)
Roof					
5th					
4th					
3rd					
2nd					
1st					
		Σ			

*Height from ground (ft).

TABLE 1.A3   Vertical Force Distribution of Base Shear

Total base shear, V, is to be calculated as 5 percent of the effective weight of the building, W, in N-S direction (for a true determination of lateral loads, please refer to Shen et al., 2020):

$$V = 0.05W \text{ kips}$$

The base shear needs to be distributed along with the building height in an inverted triangular pattern (Table 1.A3) (for a true determination of lateral loads, please refer to Shen et al., 2020).

1.A3.3   Numerical Model of the Frame
Provide a numerical model of the frame. Assume that

- Elastic beam elements are used for beams and columns.
- Beam and columns are rigidly connected.
- Rigid end offsets are incorporated at the panel zones.
- The floor slabs are assumed to be lumped at joints.
- To account for the second-order effects (P-Δ and P-δ), a leaning column consisting of rigid truss elements is introduced and the gravity loads not supported by the moment frames are transferred to these columns at each story level.
- Columns are meshed into five equal segments to take P-δ effect into account properly, when conducting stability analysis.

1.A3.4   Strength Design
Load Combinations

1. $1.4D$
2. $1.2D + 1.6L + 0.5L_r$
3. $1.2D + 1.6L_r + 1.0L$

4. $1.4D + Q_E + 0.5L$

5. $0.7D + Q_E$

Tributary Areas   Compute the tributary areas for each uniformly distributed and concentrated load for dead and live loads.

Stability Analysis: *Direct Analysis Method (DAM)*

- Build a realistic numerical model of the lateral force–resisting system with "leaning columns" to introduce the impact of gravity frames on second-order effects.
- Reduce the stiffness of the lateral force–resisting system.
- Apply notional loads in terms of the factored gravity loads of notional displacements to account for initial imperfections (Table 1.A4).
- Conduct a second-order analysis that takes P-Δ and P-δ effects into account.
- Determine capacities of members using $K = 1.0$.
- Check the ratio of second-order drifts to first-order drifts ($\Delta_{2nd}/\Delta_{1st}$) (Table 1.A5).
- Check strength for columns (Tables 1.A6 through 1.A8).

Combination	Description
Comb#1	$1.4D \pm 1.4ND$
Comb#2	$1.2D + 1.6L + 0.5L_r \pm (1.2ND + 1.6NL + 0.5NL_r)$
Comb#3	$1.2D + 1.0L + 1.6L_r \pm (1.2ND + 1.0NL + 1.6NL_r)$
Comb#4	$1.4D \pm Q_E + 0.5L$
Comb#5	$0.7D \pm Q_E$

Notes: (a) ND = notional dead load, NL = notional live load, NL_r = notional roof live load. (b) Notional loads are only included in gravity load combinations.

TABLE 1.A4   Load Combinations Including Notional Load Effects

Combination	$\Delta_{2nd}/\Delta_{1st}$ ratio*
Comb#1	
Comb#2	
Comb#3	
Comb#4	
Comb#5	

*Max. Δ2nd/Δ1st ratio along with the height of the building.

TABLE 1.A5   Ratio of Second-Order Drifts to First-Order Drifts (Δ2nd/Δ1st)

Story	Section	P_r (kips)	P_y (kips)	P_r/P_y	τ_b
Roof					
5th					
4th					
3rd					
2nd					
1st					

TABLE 1.A6 τ_b Values for Exterior Columns

Story	Section	P_r (kips)	P_y (kips)	P_r/P_y	τ_b
Roof					
5th					
4th					
3rd					
2nd					
1st					

TABLE 1.A7 τ_b Values for Interior Columns

Comb.	Story	Section	P_r (kips)	P_c (kips)	M_r (kip-ft)	M_c (kip-ft)	P-M Ratio
Comb. #1	Roof						
	5th						
	4th						
	3rd						
	2nd						
	1st						
Comb. #2	Roof						
	5th						
	4th						
	3rd						
	2nd						
	1st						
Comb. #3	Roof						
	5th						
	4th						
	3rd						
	2nd						
	1st						

TABLE 1.A8   Strength Check for Exterior/Interior Columns Based on Second-Order Analysis Results

Comb.#4	Roof							
	5th							
	4th							
	3rd							
	2nd							
	1st							
Comb.#5	Roof							
	5th							
	4th							
	3rd							
	2nd							
	1st							

TABLE 1.A8   Strength Check for Exterior/Interior Columns Based on Second-Order Analysis Results   (*Continued*)

1.A3.5   Design of Girders

Lateral Bracing of Girders   Lateral bracing of highly ductile beams in SMF is spaced at maximum spacing of $L_b = 0.095 r_y E/(R_y F_y)$, (*Section D1.2, AISC 341*) (Table 1.A9). Check strength for girders (Table 1.A10).

Story	Section	r_y (in.)	$L_{b,max}$ (in.)
Roof			
5th			
4th			
3rd			
2nd			
1st			

TABLE 1.A9   Spacing of Lateral Bracings

Story	Girder	Z_x (in.³)	M_p (kip-ft)	M_r^{1st} (kip-ft)	M_r^{2nd} (kip-ft)	M_c (kip-ft)	Fail/Pass
Roof							
5th							
4th							
3rd							
2nd							
1st							

Note: M_r^{1st} and M_r^{2nd} = required flexural strengths obtained from first-order and second-order analyses, respectively.

TABLE 1.A10   Strength Checks for Girders

Member Requirement Check limiting width-to-thickness ratio for girders (*D1.1, AISC 341*) (Table 1.A11).

Story	Girder	$b/2t_f$	λ_{hd}
Roof			
5th			
4th			
3rd			
2nd			
1st			

TABLE 1.A11   Limiting Width-to-Thickness Ratios for Girders

Check limiting width-to-thickness ratio for columns (Table 1.A12).

	Column	$b/2t_f$	$\lambda_{hd,f}$	h/t_w	$\lambda_{hd,w}$
Exterior					
Interior					

TABLE 1.A12   Limiting Width-to-Thickness Ratios for Columns

Final Design Fill in Table 1.A13 showing the final girder and column sections.

Story	Exterior column	Interior column	Girders
Roof			
5th			
4th			
3rd			
2nd			
1st			

TABLE 1.A13   Final Design

1.A3.6   Connection Design

Use reduced beam section (RBS) design for the first-story girder with 30-ft span length in accordance with AISC 358 (Table 1.A14).

Loads on the beam: $w_D = \ldots$ kip/ft; $w_L = \ldots$ kip/ft

Materials: A992, Grade 50 for W-shapes, and A36 or A572 for plates; Electrode—$F_{exx} = 70$ ksi; MA325 bolts are available.

	Beam section	a (in.)	b (in.)	c (in.)	Z_{RBS} (in.³)	C_{pr}	M_{pr} (kip-ft)	S_h (in.)	L_h (ft)	V_{RBS} (kips)	M_f (kip-ft)	πM_{pe} (kip-ft)
Exterior bays												
Interior bays												

TABLE 1.A14   Summary of RBS Design (See AISC 358 for the Definitions in the Table)

1.A3.7 Strong-Column Weak-Beam (SCWB) Requirement

Check SCWB requirement for each beam-to-column connection (AISC 341, Chapter E) (Tables 1.A15 and 1.A16).

$$\frac{\sum M_{pc}^*}{\sum M_{pb}^*} > 1.0$$

$$\sum M_{pc}^* = \sum Z_c (F_{yc} - P_r / A_g)$$

	Joint	Column above	$P_{r,above}$ (kips)	$M_{pc,above}$ (kip-ft)	Column below	$P_{r,below}$ (kip)	$M_{pc,below}$ (kip-ft)	$\sum M_{pc}^*$ (kip-ft)	$\sum M_{pb}^*$ (kip-ft)
Exterior	Roof								
	5th								
	4th								
	3rd								
	2nd								
	1st								
Interior	Roof								
	5th								
	4th								
	3rd								
	2nd								
	1st								

TABLE 1.A15   Summary of SCWB Check

Story	Exterior column	Interior column	Girders
Roof			
5th			
4th			
3rd			
2nd			
1st			

*Check whether these sections satisfy the story drift limitation of 0.02 times story height.

TABLE 1.A16   Member Sizes That Satisfy SCWB Requirement*

1.A4   Special Concentrically Braced Frame (SCBF) Design

Basic design steps for the design of concentrically braced frame will follow:

1. Calculating gravity loads.
2. Determining seismic and structural design parameters.
3. Designing braces, beams, and columns in the braced bays.
4. Designing beams and columns in the unbraced bays.
5. Constructing numerical model of the frame.
6. Checking stiffness requirement (drift limitation).

1.A4.1   Gravity Loads
Dead Load   (The same as used in the SMF.

1.A4.2   Lateral Loads
Determine the effective weight of the structure, W, that only consists of dead load given in Table 1.A1. Total base shear, V, is to be calculated as 12 percent of the effective weight of the building, W, in E-W direction:

$$V = 0.12W \text{ kips}$$

The base shear needs to be distributed along the building height in an inverted triangular pattern (Table 1.A17).

Story	w_x (kips)	h_x* (m)	$w_x(h_x)^k$	$C_{vx} = \dfrac{w_x(h_x)^k}{\sum w_x(h_x)^k}$	$F_x = C_{vx}V$ (kips)
Roof					
5th					
4th					
3rd					
2nd					
1st					
$\sum$					

*Height from ground (ft).

TABLE 1.A17   Vertical Force Distribution of Base Shear

1.A4.3 Gravity Loads on Girders, Exterior Columns, and Interior Columns

Calculate the point gravity loads to be supported by girders, exterior columns, and interior columns using the tributary areas.

1.A4.4 Brace Design

Find the axial forces in the braces and select square or round HSS (Table 1.A18).

Story	Brace section	A_g (in.^{2})	$L_{b,c}$ (ft)	L_c/r	D/t	Q_e (kips)	$F_{b,Qe}$ (kips)	$F_{b,DL}$ (kips)	$F_{b,LL}$ (kips)	P_r (kips)	P_c (kips)
Roof											
5th											
4th											
3rd											
2nd											
1st											

Notes: (a) Braces are made of ASTM A500 Gr. C; (b) length of braces from centerline to centerline; (c) slenderness ratios of all braces should be less than 200; (d) the limiting diameter-to-wall thickness ratio (D/t) for highly ductile compression members made of circular HSS is $\lambda_{hd} = 0.053E/R_yF_y$; (e) the required axial strength, P_r, is $1.4P_D + P_{QE} + P_{LL}$; (f) $F_{b,Qe}$, $F_{b,DL}$, $F_{b,LL}$ are axial force demands due to earthquake, dead load, and live load, respectively.

TABLE 1.A18   Brace Design for SCBF

1.A4.5 Beam Design

Consider the following cases in beam design (AISC 341) (Tables 1.A19 through 1.A22):

Case I: A structural analysis in which all braces are assumed to resist forces corresponding to their expected strength in compression or tension (T_{ET} for tension or C_{EB} for compression).

Case II: A structural analysis in which all braces in tension are assumed to resist forces corresponding to their expected strength (T_{ET}) and all braces in compression are assumed to resist their expected post-buckling strength (C_{EPB}).

Story	Brace section	A_g (in.^{2})	L_b (ft)	r (in.)	L_c/r	T_{ET} (kips)	C_{EB} (or C_{EPB}) (kips)
Roof							
5th							
4th							
3rd							
2nd							
1st							

Note: L_b is assumed to be 85% of the brace length from centerline to centerline.

TABLE 1.A19   Expected Brace Strengths to Be Used in Analysis of Case I/Case II

Story	Girder Section	F_u (kips)	M_{QE} (kip-ft)	M_{DL} (kip-ft)	M_{LL} (kip-ft)	P_{QE1} (kips)	M_r (kip-ft)	P_r (kips)	$M_c = \phi_c M_n$ (kip-ft)	$P_c = \phi_c P_n$ (kips)	Strength ratio*
5th											
3rd											
1st											

*Demand-to-capacity ratio, P_r/P_c.

Notes: (a) F_u is capacity-limited seismic force at the story level, (b) check strength ratio using axial force–bending moment interaction equation.

TABLE 1.A20   Brace-Intersected Girder Sections for Structural Analysis of Case I/Case II

Story	Girder section	F_u (kips)	M_{QE1} (kip-ft)	M_{DL} (kip-ft)	M_{LL} (kip-ft)	P_{QE1} (kips)	M_r (kip-ft)	P_r (kips)	$M_c=\phi_c M_n$ (kip-ft)	$P_c=\phi_c P_n$ (kips)	Strength ratio*
Roof											
4th											
2nd											

*Demand-to-capacity ratio, P_r/P_c.

Note: Check strength ratio using axial force–bending moment interaction equation.

TABLE 1.A21   Girder Sections Not Intersected by Braces for Structural Analysis of Case I/Case II

Story	Girder section		Governing analysis case	Selected girder section
	Analysis case I	Analysis case II		
Roof				
5th				
4th				
3rd				
2nd				
1st				

TABLE 1.A22   Final Sections for the Girders

1.A4.6   Column Design

Fill in Tables 1.A23 through 1.A25 for column design.

Story	Brace section	T_{ET} (kips)	C_{EB} (or C_{EPB}) (kips)	F_u (kips)	$P_{C,Left}$ (kips)	$P_{C,Right}$ (kips)	P_{QE1} (kips)
Roof							
5th							
4th							
3rd							
2nd							
1st							

TABLE 1.A23   Column Forces for Structural Analysis of Case I/Case II

Story	$P_{QE,Case1}$ (kips)	$P_{QE,Case2}$ (kips)	Governing analysis case	P_{QE} (kips)	P_{DL} (kips)	P_{LL} (kips)	P_r (kips)
Roof							
5th							
4th							
3rd							
2nd							
1st							

TABLE 1.A24   Required Axial Strength for Columns

Story	Column section	Member requirement				Strength requirement		
		$b/2t_f$	$\lambda_{hd,flange}$	h/t_w	$\lambda_{hd,web}$	P_r (kips)	P_c (kips)	Strength ratio*
Roof								
5th								
4th								
3rd								
2nd								
1st								

*Demand-to-capacity ratio, P_r/P_c.

TABLE 1.A25   Strength and Member Requirements for Columns

Story	Braces	Girders	Columns
Roof			
5th			
4th			
3rd			
2nd			
1st			

TABLE 1.A26   Final Sections for the Members in the Braced Bays

List the final sections in the braced bays (Table 1.A26).

1.A4.7   Design of Beams and Columns in Unbraced Bays

Beams and columns in the unbraced bays will be designed for gravity loads only in a similar way.

Bibliography

AASHTO, *LRFD Bridge Design Specification*, American Association of State Highway and Transportation Officials for Bridges, Washington, 2020.

Aghayere, A. O., and J. Vigil, *Structural Steel Design: A Practice Oriented Approach*, Pearson Education, Inc., New Jersey, NY, 2015.

AISC, *Seismic Design Manual*, American Institute of Steel Construction, Chicago, IL, 2018.

AISC, *The Material Steel*, American Institute of Steel Construction, A Teaching Primer for Colleges of Architecture, Chicago, IL, 2007.

AISC 341, *Seismic Provisions for Structural Steel Buildings*, ANSI/AISC Standard 341-16, American Institute of Steel Construction, Chicago, IL, 2016.

AISC 358, *Prequalified Connections for Special and Intermediate Steel Moment Frames for Seismic Applications*, American Institute of Steel Construction, Chicago, IL, 2016.

AISC 360, *Specification for Structural Steel Buildings*, ANSI/AISC Standard 360-16, American Institute of Steel Construction, Chicago, IL, 2016.

AISC Manual, *Steel Construction Manual*, 15th ed., American Institute of Steel Construction, Chicago, IL, 2016.

ASCE 7, *Minimum Design Loads and Associated Criteria for Buildings and Other Structures*, ASCE/SEI 7-16, American Society of Civil Engineers, Reston, VA, 2016.

ASTM, *Annual Book of ASTM Standards*, American Society for Testing and Materials, Philadelphia, 2020.

AWS, *Structural Welding Code-Steel (AWS D1:1:2020)*, 24th ed., American Welding Society, Danvers, MA, 2020.

Salmon, C. G., J. E. Johnson, and F. A. Malhas, *Steel Structures: Design and Behavior—Emphasizing Load and Resistance Factor Design*, 5th ed., Pearson Education, Inc., New Jersey, NY, 2009.

Shen, J., B. Akbas, O. Seker, and C. Carter, *Structural Engineering Handbook—Chapter 8: Design of Structural Steel Members*, 5th ed., McGraw-Hill, New York, 2020.

Yescas-Gonzalez, M. A., and H. K. D. H. Bhadeshia, *Cast Irons*, University of Cambridge, www.phase-trans.msm.cam.ac.uk/2001/adi/cast.iron.html (accessed June 2020).

CHAPTER 2

Design Considerations

2.1 Introduction

A structural design process begins with identifying the types of loads on a structure. The loads on structures are due to a variety of different circumstances, such as gravity loads (dead, live, snow, etc.) and lateral loads (wind, earthquake). In a typical steel building, we will have joists, girders, and steel decking on top of the plenum area. For the required stiffness, strength, and fire protection, concrete is poured on the decking. And the floor is ready to transfer all the loads imposed upon it vertically due to gravity loads and horizontally due to earthquake or wind loads.

Gravity loads on the floor deck are distributed through the framing systems, beams, girders, and columns. Girders are defined as the main load-carrying elements, i.e., they transfer both gravity and lateral loads. The tributary width for each beam or girder is half the distance to the next beam or girder on both sides. Edge beams or girders only have tributary width on one side, while interior beams or girders have it on both sides. The tributary length is the length of the bay, and the tributary area, A_T, is the product of the tributary length and the tributary width. Columns transfer all the loads from the roof level to the foundation. For columns transferring only gravity loads, it is assumed that the gravity loads are coming from the floor deck. Thus, corner, edge, and interior columns will have different tributary areas, which is easily calculated by measuring one-quarter of the area of all of the four bays around it.

Similar to gravity loads, lateral loads due to wind and earthquake are also transmitted through the framing system. The wind loads on the building surface are divided accordingly among the floors and acted upon the building at floor levels. In contrast, earthquake loads are input to the structure through the foundation but converted into lateral loads acting at floor levels. The floor system acts, in general, as a rigid diaphragm, taking the load from where it acts to the lateral force–resisting system (LFRS). LFRS takes the load and transfers it down to the foundation.

Minimum design loads are defined in building codes. A modern building code should address the design and construction of building systems through requirements emphasizing performance. International Building Code (IBC) requires that every structure be designed and constructed to resist the effects of earthquake motions in accordance with ASCE 7, *Minimum Design Loads and Associated Criteria for Buildings and Other Structures*. It is intended that major provisions are explained as practical as possible, but the reader is strongly recommended to relate them to the fundamental understanding on the seismic performance and parameters affecting the performance, since every single parameter in the provisions is either directly or indirectly related to what we know about structural performance during the design earthquake ground motions. It is believed that not only do we need to know how to use the code provisions in design practice but also we have to understand why to do so. Please keep in mind that the seismic design requirements are indeed "*Minimum*," and we are always better off when prepared to exercise our engineering judgments in dealing with seismic design issues.

2.2   Design Loads and Design Approaches

2.2.1   Vertical Loads

Vertical loads such as dead, live, rain, and snow vary depending on the type where they are originated.

Dead Load:

All gravity loads with fixed amplitude and location, including structural self-weight, permanent equipment, etc., which are calculated by the designer during the structural design process.

Live Load:

All gravity loads with amplitude and location varying significantly, including people, movable equipment, etc. Due to its uncertain nature, the design live load is defined by standard codes such as ASCE 7.

Rain and Snow Loads:

Rain and snow loads are given by standard codes such as ASCE 7.

2.2.2   Lateral Loads

Wind:

Wind disturbance is a long period of excitation. It takes 3.0–60.0 s to finish a cycle. Lift and drag forces occur when the wind is acting on any structure, which depends upon velocity (Fig. 2.1). For short period structures, design codes convert the lift and drag forces into static forces, but this approach is not well suited for long and

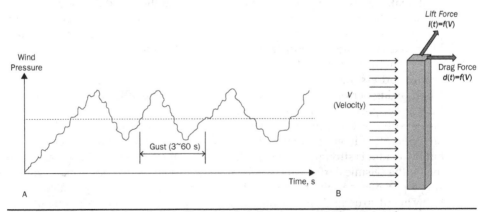

FIGURE 2.1   Characteristics of wind effect: (*a*) wind disturbance, (*b*) lift and drag forces

slender structures. Note that lift and drag forces can only be determined statistically. Wind disturbance has a dynamic effect on long-period structures such as:

- High-rise buildings with height above 400 ft
- Long-span bridges with periods longer than 3.0 s

Dynamic amplification of the wind pressure on the above structures should be introduced based on dynamic analysis and wind-tunnel tests. Wind disturbance on short-period structures such as low-rise buildings, short-span bridges, etc., has static or quasi-static nature. The wind effect on these types of structures can well be described as static pressures. The design wind load is specified by standard codes such as ASCE 7.

Earthquake Load:

Characteristics of earthquake ground motions such as the predominant period of excitations, energy input, and duration of shaking vary a lot from one event to another, from one location to another (Fig. 2.2). Thus, the seismic effect on structures has a random nature. Earthquake ground motions cause the most severe loads placed on structures. Predominant frequencies in recorded earthquake ground motions cover a range of about 1–20 Hz (0.05–1.0 s). Unfortunately, most of the structures are in this range, and the devastating collapse of structures was contributed by the significant dynamic effect of earthquakes and inadequacies in design due to technical and economic reasons. The structural system, soil–structure interactions, material, etc., have a substantial impact on the earthquake damage. In addition, the construction quality control turns out to be responsible for many unexpected collapses of well-engineered structures. Structural engineers measure ground acceleration to calculate the earthquake load in structures. The design earthquake load is given by the building code, ASCE 7.

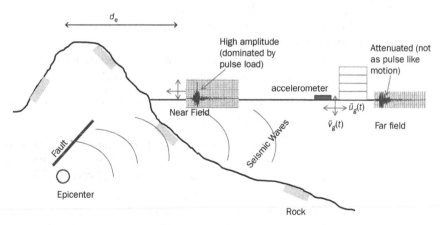

$\ddot{u}_g(t)$ = acceleration time history in horizontal direction

$\ddot{v}_g(t)$ = acceleration time history in vertical direction

Figure 2.2 Characteristics of earthquake ground motions

Impact and Blast:

These types of loads are not specified for common building structures due to their extremely uncertain nature.

2.3   Design of Structural Framing Systems

It has been observed during past earthquakes that numerous parameters are affecting structural performance, and many of them consist of uncertainties. The balance between design efforts and expected structural performance under various levels of ground motion intensity might be the best possible approach. For the given design earthquake, a typical and conventional structure may tolerate a high level of uncertainties in determining demand and capacity, and its expected seismic performance is relatively low, often constrained by cost consideration. However, if we expect a special (or critical) structure to perform well with a low level of uncertainties, a much more rigorous design procedure should be employed, and such rigorousness level should be increased for a higher design earthquake. In ASCE 7, every building structure is assigned to a seismic design category (SDC) (to decide how rigorous the design procedure should be), based on its level of impact if a failure occurs (expected performance) and response acceleration spectrum at the site (design earthquake intensity). SDCs are used to dictate every step throughout the design process, from the selection of the structural system to the determination of analytical procedure, in recognizing those factors that most significantly affect seismic performance, such as:

a. Ductility capacity (or inelastic deformation capacity) in structural systems: select the systems that have high seismic performance, indicated by ductile behavior, for the structure with high expectation, subject to strong ground motions.

b. Irregularity: More rigorous analytical procedures are required for irregular structures. Since little can be done to improve the performance of an irregular building, it is probably the best practice at present to avoid inherent poor performance of irregular structures.

c. The types of lateral force analysis that must be performed (dynamic analysis would be necessary for a structure in high SDC).

We also might have to use our "common sense" in structural engineering to ensure a sound system, which would include:

a. The structure should be designed with complete lateral force–resisting (LFR) and vertical force–resisting (VFR) elements capable of providing adequate strength, stiffness, and energy dissipation (ductility) (Fig. 2.3).

b. The structural system will perform better (more ductile in general) when LFR and VRF elements have less interaction between them. For example, when the moment-resisting frame is used as an LFR element, it will achieve its intended ductile performance when carrying no or little gravity load. The bearing wall system is an opposite example, in which the LFR and VFR elements are often combined into the same elements, resulting in low ductility.

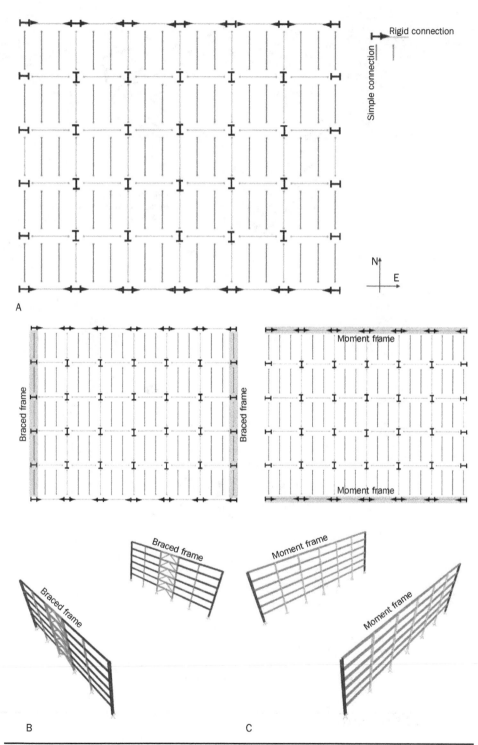

Figure 2.3 Typical structural framing of a steel building: (a) plan layout, (b) perimeter braced frames in N-S directions, (c) perimeter moment frames in E-W direction, (d) perimeter moment and braced frames, (e) interior gravity-only frames, (f) secondary beams

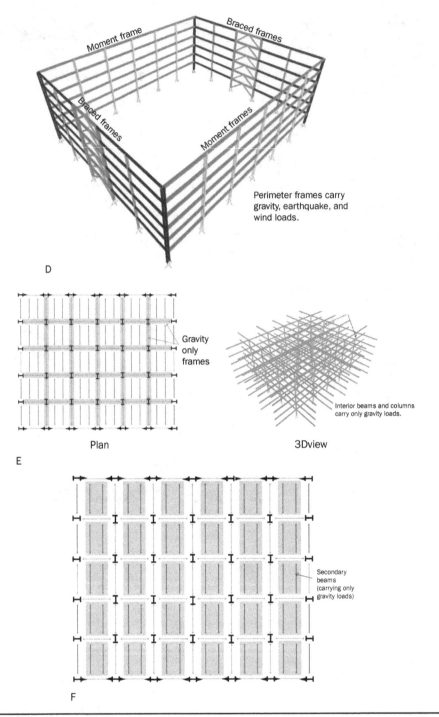

Moment frame

Braced frames

Braced frames

Moment frames

Perimeter frames carry
gravity, earthquake, and
wind loads.

D

Gravity
only
frames

Interior beams and columns
carry only gravity loads.

Plan   3Dview

E

Secondary
beams
(carrying only
gravity loads)

F

FIGURE 2.3   *(Continued)*

c. A continuous load path, or paths, with adequate strength and stiffness to transfer forces induced by the design earthquake ground motions from the points of application to the final point of resistance (often at the base), needs to be provided.

d. The LFR system with large torsional resistance is always a better choice even for a seemingly symmetric structural system. Some examples are shown in Fig. 2.4. Even if the center of mass and rigidity coincide, the codes require that 5 percent accidental eccentricity be taken into account.

e. The lateral displacement of an LFR system under earthquake or wind is generally limited to $0.02H$, where H is the building height (Fig. 2.5).

Example 2.1: Determine Seismic Forces Using Static Procedure

The typical floor plan of a six-story steel building is shown in Fig. 2.6 with elevations of perimeter frames in Fig. 2.7. All beam-to-column connections in other frames are shear connections. In International Building Code, IBC, the earthquake load and related design issues are based on *Minimum Design Loads and Associated Criteria for Buildings and Other Structures, ASCE 7*. In a seismic force–resisting system (SFRS), special moment frames are arranged in one direction and braced frames in the other orthogonal direction. Eighty psf for floors and roof, calculated based on the information for this building, including concrete decking (40 psf), ceiling, floor finishing, mechanical and electrical, exterior walls, and estimated steel member weight (30 psf), and partition (half of 20 psf per ASCE 7.) will be used for the seismic weight (calculated from all dead loads and half of the partition). See the Appendix at the end of the book for basics on seismic design of steel structures. Assume that

$$T = \begin{cases} 0.89\,\text{s for Special MRF} \\ 0.51\,\text{s for Special CBF} \end{cases}$$

$$C_s = \begin{cases} 0.14\,\text{for Special MRF} \\ 0.33\,\text{for Special CBF} \end{cases}$$

Determine the seismic lateral force distribution on the perimeter moment and braced frames.

Solution:
a. Design earthquake load
Seismic force–resisting systems:

Two special moment frames (SMFs) on Lines A and E are arranged to take the lateral forces (seismic or wind forces) in the East-West direction, and two special concentrically braced frames (SCBF) are arranged to resist lateral forces in the North-South direction. Effective seismic weight (or total building weight):

Weight on each floor/roof:
$w = (80^{psf})(120')(100') = 960,000\,\text{lb} = 960\,\text{kips}$ (floor levels and the roof)
Total building weight: $W = \sum w = 6w = 5760\,\text{kips}$

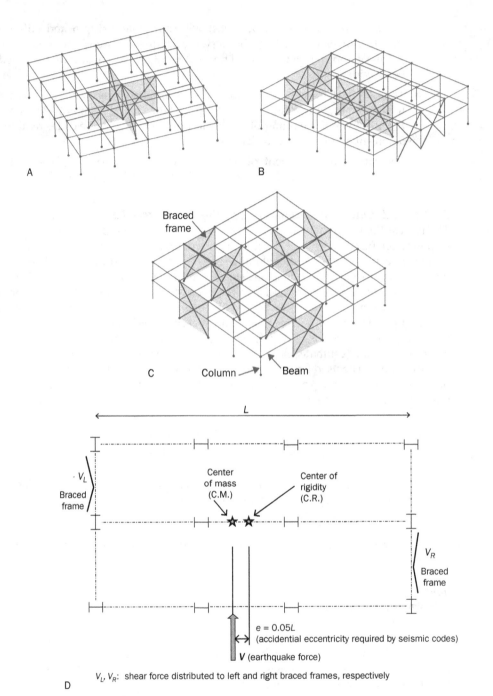

FIGURE 2.4   Structural framings: (*a*) no torsional resistance, (*b*) torsionally unbalanced system, (*c*) excellent torsional resistance with braced frames on the perimeter as well as in the middle, (*d*) accidental eccentricity

Story drift limitation ≈ 0.02H under lateral forces (earthquake and wind)

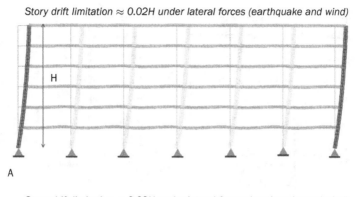

A

Story drift limitation ≈ 0.02H under lateral forces (earthquake and wind)

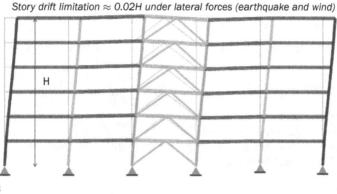

B

FIGURE 2.5 Story drift limitation under wind- or earthquake-induced lateral forces: (*a*) moment frame, (*b*) braced frame

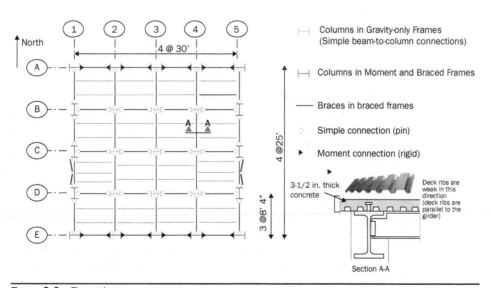

FIGURE 2.6 Floor plan

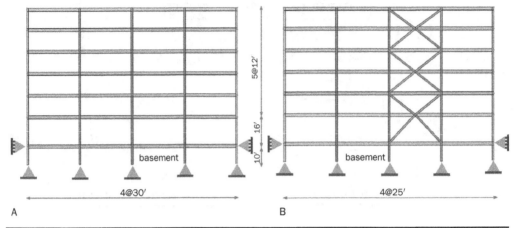

FIGURE 2.7   Elevation for the frame: (a) on Line A or Line E, (b) on Line 1 or Line 5

b. Equivalent lateral force procedure

Vertical distribution of seismic forces:

Per Section 12.8.3, ASCE 7, we can distribute the total base shear to each moment and braced frames and each floor as follows:

$k = 1.195$ for SMRF
$ = 1.101$ for SCBF

The lateral force distribution is given in Table 2.1 for the moment frame and in Table 2.2 for the braced frame. Note that story shear force V_x is equal to the sum of all lateral forces above that floor, i.e.,

$$V_x = \sum_x^n F_i$$

c. Horizontal distribution of seismic forces at each floor

We have assumed that the seismic weight (dead load) on the floors and roof are uniformly distributed. If it is the case, the total seismic forces in each direction will be shared by two moment frames and two braced frames, respectively. But it is very unlikely to be the case since the distribution of the dead load is rarely uniform in reality. To consider the possible nonuniform distribution of the weight at each floor/roof, ASCE 7 requires an accidental eccentricity of 5 percent times dimension perpendicular to the seismic forces to be included in distributing the seismic forces among lateral load–resisting frames. The seismic forces among the seismic force–resisting elements can be easily distributed in any orthogonal direction on the basis of the relative lateral stiffness of all seismic force–resisting elements in both orthogonal directions when they are available. In general, when the braced frames are used in the building, such small eccentricity does not affect moment frames since they are more flexible than the braced frames, but each braced frame might take more than half of the load. Approximately,

Floor (level)	h_x (ft)	w_x (kips)	$w_x(h_x)^k$ (kip-ft)	C_{vx}	F_x (kips)	V_x (kips)
6	76	960	169,763	0.2965	239	–
5	64	960	138,247	0.2414	195	239
4	52	960	107,869	0.1884	152	434
3	40	960	78,838	0.1377	110	586
2	28	960	51,478	0.0900	73	696
1	16	960	26,375	0.0460	37	769
Total	–	–	572,570	1.0000	806	806

TABLE 2.1 Distribution of Lateral Forces for Two Moment Frames

Floor (level)	h_x (ft)	w_x (kips)	$w_x(h_x)^k$ (kip-ft)	C_{vx}	F_x (kips)	V_x (kips)
6	76	960	72,960	0.2754	523	–
5	64	960	61,440	0.2319	441	523
4	52	960	49,920	0.1884	358	964
3	40	960	38,400	0.1449	275	1322
2	28	960	26,880	0.1014	193	1597
1	16	960	15,360	0.0580	110	1790
Total	–	–	264,960	1.0000	1900	1900

TABLE 2.2 Distribution of Lateral Forces for Two Braced Frames

we use half of the total forces in Table 2.1 for one moment frame, and 0.55 of the total forces in Table 2.2 for one braced frame, as shown in Figs. 2.8 and 2.9.

Additional discussions on the issues related to the example:

- The example building includes a basement, and the building is thus supported at two levels: the column bases at the basement level and the grade level (second floor). The total base shear at the ground level would be transferred to the perimeter basement walls at the grade level. Three-dimensional building model with proper supports (working with geotechnical engineers) at the grade level would provide the design forces for the walls.

- If this frame is modeled with computer software and a rigid horizontal restraint is placed at the ground level, there will be a large reverse shear (lever action) between the basement and second floor. It is obviously due to the two-level support conditions (at the basement floor and on the grade level), which can be seen clearly from the analysis.

- The basement issue related to the definition of the "base" and participation of effective seismic weight: The "base" in this case is the interface between the

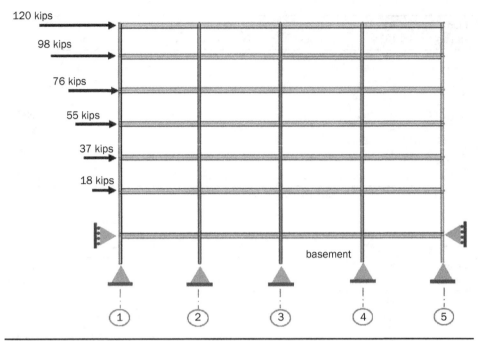

120 kips

98 kips

76 kips

55 kips

37 kips

18 kips

basement

1   2   3   4   5

FIGURE 2.8   Seismic forces along the height (one moment frame)

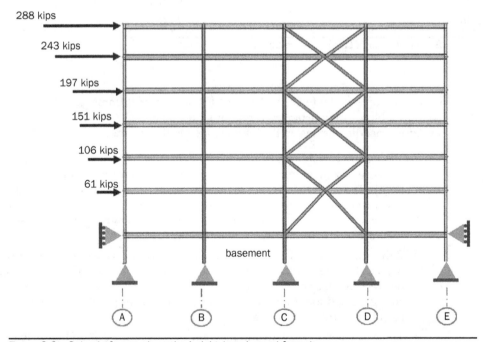

288 kips

243 kips

197 kips

151 kips

106 kips

61 kips

basement

A   B   C   D   E

FIGURE 2.9   Seismic forces along the height (one braced frame) "moving" part of the structure having relative displacement with respect to the ground and the supports (the basement and walls in this example). The effective seismic weight should include all weights associated with the moving part of the structure.

2.4   Load Paths in Steel Structures

In general, the simple and straightforward load path is desirable for economic and safety reasons. The approximate approach is often used in determining load distribution since "accurate" analysis is either too complex or unnecessary. Under gravity loads, loads acting on the floor slab are first transmitted to the secondary beams. These loads are then transferred to the girders, which also transfer them to the columns. If it is a braced frame, braces also take a smaller portion of the gravity loads. Finally, columns transfer all the loads to the foundation (Figs. 2.10a and 2.11a). However, under lateral forces such as earthquake or wind, secondary beams or beams and columns in gravity-only frames are assumed not to contribute to carry lateral forces. Thus, lateral force transfer is accomplished first through the girders and braces, which transfer the imposed loads to the columns in the LFR. Load transfer is completed when all the loads are transmitted to the foundation by the LFR frame's columns (Figs. 2.10b and 2.11b).

Example 2.2

The figures show the typical plan and elevation of a six-story steel office building (Figs. 2.12 and 2.13). Lateral force–resisting frames for wind or earthquake are on Lines 1 and 6 in N-S direction and on Lines A and F in E-W direction. On the floors, the dead load is 80 psf and the live load 50 psf. On the roof, the dead load is 65 psf and the snow load 20 psf. Note that the snow load (S) is larger than the live load on the roof (L_r).

a. Determine the distributed dead and live loads on beam B1, in kip/ft.

b. Determine the concentrated dead and live loads on beam G1, in kips.

c. Determine the factored axial loads on Column C1 on the first floor due to dead and live loads.

d. When wind load is given as 25 psf uniform pressure in all horizontal directions, determine the wind load, in kips, at each level of Frame A due to the wind pressure.

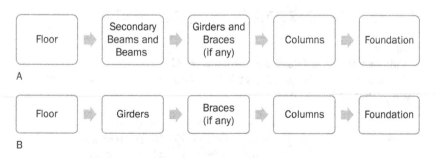

Figure 2.10   Load distribution: (a) under gravity loads, (b) under lateral loads

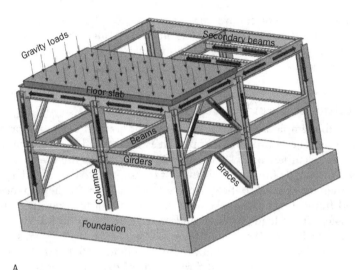

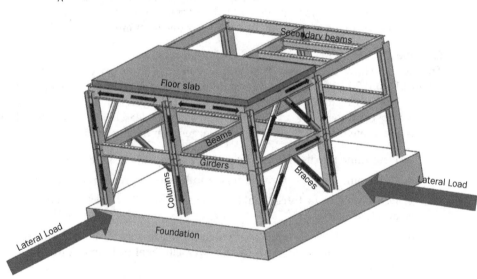

FIGURE 2.11   Illustration of load path: (a) gravity loads, (b) lateral loads

Solution:
a. Uniformly distributed dead and live loads on B1

Tributary area of B1 is 25′×10′ (Fig. 2.14a). Thus, the uniformly distributed dead and live loads acting on B1 can be found as

$$q_{D,\text{floor}} = 10' \times 80^{\text{psf}} = 800 \text{ lbs/ft} = 0.8 \text{ kip/ft}$$
$$q_{D,\text{roof}} = 10' \times 65^{\text{psf}} = 650 \text{ lbs/ft} = 0.65 \text{ kip/ft}$$
$$q_{L,\text{floor}} = 10' \times 50^{\text{psf}} = 500 \text{ lbs/ft} = 0.5 \text{ kip/ft}$$
$$q_{\text{Snow}} = 10' \times 20^{\text{psf}} = 200 \text{ lbs/ft} = 0.2 \text{ kip/ft}$$

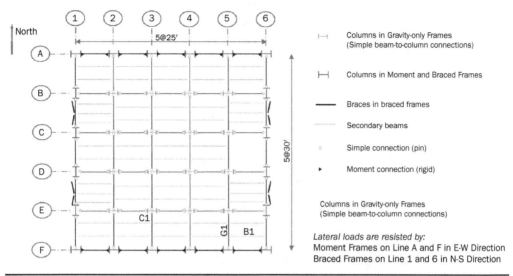

Figure 2.12 Plan view of the perimeter building

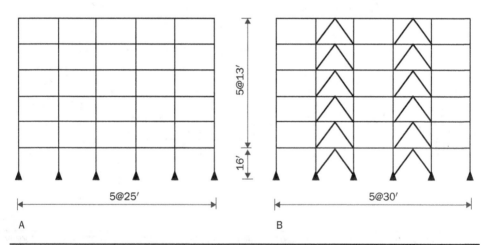

Figure 2.13 Elevation of the: (*a*) moment frames on Lines A and F, (*b*) braced frames on Lines 1 and 6

Support reaction of B1: (Fig. 2.14b)

$$R_{B1} = q \times L \ / \ 2$$

Note that live load reduction is not included in the above calculation. The load combinations to be used are (see Sec. 1.5):

1. $1.4D$ (1.7a)
2. $1.2D + 1.6L + 0.5L_r$ (1.7b)
3. $1.2D + 1.6L_r + 1.0L$ (1.7c)

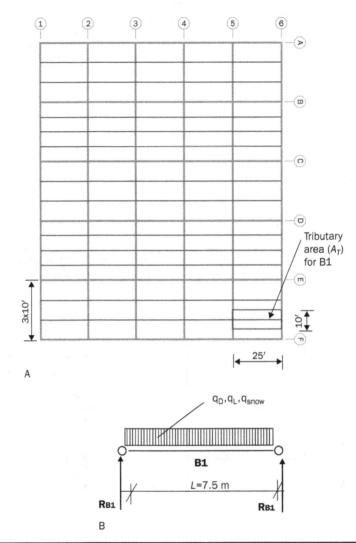

FIGURE 2.14 Dead and live loads on B1: (*a*) plan view and tributary area of B1, (*b*) uniformly distributed dead and live loads on B1

For Floors $\begin{cases} \text{Combination \#1: } 1.4D \\ q_{\text{required}} = q_u = 1.4 q_D = 1.4 \times (0.8^{\text{kip/ft}}) = 1.12 \text{ kip/ft} \\ \text{Combination \#2: } 1.2D + 1.6L \\ q_{\text{required}} = q_u = 1.2 \times (0.8^{\text{kip/ft}}) + 1.6 \times (0.5^{\text{kip/ft}}) = 1.76 \text{ kip/ft} \leftarrow \text{Controls} \end{cases}$

For Roof $\begin{cases} \text{Combination \#1: } 1.4D \\ q_{\text{required}} = q_u = 1.4 q_D = 1.4 \times (0.65^{\text{kip/ft}}) = 0.91 \text{ kip/ft} \\ \text{Combination \#3: } 1.2D + 1.6(S \text{ or } L_r) \\ q_{\text{required}} = q_u = 1.2 \times (0.65^{\text{kip/ft}}) + 1.6 \times (0.2^{\text{kip/ft}}) = 1.1 \text{ kip/ft} \leftarrow \text{Controls} \end{cases}$

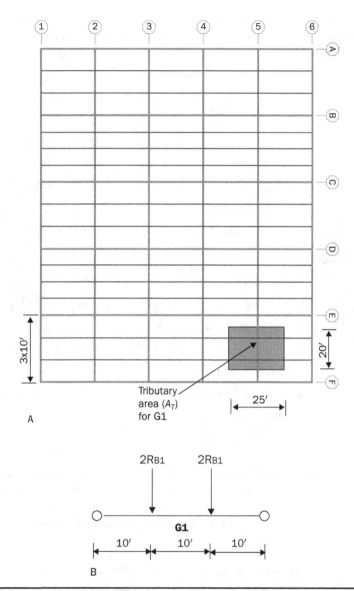

Figure 2.15 Dead and live loads on G1: (*a*) plan view and tributary area of G1, (*b*) concentrated dead and live loads on G1

Note that since floor beams (B1 on floors) solely support dead and live loads, the combinations with load types other than dead and live loads are not included.

b. Concentrated dead and live loads on G1

G1 takes loads from the secondary beams located at the left and right sides of the G1 girder. Concentrated loads, R_D and R_L, represent the support reactions of the secondary beams supported by G1.

Tributary area of G1 is 25′×20′ (Fig. 2.15a). Thus, the concentrated dead and live loads acting on G1 can be found by two alternative methods: (Fig. 2.15b)
Method 1:

R_{B1} due to Floor Dead Load:

$$R^{B1}_{D,floor} = q_{D,floor} \times L/2 = 0.8^{kip/ft} \times 25'/2 = 10 \text{ kips}$$

R_{B1} due to Roof Dead Load:

$$R^{B1}_{D,roof} = q_{D,roof} \times L/2 = 0.65^{kip/ft} \times 25'/2 = 8.13 \text{ kips}$$

R_{B1} due to Live Load:

$$R^{B1}_{L,floor} = q_{L,floor} \times L/2 = 0.50^{kip/ft} \times 25'/2 = 6.25 \text{ kips}$$

R_{B1} due to Snow Load:

$$R^{B1}_{snow} = q_{snow} \times L/2 = 0.20^{kip/ft} \times 25'/2 = 2.5 \text{ kips}$$

Method 2:

R_{B1} due to Floor Dead Load:

$$R^{B1}_{D,floor} = DL_{floor} \times A^{G1}_T/4 = 80^{psf} \times (25' \times 20')/4 = 10 \text{ kips}$$

R_{B1} due to Roof Dead Load:

$$R^{B1}_{D,roof} = DL_{roof} \times A^{G1}_T/4 = 65^{psf} \times (25' \times 20')/4 = 8.13 \text{ kips}$$

R_{B1} due to Live Load:

$$R^{B1}_{L,floor} = LL_{floor} \times A^{G1}_T/4 = 50^{psf} \times (25' \times 20')/4 = 6.25 \text{ kips}$$

R_{B1} due to Snow Load:

$$R^{B1}_{snow} = SL_{roof} \times A^{G1}_T/4 = 20^{psf} \times (25' \times 20')/4 = 2.5 \text{ kips}$$

c. Factored axial loads on Column C1 on the first floor due to dead and live loads

Column C1 on the first story supports five floors and the roof above (Fig. 2.16b).

$A^{Floor}_T = A^{Roof}_T$: Tributary Area of C1 at each floor and the roof (Fig. 2.16a)

$A^{Floor}_T = A^{Roof}_T = 25' \times 30' = 750 \text{ ft}^2$

Floor $\begin{cases} \text{Dead Load} \rightarrow P_D = (5 \times A^{Floor}_T) \times DL^{Floor} = (5 \times 750^{ft^2}) \times 80^{psf} = 300 \text{ kips} \\ \text{Live Load} \rightarrow P_L = (5 \times A^{Floor}_T) \times LL^{Floor} = (5 \times 750^{ft^2}) \times 50^{psf} = 187.5 \text{ kips} \end{cases}$

Roof $\begin{cases} \text{Dead Load} \rightarrow P_D = (A^{Roof}_T) \times DL^{Roof} = (750^{ft^2}) \times 65^{psf} = 48.8 \text{ kips} \\ \text{Snow Load} \rightarrow P_S = (A^{Roof}_T) \times S^{Roof} = (750^{ft^2}) \times 20^{psf} = 15 \text{ kips} \end{cases}$

LRFD $\begin{cases} \text{Combination \#1: } 1.4D \\ P_{required} = P_u = 1.4 \times (300^{kips} + 48.8^{kips}) = 488.3 \text{ kips} \\ \text{Combination \#2: } 1.2D + 1.6L + 0.5(L_r \text{ or } S) \\ P_{required} = P_u = 1.2 \times (300^{kips} + 48.8^{kips}) + 1.6 \times (187.5^{kips}) + 0.5 \times (15^{kips}) \\ \qquad = 726.1 \text{ kips} \leftarrow \text{Controls} \end{cases}$

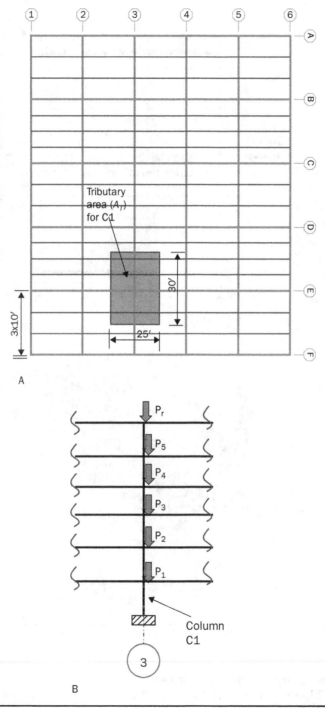

Figure 2.16 Dead and live loads on C1: (*a*) plan view and tributary area of C1, (*b*) side view of C1

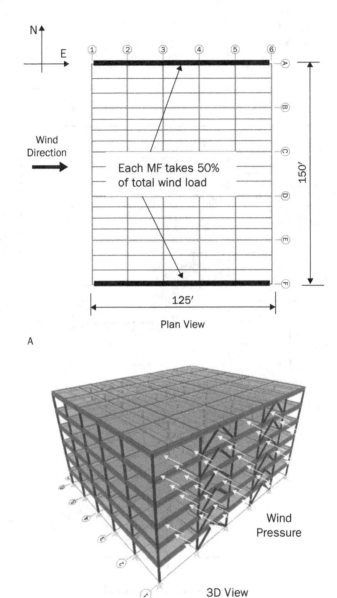

FIGURE 2.17   Wind loads acting on Frame A: (*a*) plan view, (*b*) 3D view

d. Wind load acting at each level of Frame A

(Fig. 2.17) Moment frames (MFs) support wind pressure acting on the exterior walls of the braced bays on Lines A and F. By assuming that wind pressure is uniformly distributed over the area, the diaphragm loads at each story level of MF can be calculated based on the tributary areas (Fig. 2.18).

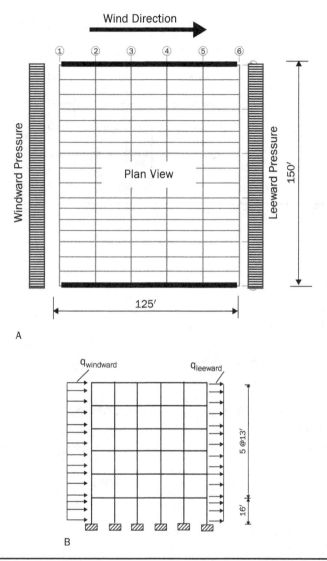

Figure 2.18 Windward and leeward pressures on Frame A: (*a*) plan view, (*b*) elevation of the identical MFs on Lines A and F

Total wind load, q_w, including windward pressure + leeward pressure acting at each story level on each MF in E-W direction can be found by multiplying the total wind load, 25 psf, with the area at which it is acting (Fig. 2.19).

$$q_w = w_{\text{total}} \times \frac{150'}{2} = 25^{\text{psf}} \times \frac{150'}{2} = 1.88 \text{ kip/ft}$$

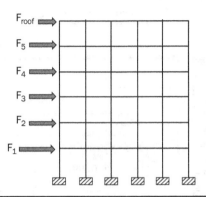

FIGURE 2.19   Wind load at each story level of Frame A

$$F_1 = q_w \times \left(\frac{16'}{2} + \frac{13'}{2} \right)$$

$$F_2 = F_3 = F_4 = F_5 = q_w \times \left(\frac{13'}{2} + \frac{13'}{2} \right)$$

$$F_{roof} = q_w \times \left(\frac{13'}{2} \right)$$

$$F_1 = 1.88^{kip/ft} \times \left(\frac{16'}{2} + \frac{13'}{2} \right) = 27.3 \text{ kips}$$

$$F_2 = F_3 = F_4 = F_5 = 1.88^{kip/ft} \times \left(\frac{13'}{2} + \frac{13'}{2} \right) = 24.4 \text{ kips}$$

$$F_{roof} = 1.88^{kip/ft} \times \left(\frac{13'}{2} \right) = 12.2 \text{ kips}$$

2.5   Problems

2.1   The typical floor plan of a four-story steel building is shown in Fig. 2.20 with elevations of perimeter frames in Fig. 2.21. All beam-to-column connections in other frames are shear connections. In International Building Code, IBC, the earthquake load and related design issues are based on *Minimum Design Loads and Associated Criteria for Buildings and Other Structures*, ASCE 7. In seismic force–resisting system, special moment frames are arranged in one direction and braced frames in the other orthogonal direction. One hundred psf for floors and roof, calculated based on the information for this building, including concrete decking (40 psf), ceiling, floor finishing, mechanical and electrical, exterior walls, and estimated steel member weight (40 psf), and partition (half of 20 psf per ASCE 7) will be used for the seismic weight (calculated from all dead loads and half of the partition). Assume that

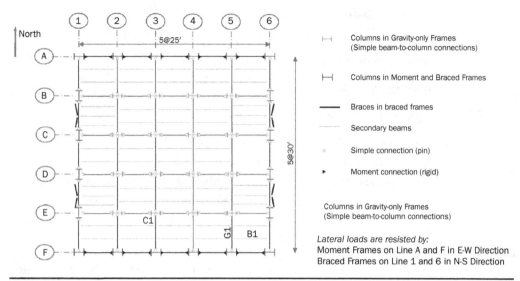

Figure 2.20 Floor plan

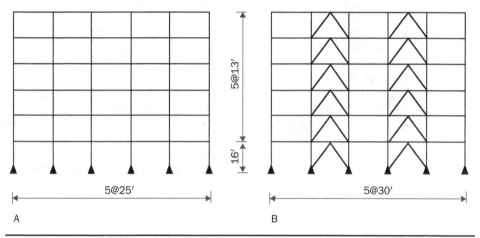

Figure 2.21 Elevation for the frame: (a) on Line A or Line F, (b) on Line 1 or Line 6

$$T = \begin{cases} 0.70 \text{ s for Special MRF} \\ 0.40 \text{ s for Special CBF} \end{cases}$$

$$C_s = \begin{cases} 0.10 \text{ for Special MRF} \\ 0.25 \text{ for Special CBF} \end{cases}$$

Determine the seismic lateral force distribution on the perimeter moment and braced frames.

2.2   Write a python code that computes

   a. Total base shear of a multi-story frame.

   b. Lateral force distribution at floor level of a building.

Related to Building Project in Chapter 1:

2.3   Please compute the following in the building project:

   a. Gravity loads acting on moment frames on Lines A and F.

   b. Gravity loads acting on a braced frame on Line 2.

   c. The design base shear and lateral distribution for the moment frames.

   d. The design base shear and lateral distribution for the braced frames.

Bibliography

AISC, *Seismic Design Manual,* American Institute of Steel Construction, Chicago, IL, 2018.

AISC 341, *Seismic Provisions for Structural Steel Buildings,* ANSI/AISC Standard 341–16, American Institute of Steel Construction, Chicago, IL, 2016.

AISC 358, *Prequalified Connections for Special and Intermediate Steel Moment Frames for Seismic Applications,* American Institute of Steel Construction, Chicago, IL, 2016.

AISC 360, *Specification for Structural Steel Buildings,* ANSI/AISC Standard 360–16, American Institute of Steel Construction, Chicago, IL, 2016.

AISC Manual, *Steel Construction Manual,* 15th ed., American Institute of Steel, Chicago, IL, 2016.

ASCE 7, *Minimum Design Loads and Associated Criteria for Buildings and Other Structures,* ASCE/SEI 7–16, American Society of Civil Engineers, Reston, VA, 2016.

Charleson, A., *Seismic Design for Earthquakes: Outwitting the Earthquake,* Architectural Press for Elsevier Ltd., UK, 2008.

Shen, J., B. Akbas, O. Seker, and C. Carter, *Structural Engineering Handbook—Chapter 8: Design of Structural Steel Members,* 5th ed., McGraw-Hill, New York, 2020.

CHAPTER 3

Tension Members

3.1 Introduction

Tension members are those structure components that primarily carry tension load. Tension members can be found in many structures, such as truss members in buildings, bracings, uplifting columns, cables, eyebars, etc. (Fig. 3.1). Typical cross-sections for tension members are given in Fig. 3.2.

Tension members can be divided into the following categories (Fig. 3.3):

- Rolled shapes and built-up sections
- Pin-connected members and eye bars
- Tie bars and rods
- Cables

There are three possible failure modes (or strength limit states) for a tension member subjected to a tension force. The member behaves elastically until reaching one of the limit states (or called failure modes) when subject to a tension force. The possible limit states are:

(a) *Limit state 1*: yielding of the gross cross-section of the member away from the connection (tensile yielding in the gross section).

(b) *Limit state 2:* fracture of the effective net area through the holes (tensile rupture in the net section).

(c) *Limit state 3*: block shear fracture through the bolts at the connection and in some cases,

(d) failure of connection elements.

Design of a tension member is to identify the controlling failure mode (the limit state with the smallest capacity) and to select the section or connection to ensure safety.

3.2 General Strength Requirement for Tension Members

General strength requirement for tension members is stated as follows:

$$P_u \leq \phi_t P_n$$

(3.1)

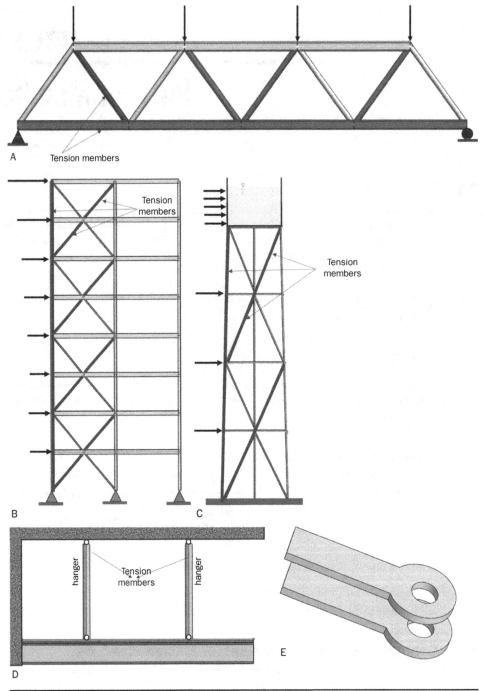

FIGURE 3.1 Tension members: (a) plane truss, (b) concentrically braced frame, (c) elevated tank, (d) hangers in buildings, (e) eye bars, (f) cables in cable-stayed bridges.

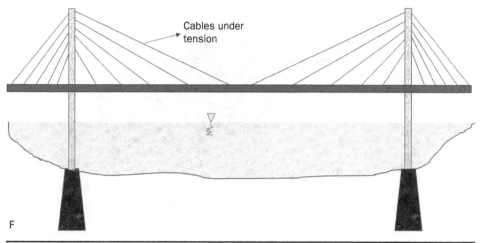

Cables under
tension

F

FIGURE 3.1 *(Continued)*

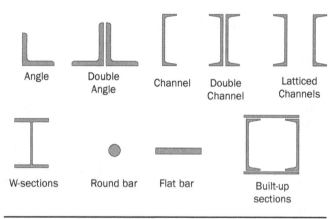

| Angle | Double Angle | Channel | Double Channel | Latticed Channels |

| W-sections | Round bar | Flat bar | Built-up sections |

FIGURE 3.2 Typical cross-sections for tension members.

where P_n = nominal tensile strength (kips)

P_u = required strength using LRFD (Load and Resistance Factor Design) load combinations (kips)

ϕ_t = resistance factor for tension members

$\phi_t P_n$ = design tensile strength (kips)

3.3 Tension Member Strength

3.3.1 Nominal Tension Strength from Yielding of Gross Area, P_n

This limit state defines the yielding of the gross section of a tension member without holes (i.e., with welded connection), and excessive deformation of the member is prevented (Fig. 3.3).

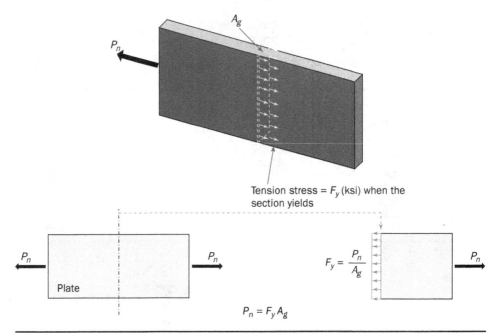

FIGURE 3.3   Yielding of the gross section.

According to AISC360, Chapter D, Eq. (D2-1), nominal tensile strength, P_n, for limit state 1: yielding in the gross section is

$$P_n = F_y A_g \tag{3.2}$$

where F_y = the specified minimum yield stress of the steel type being used, ksi
A_g = gross area of the cross-section, in.²
$\phi_t = 0.9$

3.3.2   Nominal Strength from the Fracture of Effective Net Area

Net area, A_n, is defined as the reduced cross-section of a member having holes (i.e., bolt holes) (Fig. 3.4).

Nominal tensile strength, P_n, for *limit state 2: fracture of effective net area* is

$$P_n = F_u A_e \; \text{AISC}\,360 \tag{3.3}$$

where F_u = the specified minimum tensile strength of the steel type being used, ksi
A_e = Effective net area, in.²
$\phi_t = 0.75$

F_u is considered at the fracture stage over the entire net area. The fracturing process typically starts from the edges of holes and progresses to a complete separation.

i. Definition of reduction in gross area
$d_h = d_b + 1/16''$ is the standard hole size, whereas an additional $1/16''$ in w is to account for possible damage when the hole is punched (Fig. 3.5). In other words, the difference in w and d_h is to account for damage due to the fabrication process.

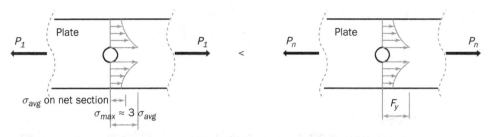

FIGURE 3.4 Stress distribution in a cross-section with holes present. (*a*) Elastic stresses (under service loads). (*b*) Nominal strength condition.

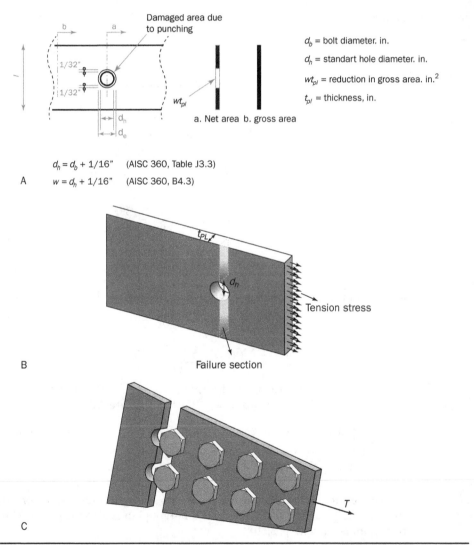

d_b = bolt diameter. in.

d_h = standart hole diameter. in.

wt_{pl} = reduction in gross area. in.^{2}

t_{pl} = thickness, in.

a. Net area b. gross area

$d_h = d_b + 1/16''$ (AISC 360, Table J3.3)

A $w = d_h + 1/16''$ (AISC 360, B4.3)

B

C

FIGURE 3.5 Reduction in gross area: (*a*) net area, (*b*) failure section, (*c*) failure through net section.

ii. Net area without staggered holes (A_n) (Fig. 3.6a)

$$A_n = A_g - \sum wt_{pl} \qquad (3.4)$$

where

$$w = d_b + \frac{1''}{8}$$

Example 3.1

The tension member in Fig. 3.7 consists of a PL $4 \times 1/2$ bolted to a $7/8''$ thick gusset plate by four (4) 5/8-in.-diameter bolts in standard holes (PL: plate). Determine the net area. (Note: the holes are made by punching.)

Solution:

$$A_n = A_g - \sum wt_{pl}$$

$$A_g = 4'' \times 1/2'' = 2.0 \text{ in.}^2$$

$$d_h = d_b + \frac{1''}{16} = \frac{5''}{8} + \frac{1''}{16} = \frac{11''}{16}$$

$$w = d_h + \frac{1''}{16} = \frac{11''}{16} + \frac{1''}{16} = \frac{3''}{4} \text{ or}$$

$$d_b + \frac{1''}{8} = \frac{5''}{8} + \frac{1''}{8} = \frac{3''}{4}$$

$$A_n = 2.0^{\text{in.}^2} - 2 \times \left(\frac{3}{4}\right)'' \left(\frac{1}{2}\right)'' = 1.25 \text{ in.}^2$$

iii. Net area with staggered holes (A_n) (might be used when bolt spacing is limited)

$$A_n = \text{smaller of} \begin{cases} A_g - \sum(wt_{pl}) & \text{(along A–B)} \\ A_g - \sum(wt_{pl}) + \sum\left(\frac{s^2 t_{pl}}{4g}\right) & \text{(along A– C)} \end{cases} \qquad (3.5)$$

where s = stagger or spacing of adjacent holes parallel to the loading direction, in.
g = gage distance transverse to the loading direction, in.

$\sum\left(\dfrac{s^2 t_{pl}}{4g}\right)$: accounts for the increase in strength due to the cross-sectional area at an angle in the failure plane

The controlling failure line is the one that gives the largest stress on an effective net area. In many cases, the critical failure path is also the path that has the minimum net area (Fig. 3.6b).

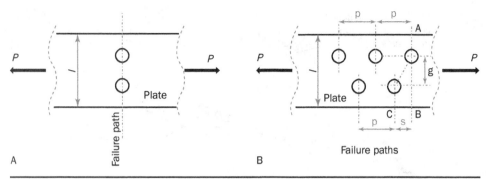

Figure 3.6 Failure path in net area: (a) without staggered holes, (b) with staggered holes.

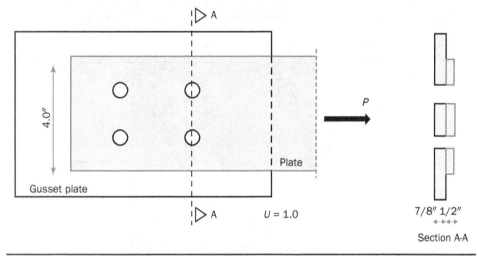

Figure 3.7 Determining net area for a tension member.

According to ANSI/AISC 360-16, when the load is transmitted to the tension member by fasteners or welds through some but not all the elements of the cross-section (e.g., only one of the legs of an angle, the flanges of a W shape but not the web, etc.) an effective, area, A_e, must be used in the design which is defined as follows:

$$A_e = A_n U \tag{3.6}$$

$$U = 1 - \frac{\overline{x}}{l} \leq 0.9 \tag{3.7}$$

U is a reduction coefficient for the concentration of shear stress (shear lag factor) (>0.60, the connection should be configured to satisfy this criterion) (Fig. 3.8),

where $\overline{x}$ = distance from the centroid of the connected part to the connection plane, in.
l = connection length (length of connection in the direction of loading between the first and last connecting points), in.

U factors for commonly used tension member with connections are given in Fig. 3.9 (AISC 360).

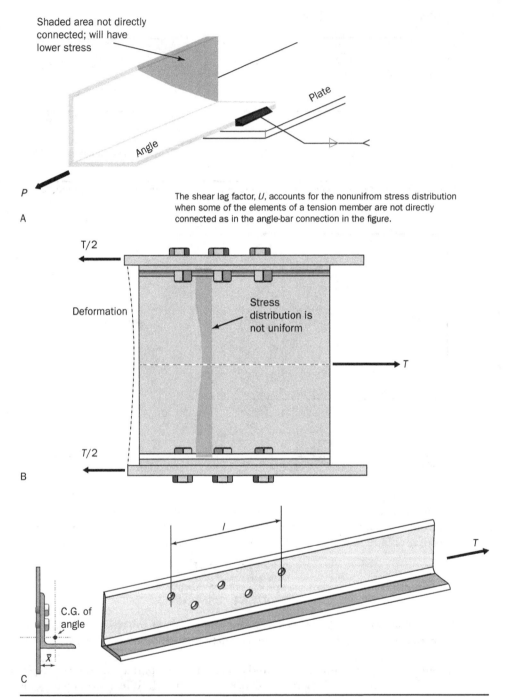

Shaded area not directly connected; will have lower stress

Plate

Angle

P

A

The shear lag factor, U, accounts for the nonunifrom stress distribution when some of the elements of a tension member are not directly connected as in the angle-bar connection in the figure.

$T/2$

Deformation

Stress distribution is not uniform

T

$T/2$

B

l

T

C.G. of angle

$\bar{x}$

C

Figure 3.8 Nonuniform stress distribution: (a) angle-bar welded connection, (b) I-shaped member with fasteners in the flanges, (c) angle-plate connection with multiple line of fasteners.

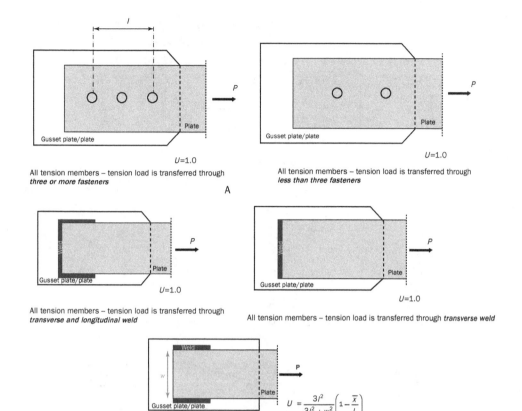

FIGURE 3.9   *U* factors: (*a*) transfer of tension loads through fasteners, (*b*) transfer of loads through transverse and longitudinal welds, (*c*) transfer of tension force in a gusset plate—HSS connections throguh welds, (*d*) W, M, S, or HP or Tees cut from these shapes-flange connected with three of more fasteners per line in the direction of loading, (*e*) W, M, S, or HP or Tees cut from these shapes-web connected with three of more fasteners pre line in the direction of loading, (*f*) single angle with four or more fasteners per line in the directions of loading, (*g*) single angle with three fasteners per line in the direcion of loading.

3.3.3   Block Shear Failure

Failure might occur by shear along a plane through the fasteners plus tension along a perpendicular plane on the area effective in resisting tearing failure. This type of failure is called "block shear failure." Nominal tensile strength, P_n, for *limit state 3: block shear rupture* is

$$P_n = (0.6F_u A_{nv} + U_{bs}F_u A_{nt}) \leq (0.6F_y A_{gv} + U_{bs}F_u A_{nt}) \tag{3.8}$$

Eq. (3.8) means that

$$P_n = \text{smaller of } (0.6F_u A_{nv} + U_{bs}F_u A_{nt}) \text{ and } (0.6F_y A_{gv} + U_{bs}F_u A_{nt}) \tag{3.9}$$

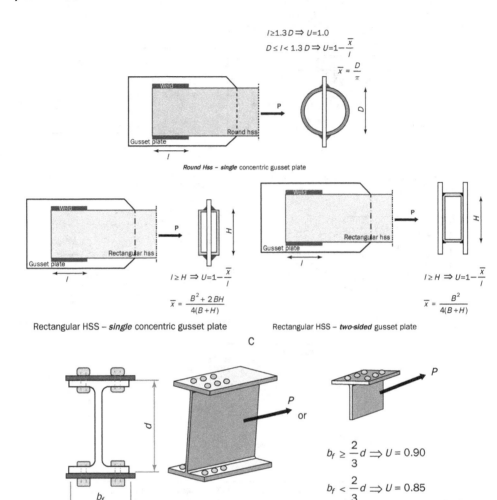

$$l \geq 1.3D \Rightarrow U=1.0$$
$$D \leq l < 1.3D \Rightarrow U=1-\frac{\overline{x}}{l}$$
$$\overline{x} = \frac{D}{\pi}$$

Round Hss – single concentric gusset plate

$$l \geq H \Rightarrow U=1-\frac{\overline{x}}{l}$$
$$\overline{x} = \frac{B^2 + 2BH}{4(B+H)}$$

Rectangular HSS – *single* concentric gusset plate

$$l \geq H \Rightarrow U=1-\frac{\overline{x}}{l}$$
$$\overline{x} = \frac{B^2}{4(B+H)}$$

Rectangular HSS – *two-sided* gusset plate

C

or

$$b_f \geq \frac{2}{3}d \Rightarrow U = 0.90$$
$$b_f < \frac{2}{3}d \Rightarrow U = 0.85$$

W, M, S or HP, or Tees cut from these shapes- *flange connected with three or more fasteners* per line in the direction of loading

D

FIGURE 3.9 *(Continued)*

where A_{nt} = net area subject to tension, in.²
A_{nv} = net area subject to shear, in.²
A_{gv} = gross area subject to shear, in.²
U_{bs} = 1.0, if tension stress is uniform
0.5, if tension stress is nonuniform

U_{bs}, in most practical cases, will be equal to 1.0 (Fig. 3.10). Therefore, Eq. (3.8) can be restated as

$$P_n = (0.6F_u A_{nv} ; 0.6F_y A_{gv})_{\min} + F_u A_{nt} \tag{3.10}$$

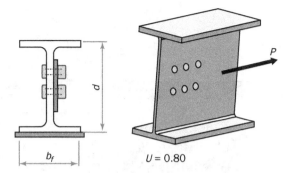

$U = 0.80$

W, M, S or HP, or Tees cut from these shapes- **web connected with three or more fasteners** per line in the direction of loading

E

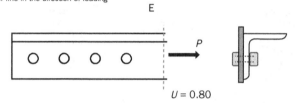

$U = 0.80$

Single **angle** with **four or more fasteners** per line in the direction of the loading

F

$U = 0.60$

Single **angle** with **three fasteners** per line in the direction of the loading

G

FIGURE 3.9   (Continued)

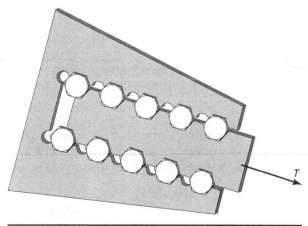

FIGURE 3.10   Block shear failure of a gusset plate.

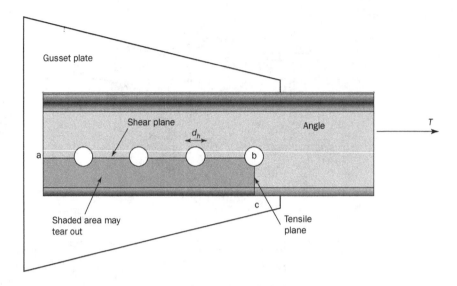

$A_{gv} = L_{ab} \times t_{angle}$

$A_{nt} = (L_{bc} - 0.5d_h) \times t_{angle}$

$A_{nv} = (L_{ab} - 3.5d_h) \times t_{angle}$

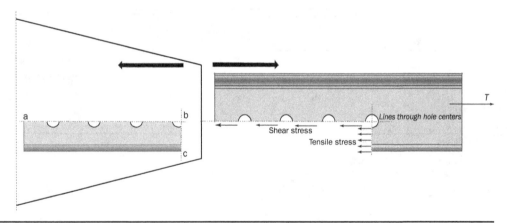

Figure 3.11   Shear and tensile stresses in an angle-gusset plate connection.

Note that block shear strength is determined by the summation of the shear and tension terms. Block shear is a rupture or tearing limit state, not a yielding limit state. The two parts are separated after the block shear failure occurs (Figs. 3.11 and 3.12).

$$A_{gv} = L_{ab} \times t_{angle}$$
$$A_{nt} = (L_{bc} - 0.5d_h) \times t_{angle}$$
$$A_{nv} = (L_{ab} - 3.5d_h) \times t_{angle}$$

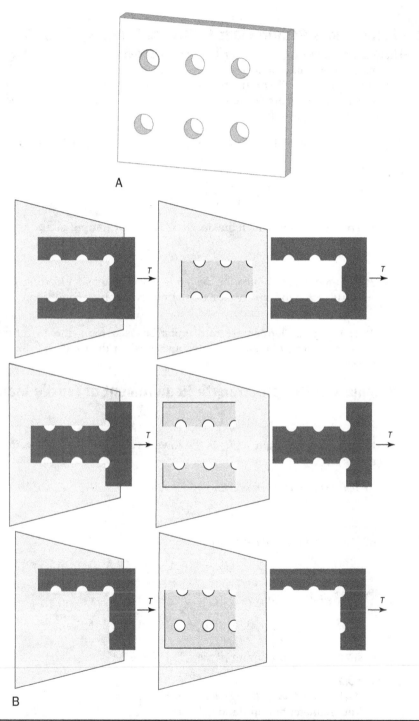

FIGURE 3.12 Block shear failure: (*a*) steel plate, (*b*) potential block shear failure modes of a tension member.

3.4   Slenderness Requirement (Stiffness Criteria)

There is no slenderness limit for the design of tension members. However, too long tension members may deflect excessively due to their own weight and vibrate when subject to wind forces. To reduce the excessive deflection or sagging and vibration problems in tension members as well as problems during construction, the following stiffness criteria is suggested:

$$\frac{L}{r} \le 300 \text{ except for steel rods and hangers in tension} (\text{AISC} 360 - 16, \text{D} - 1) \tag{3.11}$$

where $L = $ length, in.

$r = $ radius of gyration $\left(= \sqrt{\dfrac{I}{A_g}} \text{ in.} \right)$

1) For members in which the design is based on compression:

$$\frac{kL}{r} \le 200$$

2) For members in which the design is based on tension loading, but which may be subjected to some compression under other load conditions, need not satisfy the compression slenderness limit.

3) The highest slenderness ratio must be used for symmetrical members. For nonsymmetrical members, one must consider the weakest principal axis.

3.5   Summary of Design Strength Requirement of Tensile Members

$$P_u \le P_c = \phi_t P_n \tag{3.12}$$

The design tensile strength, $\phi_t P_n$, is the lowest value obtained according to the three limit states:

a) *Tensile yielding in gross area:*

$$\phi_t P_n = \phi_t F_y A_g = 0.90 F_y A_g \tag{3.13}$$

b) *Tensile rupture in the net area:*

$$\phi_t P_n = \phi_t F_u A_e = 0.75 F_u A_e \tag{3.14}$$

c) *Block shear strength:*

$$\phi P_n = 0.75 \{ (0.6 F_u A_{nv}; 0.6 F_y A_{gv})_{\min} + F_u A_{nt} \} \tag{3.15}$$

Design strength, $P_c = $ minimum $\{P_{c1}, P_{c2}, P_{c3}\}$ (Table 3.1).

Example 3.2

Given: Steel Grade A36 $\rightarrow F_y = 36$ ksi, $F_u = 58$ ksi

$d_h = 7/8$-in.-diameter bolt in standard hole

Live Load (L)/Dead Load (D) $= 3.0$

Required: Maximum tension force the angle can take.

Limit states	Nominal tension strength	Design strength
Yielding of gross area	$P_n = F_y A_g$	$P_{c1} = \phi_t P_n = \phi_t F_y A_g$ $\phi_t = 0.90$
Fracture of effective net area	$P_n = F_u A_e$ $A_e = U A_n$	$P_{c2} = \phi_t P_n = \phi_t F_u A_e$ $\phi_t = 0.75$
Block shear rupture	$P_n = (0.6 F_u A_{nv}; 0.6 F_y A_{gv})_{min} + F_u A_{nt}$	$P_{c3} = \phi_t P_n$ $\phi_t = 0.75$

TABLE 3.1 Limit States and Corresponding Strengths

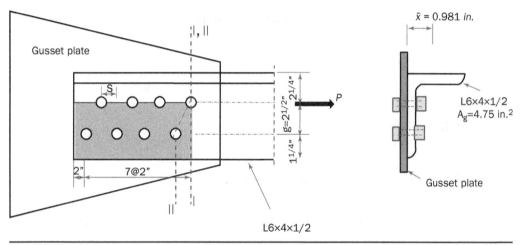

FIGURE 3.13 Gusset plate-angle connection under tension loading.

Solution:

Given: $s = 2''$, $g = 2\frac{1}{2}''$, $l = 14''$, $L = 16''$ ($= 7 \times 2'' + 2''$), $A_g = 4.75$ in.^{2}

Limit state 1: Yielding of gross section

$$P_n = F_y A_g = 36^{ksi} \times 4.75^{in.^2} = 171.0 \, kips$$

Limit state 2: Fracture of effective net area

Assume standard punched holes, so the width of the bolts will be:

$$w = \frac{7}{8} + \frac{1}{8} = 1 \text{ in.}$$

Two possible failure paths, I-I and II-II, need to be checked.

$$A_n = \text{smaller of} \begin{cases} A_g - \sum(w t_{pl}) & \text{(along I-I)} \\ A_g - \sum(w t_{pl}) + \sum \left(\dfrac{s^2 t_{pl}}{4g} \right) & \text{(along II-II)} \end{cases}$$

$$A_n = \begin{cases} 4.75^{in.^2} - 1 \times (1)''\left(\dfrac{1}{2}\right)'' = 4.25 \text{ in.}^2 \quad \text{(along I-I)} \\[2em] 4.75^{in.^2} - 2 \times (1)''\left(\dfrac{1}{2}\right)'' + 1 \times \left[\dfrac{(2'')^2 \times \left(\dfrac{1}{2}\right)''}{4 \times \left(2\dfrac{1}{2}\right)''}\right] = 3.95 \text{ in.}^2 \quad \text{(along II-II)} \end{cases}$$

$$A_n = \left\{4.25 \text{ in.}^2 ; 3.95 \text{ in.}^2\right\}_{\min} = 3.95 \text{ in.}^2$$

$$U = 1 - \frac{\bar{x}}{l} = 1 - \frac{0.981^{in.}}{14^{in.}} = 0.93 > 0.9 \Rightarrow \text{use } U = 0.9$$

$$A_e = UA_n = 0.9 \times 3.95^{in.^2} = 3.56 \text{ in.}^2$$

$$P_n = F_u A_e = 58^{ksi} \times 3.56^{in.^2} = 206.5 \text{ kips}$$

Limit state 3: Block shear rupture
The block shear area (in gray in Fig. 3.13) needs to be checked.

$$A_{gt} = \left(2\frac{1}{2}'' + 1\frac{1}{4}''\right) \times \left(\frac{1}{2}\right)'' = 1.875 \text{ in.}^2$$

$$A_{nt} = 1.875^{in.^2} - \frac{1}{2} \times (1)''\left(\frac{1}{2}\right)'' = 1.625 \text{ in.}^2$$

$$A_{gv} = 16'' \times \left(\frac{1}{2}\right)'' = 8.0 \text{ in.}^2$$

$$A_{nv} = 8.0^{in.^2} - \left(3 + \frac{1}{2}\right) \times (1)''\left(\frac{1}{2}\right)'' = 6.25 \text{ in.}^2$$

$$\begin{aligned} P_n &= (0.6F_u A_{nv} ; 0.6F_y A_{gv})_{\min} + F_u A_{nt} \\ &= \left\{\left[0.6(58^{ksi})(6.25^{in.^2})\right], \left[0.6(36^{ksi})(8.0^{in.^2})\right]\right\}_{\min} + (58^{ksi})(1.625^{in.^2}) \\ &= \left\{217.5^{kips}, 172.8^{kips}\right\}_{\min} + 94.3^{kips} \\ &= 267.1 \text{ kips} \end{aligned}$$

Thus, the design strengths for each limit states (Table 3.2) are

$$P_{c1} = \phi_t P_n = (0.9)(171^{kips}) = 153.9 \text{ kips}$$

$$P_{c2} = \phi_t P_n = (0.75)(206.5^{kips}) = 154.9 \text{ kips}$$

$$P_{c3} = \phi_t P_n = (0.75)(267.1^{kips}) = 200.3 \text{ kips}$$

Limit state	ϕ_t	P_n (kips)	$P_c = \phi_t P_n$ (kips)
Yielding of gross area	0.90	171	**153.9**
Fracture of effective net area	0.75	206.5	154.9
Block shear failure	0.75	267.1	200.3
Governing case	*Yielding of gross area*		

TABLE 3.2 Design Strengths for Each Limit State

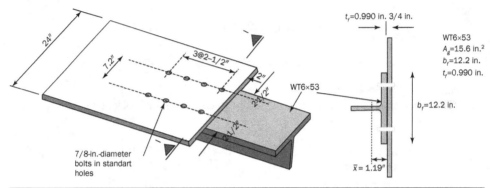

FIGURE 3.14 Tension member bolted to a plate.

Therefore, the design strength is

$$P_c = \{P_{c1}, P_{c2}, P_{c3}\}_{min} = 153.9 \text{ kips}$$

Maximum tension force that the tension member/connection can take:

$$P_u = 1.2P_D + 1.6P_L = 1.2(P) + 1.6(3P) = 6P$$

The general design equation:

$$P_u \leq P_c$$

$$6P \leq 153.9 \text{ kips}$$
$$P \leq 25.7 \text{ kips}$$

(the maximum load that the tension member/connection take)

Example 3.3

As shown in Fig. 3.14, the tension member consists of WT6 × 53 bolted to a PL24×¾ gusset plate at a joint. Eight (8) 7/8-in.-diameter bolts in standard holes are used in two lines. A572 Gr. 50 ($F_y = 50$ ksi, $F_u = 65$ ksi) steel is used for WT. Calculate the design strength ϕP_n from possible limit states in WT.

Limit state 1: Yielding of gross section

$$P_n = F_y A_g = 50^{ksi} \times 15.6^{in.^2} = 780 \text{ kips}$$

Limit state 2: Fracture of effective net area
Assume standard punched holes, so the width of the bolts will be:

$$w = \frac{7}{8} + \frac{1}{8} = 1 \text{ in.}$$

$$A_n = A_g - \sum wt_{pl} = 15.6^{\text{in.}^2} - 2 \times (0.990)'' (1)'' = 13.62 \text{ in.}^2$$

$$U = 1 - \frac{\overline{x}}{l} = 1 - \frac{1.19^{\text{in.}}}{7.5^{\text{in.}}} = 0.84 \leq 0.9 \Rightarrow \text{use } U = 0.84$$

$$A_e = UA_n = 0.84 \times 13.62^{\text{in.}^2} = 11.44 \text{ in.}^2$$

$$P_n = F_u A_e = 65^{\text{ksi}} \times 11.44^{\text{in.}^2} = 743.6 \text{ kips}$$

Limit state 3: Block shear rupture
The block shear area needs to be checked to compute the block shear strength.

$$A_{gt} = \left(2\tfrac{1}{2}\right)'' \times (0.990)'' = 2.475 \text{ in.}^2$$

$$A_{nt} = 2 \times \left[2.475^{\text{in.}^2} - \left(\frac{1}{2}\right) \times (1)''(0.990)''\right] = 3.96 \text{ in.}^2$$

$$A_{gv} = 2 \times \left[(3 \times 2\tfrac{1}{2}'' + 2'') \times (0.990)''\right] = 18.81 \text{ in.}^2$$

$$A_{nv} = 18.81^{\text{in.}^2} - 2 \times \left[3.5 \times (1)''(0.99)''\right] = 11.88 \text{ in.}^2$$

$$P_n = (0.6F_u A_{nv}; 0.6F_y A_{gv})_{\min} + F_u A_{nt}$$

$$= \left\{\left[0.6(65^{\text{ksi}})(11.88^{\text{in.}^2})\right], \left[0.6(50^{\text{ksi}})(18.81^{\text{in.}^2})\right]\right\}_{\min} + (65^{\text{ksi}})(3.96^{\text{in.}^2})$$

$$= \left\{463.3^{\text{kips}}, 564.3^{\text{kips}}\right\}_{\min} + 257.4^{\text{kips}}$$

$$= 720.7 \text{ kips}$$

Thus, the design strengths for each limit states (Table 3.3) are

$$P_{c1} = \phi_t P_n = (0.9)(780^{\text{kips}}) = 702.0 \text{ kips}$$

$$P_{c2} = \phi_t P_n = (0.75)(743.6^{\text{kips}}) = 557.7 \text{ kips}$$

$$P_{c3} = \phi_t P_n = (0.75)(720.7^{\text{kips}}) = 540.5 \text{ kips}$$

Limit state	ϕ_t	P_n (kips)	$P_c = \phi_t P_n$ (kips)
Yielding of gross area	0.90	780.0	702.0
Fracture of effective net area	0.75	743.6	557.7
Block shear failure	0.75	720.7	**540.5**
Governing case	*Block shear failure*		

TABLE 3.3   Design Strengths for Each Limit State

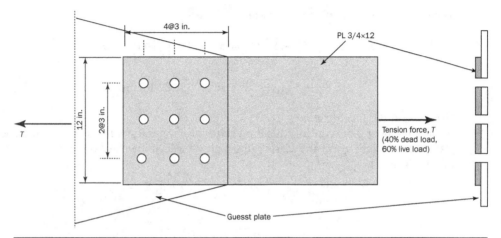

FIGURE 3.15   A plate (PL) connected to a gusset plate with fasteners.

Therefore, the design strength is

$$P_c = \{P_{c1}, P_{c2}, P_{c3}\}_{\min} = 540.5 \text{ kips}$$

Example 3.4

A PL $\frac{3}{4}$"× 12" tension member is connected to a 1-in.-thick gusset plate with nine 7/8-in.-diameter bolts, as shown in Fig. 3.15. The steel is A36 for both plates. Assume $U = 1.0$, i.e., $A_e = A_n$. Compute the following:

  a. The design strength
  b. The maximum tension force that the tension member/connection can take

Solution:

a. *Design strength*

Limit state 1: Yielding of gross section
Tension member plate controls this limit state since it has a smaller area than the gusset plate.

$$A_g = 12''\left(\frac{3}{4}\right)'' = 9 \text{in.}^2$$

$$P_n = F_y A_g = 36 \times 9 = 324 \text{ kips}$$

Limit state 2: Fracture of effective net area
Tension member plate controls this limit state since it has a smaller area than the gusset plate. Assume standard punched holes, so the width of the bolts will be

$$w = \frac{7}{8} + \frac{1}{8} = 1 \text{ in.}$$

Since $A_e = A_n$, net and effective net areas will be equal.

$$A_e = A_n = A_g - \sum wt_{pl} = 9^{\text{in.}^2} - 3 \times \left(\frac{3}{4}\right)'' (1)'' = 6.75 \text{ in.}^2$$

$$P_n = F_u A_e = 58^{\text{ksi}} \times 6.75^{\text{in.}^2} = 391.5 \text{ kips}$$

Limit state 3: Block shear rupture
Tension member plate controls this limit state since it has a smaller area than the gusset plate. Case 1 (light gray) and Case 2 (dark gray) need to be checked to get the smaller value (Fig. 3.16).

$$A_{gt} = 6'' \times \left(\frac{3}{4}\right)'' = 4.5 \text{ in.}^2$$

$$A_{nt} = 4.5^{\text{in.}^2} - 2 \times (1)'' \left(\frac{3}{4}\right)'' = 3.0 \text{ in.}^2$$

$$A_{gv} = 2 \times 9'' \times \left(\frac{3}{4}\right)'' = 13.5 \text{ in.}^2$$

$$A_{nv} = 13.5^{\text{in.}^2} - 5 \times (1)'' \left(\frac{3}{4}\right)'' = 9.75 \text{ in.}^2$$

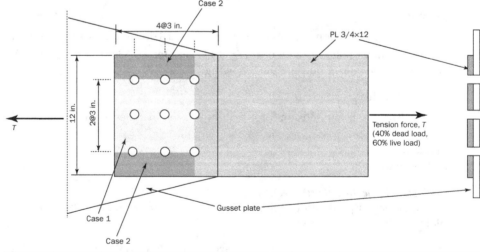

FIGURE 3.16 Potential block shear rupture areas.

Limit state	ϕ_t	P_n (kips)	$P_c = \phi_t P_n$ (kips)
Yielding of gross area	0.90	324	**291.6**
Fracture of effective net area	0.75	391.5	293.6
Block shear failure	0.75	465.6	349.2
Governing case	*Yielding of gross area*		

TABLE 3.4 Design Strengths for Each Limit State

$$P_n = (0.6F_u A_{nv}; 0.6F_y A_{gv})_{min} + F_u A_{nt}$$
$$= \left\{ \left[0.6(58^{ksi})(9.75^{in.^2}) \right], \left[0.6(36^{ksi})(13.5^{in.^2}) \right] \right\}_{min} + (58^{ksi})(3.0^{in.^2})$$
$$= \left\{ 339.3^{kips}, 291.6^{kips} \right\}_{min} + 174^{kips}$$
$$= 465.6 \text{ kips}$$

Thus, the design strengths for each limit states (Table 3.4) are

$$P_{c1} = \phi_t P_n = (0.9)(324^{kips}) = 291.6 \text{ kips}$$

$$P_{c2} = \phi_t P_n = (0.75)(391.5^{kips}) = 293.6 \text{ kips}$$

$$P_{c3} = \phi_t P_n = (0.75)(465.6^{kips}) = 349.2 \text{ kips}$$

Therefore, the design strength is

$$P_d = \left\{ P_{d1}, P_{d2}, P_{d3} \right\}_{min} = 291.6 \text{ kips}$$

b. *Maximum tension force that the tension member/connection can take*

$$P_u = 1.2P_D + 1.6P_L = 1.2(40\% \times T) + 1.6(60\% \times T) = 1.44T$$

The general design equation:

$$P_u \le P_c$$

$$1.44 \, T \le 291.6 \text{ kips}$$
$$T \le 202.5 \text{ kips}$$

(the maximum load that the tension member/connection can take)

Example 3.5   *Tension member in a simple truss*

A simple steel truss, as shown in Fig. 3.17, is loaded with 170 kips service load at joint B. Member AB consists of two plates bolted to a $1'' \times 12''$ gusset plate at joint B. Six (6) ½-in.-diameter bolts in standard holes are used in two lines at Joint B (Fig. 3.18). A36 steel ($F_y = 36$ ksi, $F_u = 58$ ksi) is used for all plates. Check if the tension member is safe. Make sure that all possible limit states in member AB and gusset plate are considered.

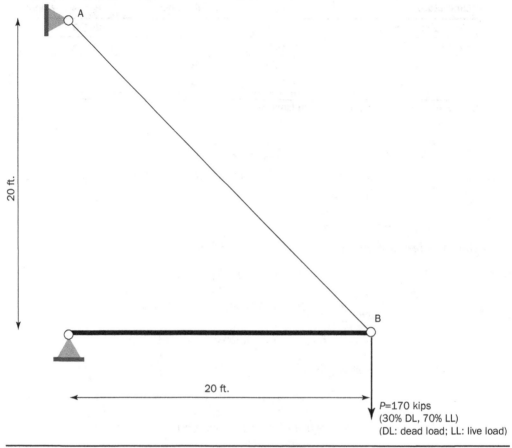

20 ft.

20 ft.

B

A

P=170 kips
(30% DL, 70% LL)
(DL: dead load; LL: live load)

FIGURE 3.17 Design of a tension member in a truss system.

Solution:

Given:

A36 steel, $F_y = 36$ ksi, $F_u = 58$ ksi.

Bolts: $d = \frac{1}{2}''$, $w = 1/2'' + 1/8'' = 5/8''$

Design strength of the member and connection:

a. Check gusset plate's design tension strength

Limit state 1: Yielding of gross section

$$A_g = 12''(1)'' = 12 \text{ in.}^2$$

$$P_n = F_y A_g = 36^{ksi} \times 12^{in.^2} = 432 \text{ kips}$$

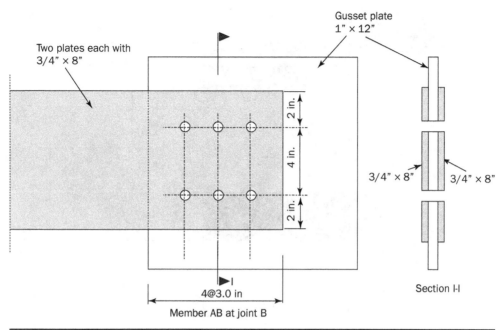

Two plates each with
3/4" × 8"

Gusset plate
1" × 12"

2 in.

4 in.

2 in.

3/4" × 8"

3/4" × 8"

4@3.0 in

Member AB at joint B

Section I-I

FIGURE 3.18   Connection detail at joint B.

Limit state 2: Fracture of effective net area

$$A_e = A_g - A_{\text{holes}} = 12^{\text{in.}^2} - 2 \times \left(\frac{5}{8}\right)'' (1)'' = 10.75 \text{ in.}^2$$

$$A_e = U A_n = 1.0 \times 10.75^{\text{in.}^2} = 10.75 \text{ in.}^2$$

$$P_n = F_u A_e = 58^{\text{ksi}} \times 10.75^{\text{in.}^2} = 623.5 \text{ kips}$$

Limit state 3: Block shear rupture
We need to discuss Case 1 and Case 2; however, the light gray area is smaller than the dark gray area, i.e., Case 1 controls. Tension member plate controls the block shear rupture since it has a smaller area than the gusset plate.

Case 1 (light gray) and Case 2 (dark gray) need to be checked to get the smaller value (Fig. 3.19).

$$A_{gt} = 4'' \times (1.0)'' = 4.0 \text{ in.}^2$$

$$A_{nt} = 4.0^{\text{in.}^2} - 1 \times \left(\frac{5}{8}\right)'' (1)'' = 3.375 \text{ in.}^2$$

$$A_{gv} = 2 \times 9'' \times (1)'' = 18.0 \text{ in.}^2$$

$$A_{nv} = 18.0^{\text{in.}^2} - 5 \times \left(\frac{5}{8}\right)'' \times (1)'' = 14.875 \text{ in.}^2$$

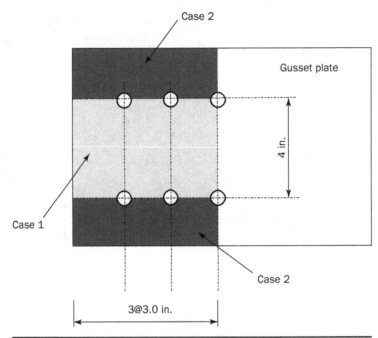

Case 2

Gusset plate

4 in.

Case 1

Case 2

3@3.0 in.

Figure 3.19 Block shear rupture in the gusset plate.

$$P_n = (0.6F_u A_{nv}; 0.6F_y A_{gv})_{min} + F_u A_{nt}$$
$$= \left\{ \left[0.6(58^{ksi})(14.875^{in.^2}) \right], \left[0.6(36^{ksi})(18.0^{in.^2}) \right] \right\}_{min} + (58^{ksi})(3.375^{in.^2})$$
$$= \left\{ 517.7^{kips}, 388.8^{kips} \right\}_{min} + 195.8^{kips}$$
$$= 584.6 \text{ kips}$$

Thus, the design strengths for each limit states are

$$P_{c1} = \phi_t P_n = (0.9)(432^{kips}) = 388.8 \text{ kips}$$

$$P_{c2} = \phi_t P_n = (0.75)(623.5^{kips}) = 467.6 \text{ kips}$$

$$P_{c3} = \phi_t P_n = (0.75)(584.6^{kips}) = 438.4 \text{ kips}$$

Therefore, the design strength is

$$P_c = \left\{ P_{d1}, P_{d2}, P_{d3} \right\}_{min} = 388.8 \text{ kips}$$

Check the two plate's design tension strength (2-3/4″×8″):
   b. Check the two plates' design tension strength

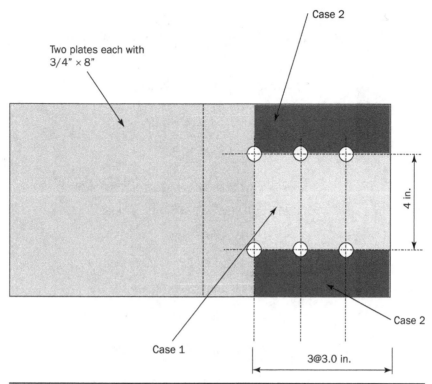

FIGURE 3.20   Block shear rupture in the two plates.

Limit state 1: Yielding *of gross section*

$$A_g = 8'' \times \left(\frac{3}{4}\right)'' \times 2 = 12 \text{ in.}^2$$

$$P_n = F_y A_g = 36^{ksi} \times 12^{in.^2} = 432 \text{ kips}$$

Limit state 2: Fracture *of effective net area*

$$A_e = A_g - A_{\text{holes}} = 12^{in.^2} - 4 \times \left(\frac{5}{8}\right)'' \times \left(\frac{3}{4}\right)'' = 10.125 \text{ in.}^2$$

$$A_e = U A_n = 1.0 \times 10.125^{in.^2} = 10.125 \text{ in.}^2$$

$$P_n = F_u A_e = 58^{ksi} \times 10.125^{in.^2} = 587.3 \text{ kips}$$

Limit state 3: Block shear rupture
  Case 1 (light gray) (Fig. 3.20):

$$A_{gt} = 4'' \times \left(\frac{3}{4}\right)'' \times 2 = 6.0 \text{ in.}^2$$

$$A_{nt} = 6.0^{in.^2} - 1 \times \left(\frac{5}{8}\right)'' \times \left(\frac{3}{4}\right)'' \times 2 = 5.0625 \text{ in.}^2$$

$$A_{gv} = 2 \times 9'' \times \left(\frac{3}{4}\right)'' \times 2 = 27.0 \text{ in.}^2$$

$$A_{nv} = 27.0^{in.^2} - 5 \times \left(\frac{5}{8}\right)'' \times \left(\frac{3}{4}\right)'' \times 2 = 22.3125 \text{ in.}^2$$

$$P_n = \{(0.6F_u A_{nv} + F_u A_{nt}), (0.6F_y A_{gv} + F_u A_{nt})\}_{min} = \{(0.6F_u A_{nv}), (0.6F_y A_{gv})\}_{min} + F_u A_{nt}$$

$$= \{[0.6(58^{ksi})(22.3125^{in.^2})], [0.6(36^{ksi})(27.0^{in.^2})]\}_{min} + (58^{ksi})(5.0625^{in.^2})$$

$$= \{776.5^{kips}, 583.2^{kips}\}_{min} + 293.6^{kips}$$

$$= 876.8 \text{ kips}$$

Thus, the design strengths for each limit states (Table 3.5) are

$$P_{c1} = \phi_t P_n = (0.9)(432^{kips}) = 388.8 \text{ kips}$$

$$P_{c2} = \phi_t P_n = (0.75)(532.2^{kips}) = 399.2 \text{ kips}$$

$$P_{c3} = \phi_t P_n = (0.75)(876.8^{kips}) = 657.6 \text{ kips}$$

Case 2 (dark gray) (Fig. 3.20):

A_{nt}, A_{gv}, A_{nv} are the same as those in case 1. Thus, P_n is the same as those in Case 1. Thus, the design strengths for each limit states are

$$P_{c1} = \phi_t P_n = (0.9)(432^{kips}) = 388.8 \text{ kips}$$

$$P_{c2} = \phi_t P_n = (0.75)(532.2^{kips}) = 399.2 \text{ kips}$$

$$P_{c3} = \phi_t P_n = (0.75)(876.8^{kips}) = 657.6 \text{ kips}$$

Therefore, the design strength is

$$P_c = \{P_{d1}, P_{d2}, P_{d3}\}_{min} = 388.8 \text{ kips}$$

Conclusion on the design strength: Checking all the limit states in member AB and gusset plate, we can conclude that the yielding of a gross area controls the design strength (with the minimum design strength).

Required design strength from structural analysis:

From the force analysis (Fig. 3.21), we can get

$$T_r = \frac{P}{\sin 45°} = \frac{1.2 \times (0.3 \times 170^{kips}) + 1.6 \times (0.70 \times 170^{kips})}{0.707} = 355.9 \text{ kips}$$

Limit state	Gusset plate			Two plates		
	ϕ_t	P_n (kips)	$P_c = \phi_t P_n$ (kips)	ϕ_t	P_n (kips)	$P_c = \phi_t P_n$ (kips)
Yielding of gross area	0.90	432	**388.8**	0.90	432	**388.8**
Fracture of effective net area	0.75	623.5	467.6	0.75	532.2	399.2
Block shear failure	0.75	584.6	438.4	0.75	876.8	657.6
Governing case	*Yielding of gross area*					

TABLE 3.5   Design Strengths for Each Limit State

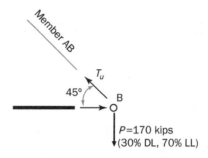

FIGURE 3.21   Free-body-diagram at joint B.

$$T_u = T_r \leq T_{\text{design}} = P_d$$

355.9 kips ≤ 388.9 kips Safe!

Example 3.6
An angle L 4×3×3/8 tension member is shown in Fig. 3.22. The angle (with its longer leg) is connected to a 5/8″×5″ gusset plate by three 7/8-in.-diameter bolts in the standard holes, using a pitch, $p = 3$ in.; end distance, $L_e = 1\frac{1}{2}$ in.; and side distance, $L_s = 1\frac{1}{2}$ in. A36 ($F_y = 36$ ksi and $F_u = 58$ ksi) steel is used. Compute the design strength.

Solution:

$$\text{A36 Steel} \rightarrow F_y = 36 \text{ ksi}; \quad F_u = 58 \text{ ksi}$$

$$A_{g,\text{angle}} = 2.49 \text{ in.}^2 (\text{AISC Manual Part 1: Dimensions})$$

Limit state 1: Yielding of gross section

$$P_n = F_y A_g \rightarrow A_{g,\text{angle}} = 2.49 \text{ in.}^2 < A_{g,\text{plate}} = (5/8)'' \times 5'' = 3.125 \text{ in.}^2 \rightarrow \text{Angle governs}$$
$$P_n = 36^{\text{ksi}} \times 2.49^{\text{in.}^2} = 89.64 \text{ kips}$$
$$P_{c1} = \phi_t P_n = 0.9 \times 89.64^{\text{kips}} = 80.6 \text{ kips}$$

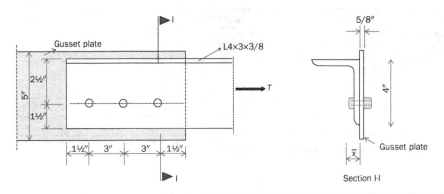

Figure 3.22   An angle tension member connected to a gusset plate.

Limit state 2: Fracture of effective net area

$$P_n = F_u A_e$$

$$w = d_b + 1/8'' = \left(7/8\right)'' + \left(1/8\right) = 1 \text{ in.}$$

$$\left.\begin{array}{l} A_{n,\text{angle}} = A_{g,\text{angle}} - t_{\text{angle}} \times w = 2.49^{\text{in.}^2} - 1 \times \left(3/8\right)'' \times (1)'' \\ \qquad = 2.115 \text{ in.}^2 \\[6pt] U_{\text{angle}} = 1 - \dfrac{\bar{x}}{L} = 1 - \dfrac{0.775''}{6''} = 0.87 \rightarrow A_{e,\text{angle}} = U A_{n,\text{angle}} \\ \qquad = 0.87 \times 2.115^{\text{in.}^2} = 1.84 \text{ in.}^2 \\[6pt] A_{n,\text{plate}} = A_{g,\text{plate}} - t_{\text{plate}} \times w = 3.125^{\text{in.}^2} - 1 \times (5/8)'' \times (1)'' \\ \qquad = 2.5 \text{ in.}^2 \\[6pt] U_{\text{plate}} = 1 \rightarrow A_{e,\text{plate}} = U A_{n,\text{plate}} = 1 \times 2.5^{\text{in.}^2} = 2.5 \text{ in.}^2 \end{array}\right\}\ \text{Angle governs!}$$

$$P_n = 58^{\text{ksi}} \times 1.84^{\text{in.}^2} = 106.7 \text{ kips}$$

$$P_{c2} = \phi_t P_n = 0.75 \times 106.7^{\text{kips}} = 80.1 \text{ kips}$$

Limit state 3: Block shear rupture (Fig. 3.23)

$$P_n = (0.6 F_u A_{nv} ; 0.6 F_y A_{gv})_{\min} + F_u A_{nt}$$

$$A_{nt} = (1.5'' - w/2) \times \left(3/8\right)'' = 0.375 \text{ in.}^2 \ (w = 1.0'')$$

$$A_{gv} = (3'' + 3'' + 1.5'') \times \left(3/8\right)'' = 2.8125 \text{ in.}^2$$

$$A_{nv} = A_{gv} - \sum A_{\text{holes}} = 2.8125^{\text{in.}^2} - 2.5 \times (w) \times (3/8)'' = 1.875 \text{ in.}^2$$

$$P_n = (0.6 \times 58^{\text{ksi}} \times 1.875^{\text{in.}} ; \ 0.6 \times 36^{\text{ksi}} \times 2.8125^{\text{in.}^2})_{\min} + 58^{\text{ksi}} \times 0.375^{\text{in.}^2} = 82.5 \text{ kips}$$

Note: Since the thickness of the plate is larger than the thickness of the leg of the angle, angle governs for block shear failure (Table 3.6):

$$P_{c3} = \phi_t P_n = 0.75 \times 82.5 = 61.9 \text{ k}$$

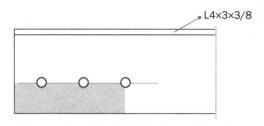

FIGURE 3.23 Block shear rupture in the angle.

Limit state	ϕ_t	P_n (kips)	$P_c = \phi_t P_n$ (kips)
Yielding of gross area	0.90	89.64	80.6
Fracture of effective net area	0.75	106.7	80.1
Block shear failure	0.75	82.5	**61.9**
Governing case	*Block shear failure*		

TABLE 3.6 Design Strengths for Each Limit State

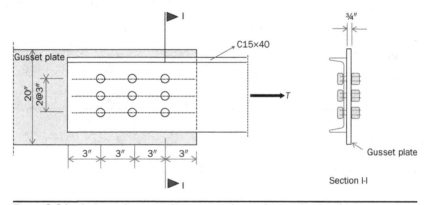

FIGURE 3.24 A channel-gusset plate connection under tension loading.

Example 3.7

The structural member shown in Fig. 3.24 consists of channel C15 × 40 and a plate with 20 in. width and ¾ in. thickness. All three limit states in both the channel and gusset plate need to be considered, and the smallest design strength from a total of six possible design strength values is the actual design strength of the member. The 3/4-in.-diameter bolts are installed in standard holes, and A36 (F_y = 36 ksi, F_u = 58 ksi) steel is used for both channel and plate. The tension force T is due to 70% dead load and 30% live load. Assume $A_e = 0.85A_n$. Determine the maximum tension force the tension member/connection can take.

Solution:

$$\text{A36 Steel} \rightarrow F_y = 36 \text{ ksi}; \; F_u = 58 \text{ ksi}$$
$$A_e = 0.85A_n$$

Limit state 1: Yielding of gross section

$$P_n = F_y A_g$$

$$A_{g,plate} = 20'' \times (3/4)'' = 15.0 \text{ in.}^2 > A_{g,channel} = 11.8 \text{ in.}^2 \, (\text{AISC Manual Part 1: Dimensions})$$

Channel controls!

$$P_n = 36^{ksi} \times 11.8^{in.^2} = 425 \text{ kips}$$

$$P_{c1} = \phi_t P_n = 0.9 \times 425 = 382.3 \text{ kips}$$

Limit state 2: Fracture of effective net area

Channel	Plate
$P_n = F_u A_e$ $w = d_b + 1/8'' = 3/4'' + 1/8''$ $\quad = 0.875 \text{ in.}$ $t_w = 0.52 \text{ in.}$ $A_n = 11.8^{in.^2} - [3 \times (0.875^{in.} \times 0.52^{in.})]$ $\quad = 10.435 \text{ in.}^2$ $A_e = U A_n = 0.85 \times 10.435^{in.^2} = 8.87 \text{ in.}^2$	$P_n = F_u A_e$ $w = d_b + 1/8'' = 3/4'' + 1/8''$ $\quad = 0.875 \text{ in.}$ $A_n = 3/4'' \times 20'' - 3 \times (0.875^{in.} \times 3/4'')$ $\quad = 13.03 \text{ in.}^2$ $A_{e,plate} = A_{n,plate} = 13.03 \text{ in.}^2 > A_{e,channel}$ $\quad = 8.87 \text{ in.}^2$ Channel Controls!

$$P_n = 58^{ksi} \times 8.87^{in.^2} = 514 \text{ kips} \qquad P_{c2} = \phi_t P_n = 0.75 \times 514 = 386 \text{ kips}$$

Limit state 3: Block shear rupture (Fig. 3.25)
For the channel:

$$P_n = (0.6 F_u A_{nv}; 0.6 F_y A_{gv})_{min} + F_u A_{nt}$$
$$A_{nt} = 6^{in} \times 0.52^{in} - 2 \times (0.875^{in} \times 0.52^{in}) = 2.21 \text{ in.}^2$$
$$A_{gv} = (9^{in} \times 0.52^{in}) \times 2 = 9.36 \text{ in.}^2$$
$$A_{nv} = 9.36^{in.^2} - 5 \times (0.875^{in} \times 0.52^{in}) = 7.09 \text{ in.}^2$$

$$P_n = (0.6 \times 36^{ksi} \times 9.36^{in.^2}; \, 0.6 \times 58^{ksi} \times 7.09^{in.^2})_{min} + 58^{ksi} \times 2.21^{in.^2}$$
$$P_n = (202.18^{kips}; 246.73^{kips})_{min} + 128.18^{kips} = 330 \text{ kips}$$

For the plate:

$$P_n = (0.6 F_u A_{nv}; 0.6 F_y A_{gv})_{min} + F_u A_{nt}$$
$$A_{nt} = 3/4'' \times 6^{in} - 2 \times (0.875^{in} \times 3/4'') = 3.1875 \text{ in.}^2$$
$$A_{gv} = (9^{in} \times 3/4'') \times 2 = 13.5 \text{ in.}^2$$
$$A_{nv} = 13.5^{in.^2} - 5 \times (0.875^{in.^2} \times 3/4'') = 10.22 \text{ in.}^2$$

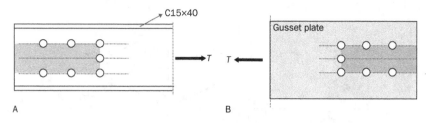

FIGURE 3.25   Block shear rupture areas: (*a*) in the channel, (*b*) in the plate.

Limit state	ϕ_t	P_n (kips)	$P_c = \phi_t P_n$ (kips)
Yielding of gross area	0.90	425	382.3
Fracture of effective net area	0.75	514	386
Block shear failure	0.75	330	**247.5**
Governing case	*Block shear failure*		

TABLE 3.7   Design Strengths for Each Limit State

$$P_n = (0.6 \times 36^{ksi} \times 13.5^{in.^2} ; 0.6 \times 58^{ksi} \times 10.22^{in.^2})_{min} + 58^{ksi} \times 3.187^{in.^2}$$
$$P_n = (291.6^{kips} ; 355.6^{kips})_{min} + 184.85^{kips} = 476.45 \text{ kips}$$

The channel controls the block shear rupture limit state. Thus, as summarized in Table 3.7, the maximum tension force the connection can take is

$$P_{c3} = \phi_t P_n = 0.75 \times 330^{kips} = 247.5 \text{ kips}$$

Example 3.8

The steel truss, as shown in Fig. 3.26, is loaded with service load P at points B, C, and D. $P = 70$ kips, including 60% dead load and 40% live load. Member BC consists of two angles (2L 6 $\times$ 4 $\times$ ½), as shown in Fig. 3.27, bolted to a 5/8-in.-thick gusset plate by six (6) 3/4-in.-diameter bolts in standard holes (Fig. 3.27). Check if member BC is safe, based on all three limit states of the angles, and block shear of the gusset plate. Assume $A_e = 0.90 A_n$. A36 steel ($F_y = 36$ ksi, $F_u = 58$ ksi) is used for all components.
   Given:

$$\text{A36 Steel} \rightarrow F_y = 36 \text{ ksi}; F_u = 58 \text{ ksi}$$
$$A_e = 0.90 A_n; P = 70 \text{ kips}$$

Find P_u for member BC using structural analysis.

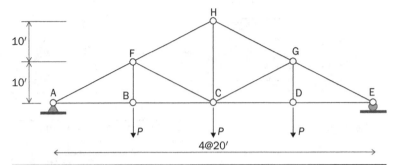

Figure 3.26 Steel truss loaded with service load P.

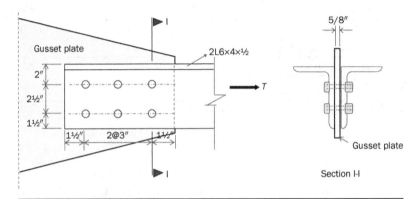

Figure 3.27 Connection detail of member BC.

Step 1: Find support reactions (Fig. 3.28).

$$\sum M_B = 0$$
$$R_A \times 80 = P \times 20' + P \times 40' + P \times 60'$$
$$R_A = 1.5P$$

Step 2: Cut from 1-1 and take the free body diagram (Fig. 3.29).

$$\sum M_F = 0$$
$$R_A \times 20' = F_{BC} \times 10'$$
$$F_{BC} = 3P$$

Step 3: Find P_u for member BC.

$$P_u = 1.2D + 1.6L = 1.2 \times (0.6 \times 3P) + 1.6 \times (0.4 \times 3P)$$
$$P_u = 4.08P = 4.08 \times 70^{kips} = 285.6 \text{ kips}$$

Step 4: Limit states of the angles.

Limit state 1: Yielding of gross section

$$P_n = F_y A_g \rightarrow A_{g,2L} = 2 \times 4.75^{in.^2} = 9.5 \text{ in.}^2$$

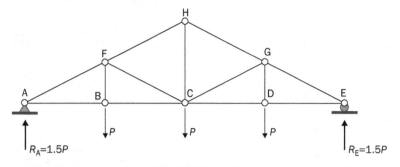

FIGURE 3.28 Finding support reactions.

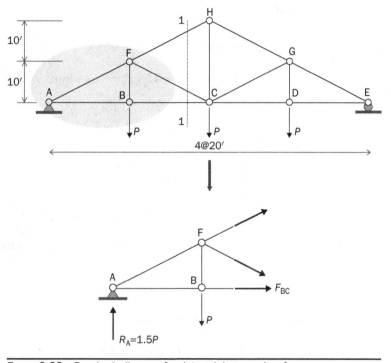

FIGURE 3.29 Free-body-diagram for determining member forces.

$$P_n = 36^{ksi} \times 9.5^{in.^2} = 342 \text{ kips}$$

$$P_{c1} = \phi_t P_n = 0.9 \times 342^{kips} = 307.8 \text{ kips}$$

Limit state 2: Fracture of effective net area

$$P_n = F_u A_e$$
$$w = d_b + 1/8'' = 3/4'' + 1/8'' = 0.875''$$
$$A_n = 9.5^{in.^2} - [4 \times (0.875^{in.} \times 0.50^{in.})] = 7.75 \text{ in.}^2$$
$$A_e = U A_n = 0.9 \times 7.75^{in.^2} = 6.975 \text{ in.}^2$$

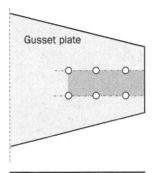

Figure 3.30 Block shear rupture in the gusset plate.

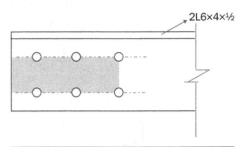

Figure 3.31 Block shear rupture mode 1 in the angles.

$$P_n = 58^{ksi} \times 6.975^{in.^2} = 404.55 \text{ kips}$$

$$P_{c2} = \phi_t P_n = 0.75 \times 404.55^{kips} = 303.4 \text{ kips}$$

Limit state 3: Block shear rupture

- Block shear of gusset plate (Fig. 3.30)

$$P_n = (0.6F_u A_{nv}; 0.6F_y A_{gv})_{min} + F_u A_{nt}$$
$$A_{nt} = 2.5^{in.} \times 5/8'' - 0.875^{in.} \times 5/8'' = 1.016 \text{ in.}^2$$
$$A_{gv} = 7.5^{in.} \times 5/8'' \times 2 = 9.375 \text{ in.}^2$$
$$A_{nv} = 9.375^{in.^2} - 5 \times 0.875^{in.} \times 5/8'' = 6.641 \text{ in.}^2$$

$$P_n = (0.6 \times 36^{ksi} \times 9.375^{in.^2}; 0.6 \times 58^{ksi} \times 6.641^{in.^2})_{min} + 58^{ksi} \times 1.016^{in.^2}$$
$$P_n = (202.5^{kips}; 231.11^{kips})_{min} + 58.93^{kips} = 261.43 \text{ kips}$$

$$P_{c3} = \phi P_n = 0.75 \times 261.43^{kips} = 196.07 \text{ kips}$$

- Block shear rupture mode 1 of angles (2L) (Fig. 3.31)

$$t_{gusset} = 5/8'' < t_{angles} = 2 \times 1/2'' = 1''$$

Note: Since the total thickness of two angles is larger than gusset plate's thickness, the block shear failure mode for the gusset plate governs.

- Block shear rupture mode 2 of angles (2L) (Fig. 3.32)

$$P_n = (0.6F_u A_{nv}; 0.6F_y A_{gv})_{min} + F_u A_{nt}$$
$$A_{nt} = 2 \times (4^{in.} \times 1/2'' - 1.5 \times 0.875^{in.} \times 1/2'') = 2.6875 \text{ in.}^2$$
$$A_{gv} = 2 \times (7.5^{in.} \times 1/2'') = 7.5 \text{ in.}^2$$
$$A_{nv} = 7.5^{in.^2} - 2 \times (2.5 \times 0.875^{in.} \times 1/2'') = 5.3125 \text{ in.}^2$$

$$P_n = (0.6 \times 36^{ksi} \times 7.5^{in.^2}; 0.6 \times 58^{ksi} \times 5.3125^{in.^2})_{min} + 58 \times 2.6875^{in.^2}$$
$$P_n = (162^{kips}; 184.88^{kips})_{min} + 155.88^{kips} = 317.88 \text{ kips}$$

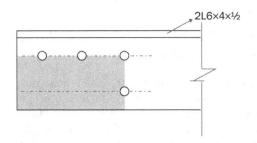

Figure 3.32 Block shear rupture mode 2 in the angles.

Limit state	ϕ_t	P_n (kips)	$P_c = \phi_t P_n$ (kips)
Yielding of gross area	0.90	307.8	307.8
Fracture of effective net area	0.75	404.55	303.4
Block shear failure	0.75	261.43	**196.07**
Governing case	*Block shear failure*		

Table 3.8 Design Strengths for Each Limit State

A summary of design strengths based on each limit state is given in Table 3.8. Thus,

$$P_{c4} = \phi P_n = 0.75 \times 317.88^{\text{kips}} = 238.41 \text{ kips}$$

Step 5: Check capacity/demand ratio.

$$\phi P_n = 196.07 \text{ kips} < P_u = 285.6 \text{ kips No Good!}$$

3.6 Problems

3.1 An angle L6 × 4 × ½ tension member is shown in Fig. 3.33. The angle (with its longer leg) is connected to a 9/16″ × 8″ gusset plate by three 7/8-in.-diameter bolts in the standard holes, using a pitch, $p = 3$ in.; end distance, $L_e = 1½$ in.; and side distance,

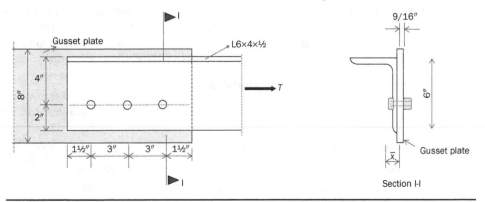

Figure 3.33 An angle tension member connected to a gusset plate.

$L_s = 1½$ in. A36 ($F_y = 36$ ksi and $F_u = 58$ ksi) steel is used. Compute design strength. Include U in the calculation $U = 1 - \overline{x}/L$ and $A_e = UA_n$ (Note: A_g of L6 × 4×½ = A in AISC Manual Part 1: Dimensions. You can also find $\overline{x}$ there.)

3.2   The structural member shown in Fig. 3.34 consists of channel C10×30 and a plate with 12 in. width and ½ in. thickness. All three limit states in both the channel and gusset plate need to be considered, and the smallest design strength from a total of six possible design strength values is the actual design strength of the member. The 3/4-in.-diameter bolts are installed in standard holes, and A36 ($F_y = 36$ ksi, $F_u = 58$ ksi) steel is used for both channel and plate. The tension force T is due to a 60% dead load and 40% live load. Determine the maximum tension force the tension member/connection can take.

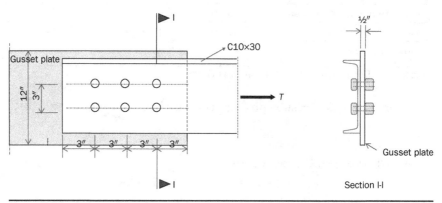

FIGURE 3.34   A channel tension member connected to a gusset plate.

3.3   The steel truss, as shown in Fig. 3.35 (right), is loaded with service load P at points B, C, and D. $P = 50$ kips, including 70% dead load and 30% live load. Member BC consists of two angles (2L8×6×7/8), as shown in Fig. 3.35 bolted to a 1-in.-thick gusset plate by six (6) 7/8-in.-diameter bolts in standard holes. Check if member BC is safe, based on all three limit states of the angles, and block shear of the gusset plate. Assume $A_e = 0.90 A_n$. A36 steel ($F_y = 36$ ksi, $F_u = 58$ ksi) is used for all components.

3.4   Please answer the following questions:

    a. What are the strength limit states that should be considered for tension members?

    b. Why is the slenderness ratio limited for common tension members?

    c. Why should we introduce reduction factor U for the net area?

    d. Why should you choose the smallest strength from all limit states as design strength?

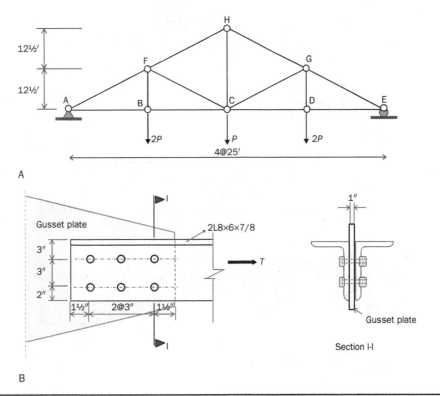

A

B

Figure 3.35 Steel truss: (a) loading, (b) connection detail of member BC.

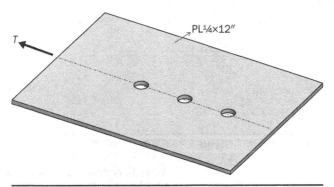

Figure 3.36 A plate tension member with single lines of holes in the direction of loading.

3.5 Compute the maximum tension force (*T*) that the plate in Fig. 3.36 can take. Steel grade is A36 with $F_y=36$ ksi and $F_u=58$ ksi. Dead load (*D*) and live load (*L*) is 0.25*T* and 0.75*T*, respectively. The 7/8-in.-diameter bolts are installed in standard single lines of holes in the direction of loading. Use load combinations of 1.4*D* and 1.2*D*+1.6*L*.

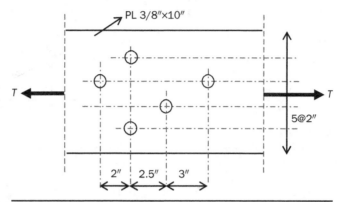

FIGURE 3.37   A plate tension member.

3.6   Compute the maximum tension force (T) that the plate in Fig. 3.37 can take. Steel grade is A36 with $F_y = 36$ ksi and $F_u = 58$ ksi. Dead load (D) to live load (L) ratio (D/L) is 4. The 8/9-in.-diameter bolts are installed in standard holes. Use load combinations of $1.4D$ and $1.2D+1.6L$.

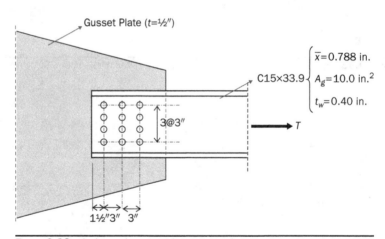

FIGURE 3.38   A channel-gusset plate connection.

3.7   A channel C15×33.9 tension member is connected to a gusset plate (Fig. 3.38). Steel grade is A36 with $F_y = 36$ ksi and $F_u = 58$ ksi. Dead load (D) and live load (L) are $0.2T$ and $0.8T$ (kips), respectively. The 7/8-in.-diameter bolts are installed in standard holes. The load combinations are $1.4D$ and $1.2D+1.6L$. Compute the maximum tension force (T) the connection can take.

3.8   An angle L5 × 3½ × ½ tension member is connected to a gusset plate (Fig. 3.39). Steel grade is A572 Grade 50 with $F_y = 50$ ksi and $F_u = 65$ ksi. Dead load (D) and live load (L) are 20 kips and 70 kips, respectively. The 7/8-in.-diameter bolts are installed in standard holes. The load combinations are $1.4D$ and $1.2D+1.6L$. Assume $U = 0.85$

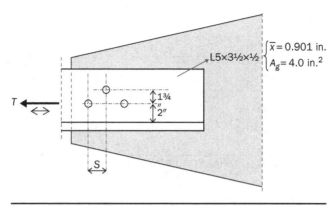

FIGURE 3.39 An angle-gusset plate connection under tension loading.

(needs to be verified later). All bolts are not shown (assume at least 3 bolts per line). Compute the pitch (*s*) (both theoretical and specified in 0.5 in. multiples) in the 5 in. leg.

3.9 Write a python code that can compute the design tension strength of a member based on three limit states and maximum tension force that the connection/member can take.

Related to Building Project in Chapter 1:

3.10 Please compute the following for the building given in the appendix in Chapter 1:

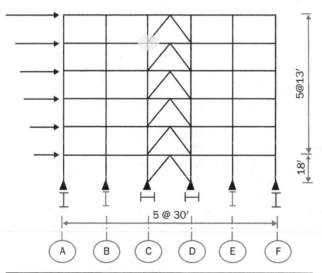

FIGURE 3.40 Elevation of the SCBFs on Line 1 and 6 with inverted V-bracing in the six-story steel building in the Building Project.

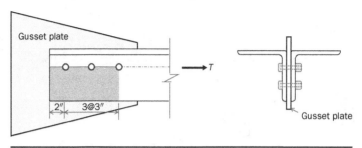

Figure 3.41   Brace-gusset plate connection.

Design the inverted V-braces at the upper story of the braced frame using the lateral forces you determined (light gray area in Fig. 3.40). Use double-angles and assume a single line of three 7/8-in.-diameter bolts at the ends (Fig. 3.41). Assume inverted V-braces are tension-only members and check the limit states for tension on gross and net area and block shear (use A500 Gr. C steel).

Bibliography

AASHTO, *LRFD Bridge Design Specification*, American Association of State Highway and Transportation Officials for Bridges, Washington, DC, 2020.

Aghayere, A. O., and J. Vigil, *Structural Steel Design: A Practice Oriented Approach*, Pearson Education, Inc., New Jersey 2015.

AISC, *Seismic Design Manual*, American Institute of Steel Construction, Chicago, IL, 2018.

AISC, *The Material Steel*, American Institute of Steel Construction, A Teaching Primer for Colleges of Architecture, Chicago, IL, 2007.

AISC 341, *Seismic Provisions for Structural Steel Buildings*, ANSI/AISC Standard 341-16, American Institute of Steel Construction, Chicago, IL, 2016.

AISC 358, *Prequalified Connections for Special and Intermediate Steel Moment Frames for Seismic Applications*, American Institute of Steel Construction, Chicago, IL, 2016.

AISC 360, *Specification for Structural Steel Buildings*, ANSI/AISC Standard 360-16, American Institute of Steel Construction, Chicago, IL, 2016.

AISC Manual, *Steel Construction Manual*, 15th ed., American Institute of Steel Construction, Chicago, IL, 2016.

ASCE 7, *Minimum Design Loads for Buildings and Other Structures*, ASCE/SEI 7-16, American Society of Civil Engineers, Reston, VA, 2016.

Salmon, C. G., J. E. Johnson, and F. A. Malhas, *Steel Structures: Design and Behavior—Emphasizing Load and Resistance Factor Design*, 5th ed., Pearson Education, Inc., New Jersey 2009.

Shen, J., B. Akbas, O. Seker, and C. Carter, *Structural Engineering Handbook—Chapter 8: Design of Structural Steel Members*, 5th ed., McGraw-Hill, New York, 2020.

CHAPTER 4

Design of Columns

4.1 Introduction

Columns are the structural members, primarily supporting concentrically applied axial compressive loads. In other words, column members are idealized to be subjected to a uniform compressive stress ($\sigma_c = P/A_g$ where P is the compressive force acting along the centerline and A_g is the gross area of the cross-section), as shown in Fig. 4.1. Although flexural and/or torsional deformations generally exist along with axial compression in an actual axially loaded structural member, the impact of these deformations on the column behavior is considered to be negligible and thus omitted from the discussions and examples covered in this chapter. It should be noted that flexural and torsional demands on a column member cannot be disregarded in the design process unless they are due to certain instances, such as bending moment transferred from a shear connection (pinned connection). In such cases, the demand is considered relatively small and usually uncertain as opposed to the well-quantified flexural demand on "beam-columns." The interaction between the axial and flexural deformations will be described in Chapter 7.

Column members can be found in nearly all structural systems, including gravity load–resisting systems, lateral load–resisting systems, and bridges to transfer axial loads from one member to another and, eventually, to the foundation. Typical examples of column members are demonstrated in Fig. 4.2. Given their deformed shapes, the top chord and the vertical members of the truss shown in Fig. 4.2a, one of the bracings at each story level of the braced frame shown in Fig. 4.2c, and the towers of the suspension bridge shown in Fig. 4.2d tend to shorten under the given loading conditions; therefore, these can be deemed representative column members. Numerous steel shapes can be employed in a steel building as a column member, including prismatic (i.e., a bar of constant cross-section along its length) and, despite their rare utilization, nonprismatic (e.g., tapered or stepped) sections. As shown in Fig. 4.3, columns can be made of plates (flat bars), single or double channels, I-shapes, single or double angles, tee shapes, tubes (circular or rectangular), latticed or battened channels, and so on. Steel fabrication process might as well differ from shape to shape and size to size. As such, columns can be manufactured by hot-rolling and cold-forming. It is essential for a design engineer to consider the pros and cons of each fabrication process that might affect the performance and economy of the structural members. For instance, cold-formed structural steel products are superior to their hot-rolled counterparts in terms of dimensional tolerances, straightness, and surface smoothness. Hot-rolled shapes, on the other hand, are more economical and weldable, and also fabrication of much larger sizes is possible

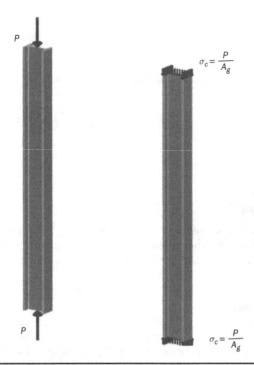

FIGURE 4.1   A centrally loaded I-shaped steel column member (left) and uniform compressive stress distribution over the entire cross-section (right).

owing to ease of hot forming. Therefore, considering the desired level of ductility, tolerance, corrosion resistance, weldability, fatigue life, etc., selection of an optimal material-shape couple among various options usually stands somewhere between economy and performance, which is highly dependent on the constraints/needs/requirements based on structure type (e.g., an industrial or multistory building), type (e.g., gravity load, wind load, and seismic load), and magnitude of the computed loads acting on the member, structural member type (e.g., bracings, truss members, and columns), and connection type.

To recognize the design process as well as the parameters affecting the behavior, first, the limit states that are applicable to column members should be well understood. The primary failure mode of a column member, unlike tension members, is global (or overall) buckling. Buckling is an instability condition that occurs when an axially loaded member reaches the critical load (i.e., buckling load) at which the member can no longer support increasing loads and consequently begins to deform laterally as the axial deformation becomes greater. For a basic comprehension of such instability, the finite element (FE) simulation of a column given in Movie 4.1, which clearly demonstrates the alter in the behavior subsequent to the global buckling, and the load-deformation relationship given in Fig. 4.4 can be examined.

1. In the initial stage, for a load smaller than the critical load (i.e., $P < P_{cr}$), only axial deformation exists and the column remains straight and stable.

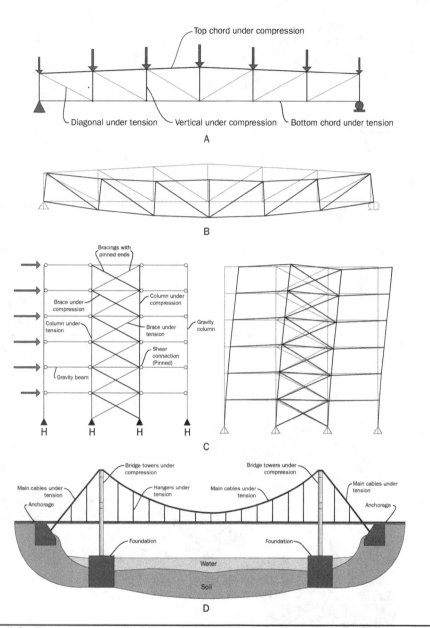

FIGURE 4.2 Typical examples of column members. (*a*) Truss system subject to gravity loads. (*b*) Deformed shape of the given truss under gravity loads. (*c*) Braced frame subject to wind load (left) and the deformed shape under wind load (right). (*d*) Bridge towers.

2. Instantaneously, the bar with pinned ends becomes unstable at a load typically lower than its yielding capacity, P_y, and subsequently undergoes an increasing bending deformation similar to that of a beam subject to flexural deformation.

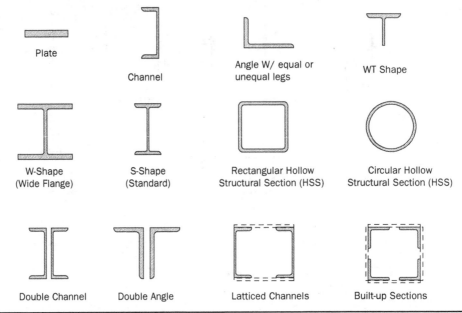

FIGURE 4.3 Typical column cross-sections.

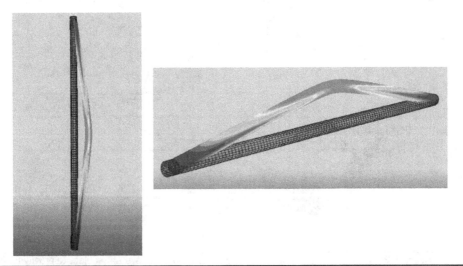

MOVIE 4.1 Global buckling of a pinned bar. (To view the entire movie, go to www.mhprofessional. com/design-of-steel-structures.)

This post buckling-induced second-order bending moment grows as the lateral displacement at the mid-length increases and accordingly the flexural deformation becomes more dominant. Therefore, as illustrated with the load-deformation curve in Fig. 4.4, once the bifurcation point (i.e., the critical load) is attained, the load-carrying capacity of a column gradually reduces from its peak value, P_{cr}, due to the fact that the

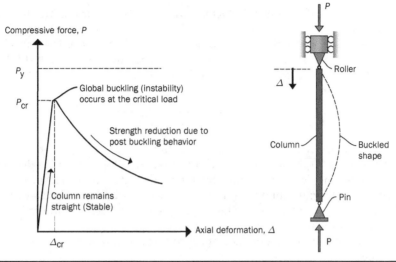

Figure 4.4 Force-deformation relationship of a single column.

behavior changes abruptly from "column" to "beam-column" when the applied axial compressive force exceeds the critical load.

Prior to attending the actual behavior further as well as the design procedure, a brief summary of the historical development of column theory, potential failure modes, and uncertainties that affect column strength will be presented by means of simulation-based discussions in the following sections. Subsequently, the design procedure considering each limit state will be demonstrated through worked examples. Finally, design examples based on possible real-life applications will be demonstrated to better explain the previously described design process and basic concepts.

4.1.1 Brief Summary of Column Theory

For design purposes, estimating the magnitude of buckling load is of concern and thus establishing a buckling load formula that easily and accurately quantifies the load-carrying capacity of a column is essential for the convenience of design. The very theoretical foundation of column theory is provided by mathematician *Leonhard Euler* (Euler, 1744). He derived *the Euler formula* given in Eq. (4.1) from an analytical solution to global buckling of an "ideal column" with perfectly pinned end restraints:

$$P_E = P_{cr} = \frac{\pi^2 EI}{L^2} \tag{4.1}$$

where P_E is the critical buckling load of the Euler column (i.e., ideal column), E is the modulus of elasticity, I is the moment of inertia of the cross-section about the axis of buckling, and L is the length of the column. If a column is supported at the ends with any means other than simple support, a general formula including various support conditions at member ends can be expressed as follows:

$$P_{cr} = \frac{\pi^2 EI}{(KL)^2} \tag{4.2}$$

where K is the effective length factor that accounts for various end restraints. Further discussions on effective length can be found in the following sections. As indicated in Eq. (4.3), Euler's equation can also be expressed in terms of the critical compressive stress, F_{cr} by dividing P_{cr} by gross cross-sectional area, A_g and substituting r^2 for I/A. Thus, the elastic critical stress formula becomes

$$F_{cr} = \frac{P_{cr}}{A_g} = \frac{\pi^2 E}{(KL/r)^2} \tag{4.3}$$

where F_{cr} is the average compressive stress when buckling occurs, r is the radius of gyration ($r = \sqrt{I/A}$) about the governing axis, and the term in the denominator, KL/r, is referred to as slenderness ratio. Such transformation of the elastic buckling formula is considered significant owing to the fact that associating the buckling stress, F_{cr}, with slenderness ratio, KL/r, allows the buckling tendency of a column to be recognized by inspecting a distinctive term.

An ideal column is a mathematical concept such that the compression member is assumed to be:

1. Elastic (i.e., independent of the strength of the material)

2. Prismatic

3. Perfect (i.e., initially straight and residual stresses do not exist)

4. Loaded through the centroidal axis of the member (no eccentricity)

With regard to material nonlinearity, Euler's approach might only be applicable to certain cases where buckling occurs before the fibers in the cross-section reaches the proportional limit ($\sigma_c < F_{prop}$), and thus it was regarded insufficient to estimate the critical loads of stocky columns (i.e., columns that possess small KL/r ratios). Many researchers had investigated perfect inelastic column behavior in the following years after Euler's work. Based on the succeeding work (Engesser, 1891), it had been demonstrated that the critical stress, F_{cr}, of a stubby, or intermediate, column can be more accurately evaluated by substituting the elastic modulus (i.e., the slope of the linear portion), E, with the tangent modulus (the slope of the nonlinear portion in stress–strain curve given in Fig. 4.5), E_t,

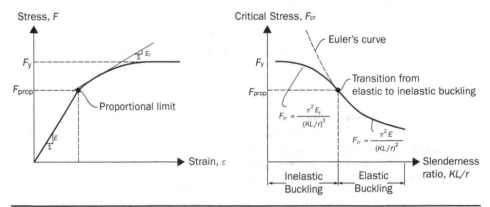

FIGURE 4.5 Stress–strain relationship (left) and column strength curves based on Euler's and tangent modulus theories (right).

in Euler's equation. As can be interpreted from Eq. (4.3), on account of the inversely proportional relationship between the average compressive stress over the cross-section and the slenderness ratio, the buckling stress grows larger as the column becomes stockier. Therefore, less slender columns tend to buckle inelastically against the most prominent assumption of Euler's theory. The buckling load of a perfect inelastic column based on the basic tangent modulus theory (Engesser, 1891) is expressed as follows:

$$F_{cr,t} = \frac{\pi^2 E_t}{(KL/r)^2} \tag{4.4}$$

Needless to say, the two expressions given in Eqs. (4.3) and (4.4) are identical except that the latter reflects material nonlinearity. Hence, the two curves shown in Fig. 4.5 coincide with each other at the proportional limit, F_{prop}, where the buckling transitioned from elastic to inelastic or vice versa. Furthermore, by comparing the two curves, one can notice that Euler's curve impractically tends to infinity whereas the tangent curve asymptotically approaches to yielding stress, F_y, as slenderness ratio reduces.

Besides its improved accuracy compared to Euler's formula, both basic tangent modulus theory and its modified version, double-tangent theory, lacked explaining the true inelastic column behavior. It had been later shown that the maximum load that can be supported by a perfect inelastic column is between the critical loads obtained from tangent modulus theory and Euler's formula (Shanley, 1947). Further details of the historical development of column theory can be found in Euler (1744), Engesser (1891), Shanley (1947), and Ziemian (2010).

Although early approaches based on the critical load of a perfect column (Euler, 1744; Engesser, 1891; Shanley, 1947), collectively, lay a basis for initiation of the modern column design formulas, we now know that columns are imperfect due to several sources, such as initial out-of-straightness and residual stresses on account of differential cooling after hot-rolling or welding. It is, therefore, usually not possible to estimate the buckling load of an actual column by directly using the critical load concept originated from Euler's approach, which sets an upper bound for the actual buckling load. Thus, in order to embrace the state of the art in column design formulas, the effect of imperfections on column behavior and potential failure modes of a compression member are discussed in the subsequent sections of this chapter.

4.2 Column Behavior

In general, behavior of a structural member can be associated with the limit state that controls its capacity. Fundamentally, failure modes of compression members can be divided into two major groups: (1) overall (global) buckling and (2) local buckling. Overall buckling may also differ from one cross-section to another. As depicted in Fig. 4.6, global buckling, which is typically the desired failure mode for a column, is divided into three subcategories. It is noteworthy that the type of the postbuckling deformation defines this characterization, which will be explained in detail later in this section utilizing FEM-based simulations (Smith, 2009).

Besides the foregoing classification involving global and local buckling, a further categorization can be based on the material behavior. Regardless of the buckling type (i.e., flexural, torsional, and flexural-torsional), buckling might occur elastically (under a small load) in a slender column or inelastically (under a significantly larger load) in a

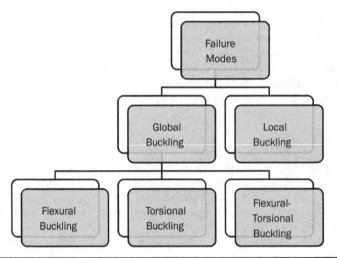

Figure 4.6   Classification of failure modes for columns.

stocky column. The magnitude of the governing slenderness ratio, $(KL/r)_{max}$, determines whether a compression member buckles elastically or inelastically. As previously mentioned, slenderness ratio is a property that reflects the interrelation between boundary conditions, length, and cross-sectional properties of a column. Slender columns, which would buckle elastically, are either very long or possess a small cross-section, or both. If we reduce the length and/or increase the second moment of the cross-sectional area, the same column might buckle inelastically. Figure 4.7 presents a schematic comparison between buckling loads of two bars of identical cross-section. One of the columns is longer (more slender) than the other one, which leads to elastic buckling under a relatively small axial load of $P_{cr,\,elastic}$. The intermediate column (stockier), on the other hand, buckles under a larger load, $P_{cr,\,inelastic}$. Since some of the fibers yielded prior to inelastic buckling, the stocky column had a plateau-like segment as seen in the load-deformation relationship given in Fig. 4.7, while the ultimate load-bearing capacity of the slender column deteriorated suddenly without having a plateau after buckling.

An improved understanding of elastic and inelastic buckling depicted in Fig. 4.7 can be achieved by investigating the FE simulation given in Movie 4.2. The simulation compares the buckling and postbuckling behavior of two circular tubing cut from the same HSS10×0.500 (ASTM A500, Gr. B, $F_y = 42$ ksi) with two different lengths corresponding to slenderness ratios of 80 and 150, respectively. Note that red and gray colors in the legend of stress contours indicate the yield stress ($F_y = 42$ ksi) and any stress greater than yield stress, respectively. By investigating the simulation (Movie 4.2), the following can be discussed:

1. As illustrated in the simulation, the slender (longer) column buckles earlier than the intermediate column, even though both specimens have the same cross-section and were subjected to the exact same ever-increasing loading.

2. By comparing the stress distribution along the length of the two specimens, it can be observed that global buckling is initiated before any fiber of the slender column reached the elastic limit while some of the fibers in the intermediate

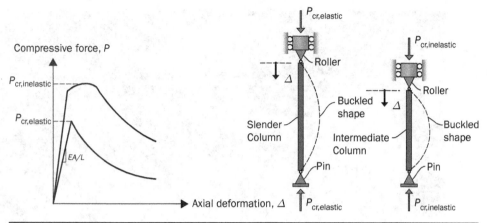

Figure 4.7   Elastic vs. inelastic global buckling (not to scale).

column experienced yielding right before buckling. In other words, due to small buckling load required to initiate buckling in slender columns, all the fibers of the slender column remained elastic. On the other hand, inelastic material behavior (i.e., yielding of some of the fibers) was observed in the stocky column before instability takes place, which led to a larger compressive load capacity with the same cross-section.

4.2.1   Global Buckling

So far, the discussions on the column behavior have been based solely on flexural global buckling. However, a column might be susceptible to torsional or flexural-torsional buckling depending upon its cross-sectional shape, such as symmetry or asymmetry of a cross-section, and boundary conditions. Nature of each buckling mode is identified and thoroughly explained by the simulation-assisted discussions presented in the following sections.

4.2.1.1   Flexural Buckling

In order for a buckling mode to be identified as flexural buckling, postbuckling deformation should not involve any deformation other than bending. Solid sections (e.g., round or flat bars), I-shapes, square or circular tubing (HSS), and single angles might be prone to flexural buckling unless their cross-sections are composed of very thin (slender) elements that might become locally unstable. Even though flexural buckling could take place about either one of the principal axes of a compression member, this type of buckling, in most cases, occurs about the weak axis (i.e., the axis having the smallest radius of gyration) of singly or doubly symmetric shapes. Symmetrical shapes, in particular, buckle about the axis with the largest slenderness ratio, since the load-carrying capacity reduces as the slenderness ratio increases. To illustrate why buckling occurs about the weak axis, deformed shapes due to the potential flexural buckling modes, which are (a) buckling about minor (weak) axis and (b) buckling about major (strong) axis, and schematic load-deformation curves of a common W-shape with pinned ends

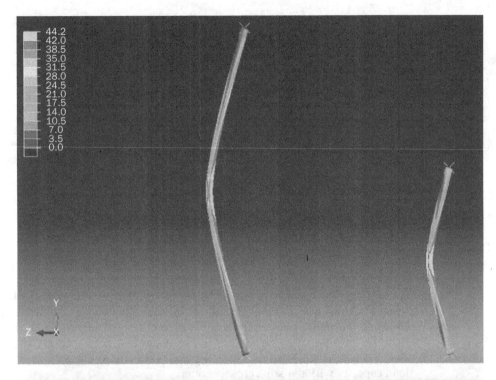

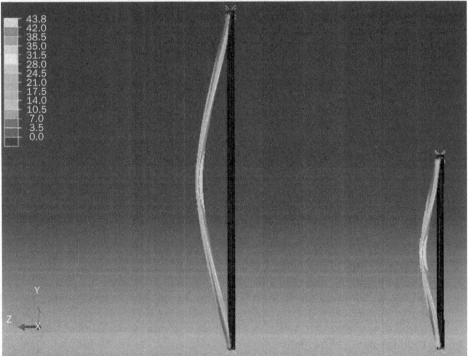

MOVIE 4.2   Elastic vs. inelastic global buckling of a circular tube. (To view the entire movie, go to www.mhprofessional.com/design-of-steel-structures.)

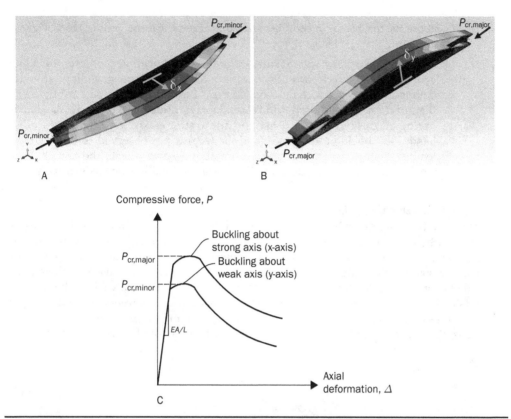

FIGURE 4.8 Potential flexural buckling modes of a W-shape with pinned end restraints. (*a*) Buckling about the weak axis (*y*-axis). (*b*) Buckling about the strong axis (*x*-axis). (*c*) Schematic load-deformation curves.

are given in Fig. 4.8. As indicated in Fig. 4.8*c*, if we increase the applied axial load until reaching the buckling load about weak axis, $P_{cr,}$ minor, the column would buckle prior to attaining the buckling load about strong axis, $P_{cr,}$ major. Hence, when the same boundary conditions are applied (e.g., free rotation about the perpendicular axes), it is very likely that buckling would occur about the minor axis of a W-shape. Note that the ratio of r_x/r_y is always greater than one for W-shapes (for common W12 or W14, the ratio is around 1.7).

Likewise, the simulation shown in Movie 4.3 demonstrates flexural buckling of a W14×132 about the weak and strong axes, respectively. By comparing the two buckling modes, one can observe the following:

1. Postbuckling deformations, in both weak- and strong-axes buckling, involve bending deformation only.

2. In both cases, lateral deformation following the buckling is initiated perpendicular to the axis that buckling takes place about.

3. If we compare the buckling loads about the two axes, in this simulation, the load obtained from the buckling about the major (*x*-axis) was roughly 15% higher than its minor counterpart, the reason being that the radius of gyration

about the major axis (r_x) is larger than that about the minor axis (r_y) for the given W-shape.

4. Postbuckling deformation results in local flange buckling soon after buckling about x-axis, which would increase the rate of postbuckling strength deterioration and reduce the axial stiffness further. This can be attributable to the direction of bending deformation. Flanges are bent about their weak axis during the postbuckling deformation due to strong-axis buckling. On the other hand, local buckling of flange (and/or web) is not observed throughout the loading history when buckling occurs about y-axis due to the fact that the weak-axis postbuckling deformation is resisted by the strong axis of flanges.

4.2.1.2 Torsional Buckling

Torsional buckling describes a failure mode in which a compression member loses its stability by twisting about the centroidal axis rather than undergoing a bending deformation as in flexural buckling. Doubly symmetric open shapes with slender elements may be susceptible to this type of buckling. Doubly symmetric closed sections (e.g., circular or rectangular tubing), on the other hand, are not conceivably vulnerable to torsional buckling owing to their high torsional resistance regardless of slenderness of their cross-sectional elements. As one of the most commonly used open column sections, the likelihood of torsional buckling failure is quite low for wide-flange sections

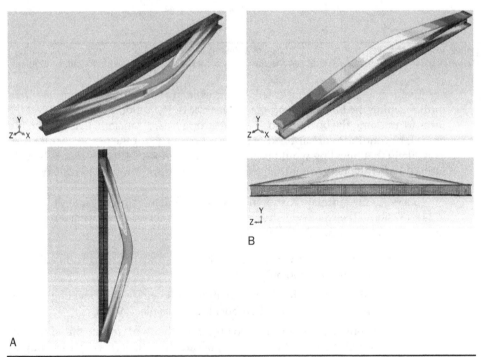

Movie 4.3 Global buckling modes about strong and weak axes. (*a*) Weak-axis buckling. (*b*) Strong-axis buckling. (To view the entire movie, go to www.mhprofessional.com/design-of-steel-structures.)

when the torsional and flexural buckling lengths are close. It is also noteworthy that even with low torsional resistance (e.g., cruciform shape with slender elements), flexural buckling might still govern for the doubly symmetric shapes possessing relatively large overall slenderness ratios (KL/r). In other words, the governing failure mode of doubly symmetric sections (e.g., I-shape, cruciform or flanged cruciform shapes, and two or four angles in cruciform shape) is determined by the interaction between slenderness ratio, KL/r, resistance against torsion, and width-to-thickness ratio of the cross-sectional elements.

Buckling of a cruciform-shaped steel column with slender elements, shown in Fig. 4.9, can be a typical example of such buckling. As seen in the section view, the legs are twisted (θ_{twist}) about the centroid (also shear center for doubly symmetric shapes) after buckling. To better visualize this failure mode, buckling of the same cruciform section build up from thin plates having an aspect ratio (i.e., width-to-thickness ratio) of 15 is simulated in Movie 4.4. By inspecting the section, side and isometric views given in Movie 4.4, the following can be observed:

1. As seen in the section view of Movie 4.4, the cross-section symmetrically rotates about the longitudinal axis without translating laterally after buckling. Postbuckling deformation is, therefore, purely torsional.

2. As seen in the side and isometric views (Movie 4.4), similar to the postbuckling lateral deformation observed in the flexural buckling of a W-shape (see simulation given in Movie 4.3), applied axial deformation is mainly compensated by the torsional deformation concentrated at the midlength of the member.

3. Each leg behaves in the same manner as a cantilever. Therefore, because of the torsional restraint at both ends, the stress due to twisting at the legs increases from free end to the centroid with increasing twist angle, θ_{twist}.

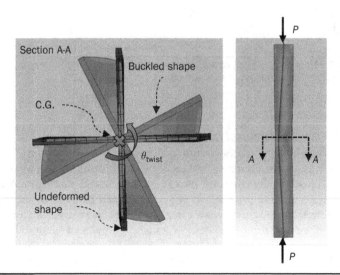

FIGURE 4.9 Torsional buckling of a cruciform shape with slender elements.

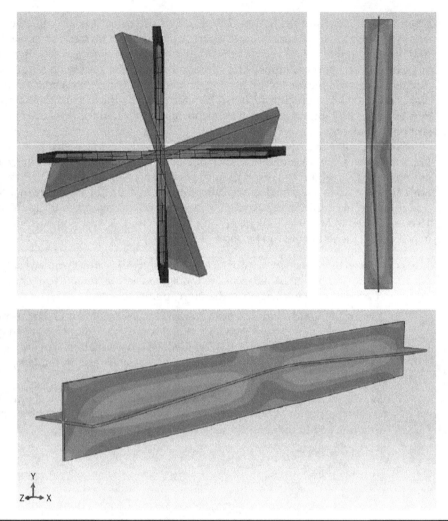

MOVIE 4.4 Torsional buckling of a cruciform shape. (To view the entire movie, go to www. mhprofessional.com/design-of-steel-structures.)

4.2.1.3 Flexural-Torsional Buckling

This combined buckling mode might occur when singly symmetric or unsymmetric sections, such as angles, channels, tee sections, double angles are subject to axial compression. Postbuckling deformation of a representative flexural-torsional buckling-prone unequal-leg angle is presented in Fig. 4.10. As can be interpreted from the figure, the section both translates and twists after buckling. Therefore, flexural-torsional buckling can be defined as a limit state during which flexural and torsional deformations occur simultaneously.

To elucidate this, postbuckling behavior of a single unequal-leg angle is loaded with an increasing axial load in compression through its centroid. Movie 4.5 presents the

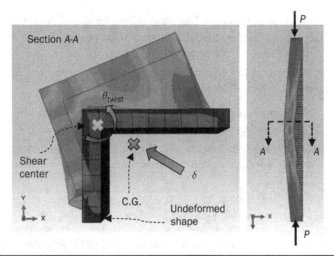

FIGURE 4.10   Flexural-torsional buckling of an unequal-leg angle.

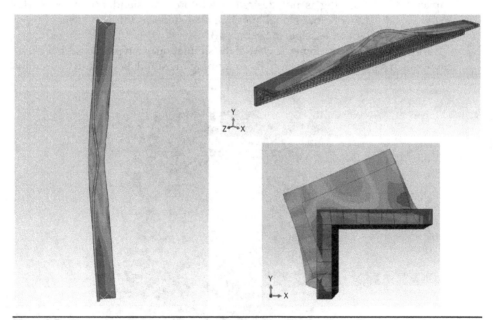

MOVIE 4.5   Flexural-torsional buckling of a single angle. (To view the entire movie, go to www.mhprofessional.com/design-of-steel-structures.)

simulation of the single angle. In what follows, the observations carried out on the simulation are summarized:

1. The exerted axial force/deformation on the member is counteracted by the combination of bending and torsional deformations primarily accumulated at the midlength. Therefore, a stress concentration due to the combined effects occurred in the vicinity of the midlength.

2. Once flexural-torsional buckling takes place, the angle begins to be bent about the minor axis (i.e., the lateral translation perpendicular to z-axis, δ) as well as to twist about the shear center (θ_{twist}).

3. By evaluating the two postbuckling deformations (i.e., lateral translation and twist) individually, it is safe to claim that

   a. No torsion would occur if the load is applied to the shear center of the cross-section, since the cross-section rotates about its shear center.

   b. The lateral deformation, δ, followed a linear path along the line between the centroid and shear center perpendicular to the minor axis.

4.2.2 Local Buckling

Local buckling occurs when a compression element in the cross-section becomes locally unstable due to its high width-to-thickness ratio. Thus, any shape, regardless of its symmetry or asymmetry, can be vulnerable to such instability depending on the width-to-thickness ratio (slenderness) and boundary conditions (e.g., stiffened or unstiffened elements) of its cross-sectional elements. Because of the resulting low load-carrying capacity, local buckling is not desired. Therefore, the slender/nonslender limits to avoid local buckling before global buckling are listed for different element types, such as flange or web of an I-shape, in Section 4.3.

Movie 4.6 shows an isometric view of the simulation of a typical, local buckling–prone thin-walled square hollow section. By examining the simulation, it can be seen that local

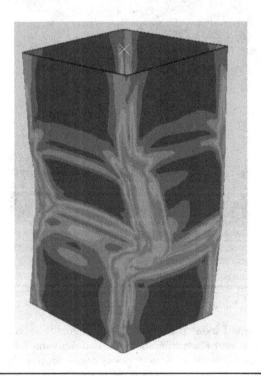

MOVIE 4.6   Local buckling of a thin-walled square tube. (To view the entire movie, go to www.mhprofessional.com/design-of-steel-structures.)

buckling occurs before the buckling of the entire column. Because of the limited capacity of the slender (i.e., too thin or too long or both) compression elements in the cross-section, the critical load due to local buckling limit state is lower than the global buckling limit states for the column composed of slender elements. This type of buckling is attributable to a combination of a large width-to-thickness ratio of the walls and low overall slenderness ratio (KL/r). It should be noted that, as previously demonstrated in Movie 4.3 by the local flange buckling after global buckling, local buckling eventually takes place either before or after global buckling, regardless of the buckling mode. Therefore, a local deformation similar to the one seen in the section view given in Fig. 4.11 can be observed near the midlength of a longer (slenderer) member after global buckling due to the imposed postbuckling bending deformation. Thus, the sequence of local and global buckling is determined by the two influential factors: (1) width-to-thickness ratio of cross-sectional elements, and (2) overall slenderness ratio of compression member (KL/r).

4.2.3   Effect of Imperfections

As previously pointed out, steel columns are not "ideal" for a variety of reasons. The compressive strength of a column is affected by the two major factors, namely, residual stresses and initial out-of-straightness as well as the interaction between them. Even though it is not straightforward to accurately include these initial imperfections (i.e., residual stresses and geometric imperfections) in structural analysis and design, their impact on stiffness and strength can be somehow incorporated when stability is of concern. As for the impact of imperfections on member strength, it has been included in the column strength curves developed by Structural Stability Research Council (SSRC), from which AISC column design formulas originated. Considering the complexity and substantial variability of each effect, this section intends to provide concise information on the subject. Therefore, the significance of the initial straightness and heat treatment–induced effects on the column strength, in particular, will be briefly illustrated

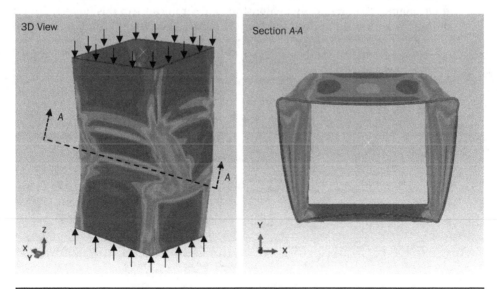

FIGURE 4.11   Local buckling of a thin-walled square tube.

in this section while the effect of some other uncertainties, such as end restraints, is not included in the discussions. A comprehensive literature survey on the influence of imperfections can be found in Ziemian (2010).

In any steel column, an initial curvature would be introduced during the process of making the column. This type of geometric imperfection is also referred to as initial out-of-straightness (or initial crookedness). To comprehend the impact, suppose that an initial deflection, δ_0, exists at a given section before axial load P is applied (initial stage), as shown in Fig. 4.12a. Due to the initial deflection (δ_0), the column is subject to a small bending moment (P times δ_0) as soon as loaded with P, and this initial second-order bending moment (P times δ) grows larger and larger as the lateral deformation due to the applied load, δ_p, increases. As indicated with the load-deformation curves given in Fig. 4.12b, on account of the inelasticity and geometric imperfection effects, the maximum load that can be applied to the column is reduced from Euler buckling load, P_E, to P_{max}, where P_{max} is the ultimate load-carrying capacity of an imperfect inelastic column. Note that P_{max} is higher than the critical load obtained from the tangent modulus theory presented earlier (Shanley, 1947). This reduction is caused by the nonuniform stress distribution that exists from the very beginning, and therefore some portions of the cross-section might have reached the yield stress ($\sigma_c = F_y$) earlier than others due to the compressive stress imposed by combined axial and flexural deformations.

Magnitude of member imperfections is often measured and defined by the ratio of δ_0/L, where δ_0 is the initial deflection at the midlength and L is the total length of the column. The acceptable initial out-of-straightness is defined as $\frac{1}{8}$ in. times the total length (in terms of feet) divided by 10 for W- and HP-shapes (ASTM A6, 2017). The permissible variation for straightness for channels, square and circular HSS, pipes, S- and M-shapes is two times the tolerance specified for W-shapes (i.e., $\frac{1}{8}$ in. times the total length divided by 5) (ASTM A6, 2017).

An experimental study carried out by Bjorhovde et al. (1972) pointed out that the variation in column strength over the changes in the magnitude of initial deflection can be substantial. To numerically demonstrate the significance of this impact, three different imperfection magnitudes, which are L/500, L/1000, and L/2000, were introduced

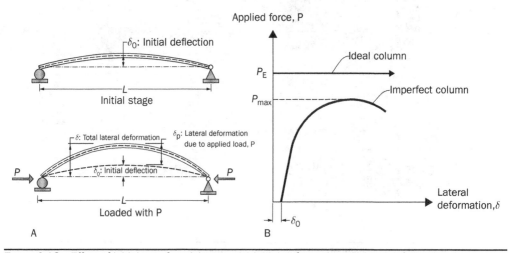

FIGURE 4.12 Effect of initial out-of-straightness. (a) Initial deformation. (b) Load-deformation curve.

to three identical circular tubes (HSS100.500) with a slenderness ratio (KL/r) of 80. The behavior of the three tubes is compared in Movie 4.7 and Fig. 4.13 in terms of buckling sequence and load-deformation curves, respectively. Note that, for convenience, the permissible limits for hot-rolled or built-up W-shapes and tubing are, typically, taken as L/1000 and L/500, respectively (Ziemian, 2010). Thus, the three initial deflection magnitudes introduced in the simulation represent the permitted maxima (i.e., L/500 and L/1000) and a larger magnitude that might stand for the actual δ_0/L value of a cold-formed HSS (i.e., L/2000). In the movie, the magnitude of initial deflection increases from left to right and Fig. 4.13 presents the schematic load-deformation curves of the three simulated specimens. Based on the observations carried out on the simulation given in Movie 4.7, the following can be discussed:

1. Under the same loading rate, buckling initiated in the specimen with the largest initial imperfection (L/500) first, then followed by the buckling of the specimens with relatively smaller initial deflections, L/1000 and L/2000, respectively.

2. The buckling load increases as the magnitude of the initial deflection decreases. Based on the simulation results, the buckling loads corresponding to the magnitudes of L/500, L/1000, and L/2000 are 377, 447 and 507 kips, respectively, which evidently reveals the considerable impact of initial conditions.

3. Besides its impact on the strength, the slight difference in the initial stiffness is also attributed to the magnitude of geometric imperfections, as seen in Fig. 4.13.

Besides geometric imperfections, structural steel is inevitably exposed to plastic deformations during fabrication. These plastic deformations remain after the fabrication processes such as hot-rolling, welding, bending during cold-forming, flame cutting, cold straightening, hole punching, and so forth. Distribution and magnitude of residual stresses due to hot-rolling process, which produces the most significant magnitudes for hot-rolled shapes, is highly dependent on the geometry of a cross-section and its cross-sectional elements (Tall, 1964). For instance, flanges of a W-shape are, usually, more vulnerable to residual stresses than web because of their thickness. Sharp corners also tend to be affected by residual stresses more than other parts. Edges of the flanges, therefore, experience the largest compressive strain amplitudes. Figure 4.14 shows the

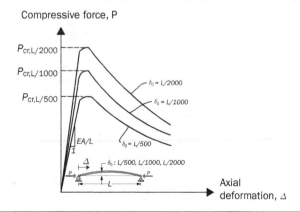

Figure 4.13   Impact of initial out-of-straightness on buckling load (not to scale).

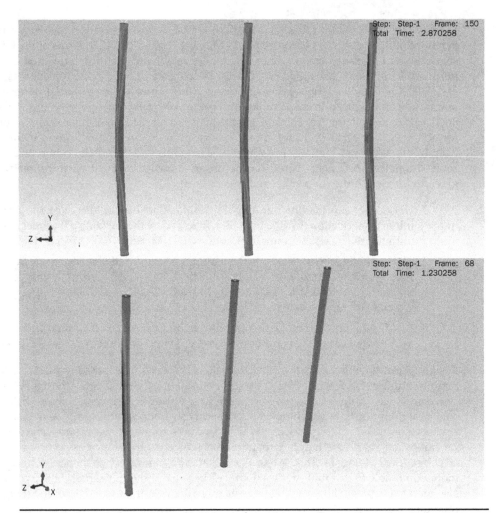

Movie 4.7 Impact of magnitude of initial out-of-straightness. (To view the entire movie, go to www.mhprofessional.com/design-of-steel-structures.)

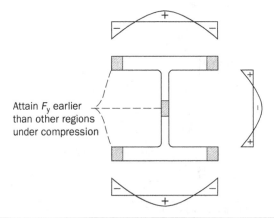

Figure 4.14 Residual stress distribution on an I-shape due to hot-rolling.

residual stress distribution of a wide-flange section. As indicated in the figure, when compressive stress is applied, the dashed regions attain yielding stress earlier than others do. Even though the tensile and compressive strains balance each other over the entire cross-section, this nonlinearity due to residual stresses results in buckling of a compression member earlier than expected.

4.2.4 Effective Length

For the sake of clarity, the discussions on the column behavior and strength have been concentrated on the behavior of isolated simply supported columns till now. A column in an actual structure, however, is often far from being isolated. Thus, the interaction between the connected members should be incorporated in the boundary conditions while designing a compression member. To be able to recognize this impact, the physical meaning of effective length factor should be described and understood first.

Effective length factor, K is defined as the distance from the points of inflection, where the bending moment is zero. Therefore, the theoretical (or elastic) values of K factors for the compression members with ideal boundary conditions can be derived mathematically. The effective length factors for isolated members with ideal supports can be found in *AISC Specification* Comm. 7.2 (AISC, 2016). The support conditions in real structures, however, are often different from ideal supports, such as pinned, fixed, etc. *AISC Specification* (AISC, 2016) suggests approximate values of the effective length factor, K for design to account for the adverse impact of the boundary conditions in real structures on column buckling load. Figure 4.15 summarizes some of the theoretical effective length factors (K) along with the K values recommended by AISC. It is apparent from Figs. 4.15a and b that with the same boundary conditions (e.g., fixed-pin), the compression members incorporated in sway frames (i.e., sidesway uninhibited) results in larger K factors than those in nonsway frames (i.e., sidesway inhibited), which is ascribable to the influence of lateral translation at joints on the effective length and accordingly on the column buckling load.

Likewise, inspection of the deformed shape given in Fig. 4.16 would better explain the influence of joint translation on the effective length. The figure represents the deformed shape of a moment frame after one of the columns buckles in the plane. The moment frame with rigid joints that are able to rotate and translate laterally deforms so that the distance between the inflection points after buckling becomes larger than its original length. Therefore, the effective length, KL is greater than L for a column in a moment frame that is pinned at base. Note that buckling of either one of the columns given in Fig. 4.16 might give rise to the same schematic deformed shape given in the figure.

Effective lengths of a column in perpendicular planes might substantially affect the column behavior as well as the compressive strength by altering the buckling mode. As pointed out in Sec. 2.1.1, columns buckle about the axis corresponding to the largest slenderness ratio, which is a function of effective length, KL, and radius of gyration, r. Therefore, the axis of buckling is controlled by the interconnection of the ratio of r_x/r_y and KL_x/KL_y. Figure 4.17 demonstrates buckled shapes of a W-shape about the two perpendicular axes. The column would buckle about either x- or y-axis. Effective length in x-direction, KL_x, equals to the entire length of the member and the buckled shape would be similar to a half-sine curve. In case of minor-axis buckling, however, the curvature of

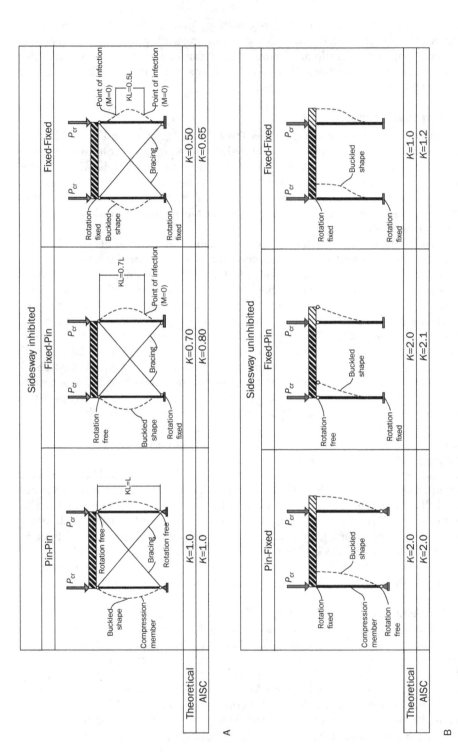

Figure 4.15 Theoretical and recommended K values. (*a*) Columns in braced frames. (*b*) Columns in unbraced frames.

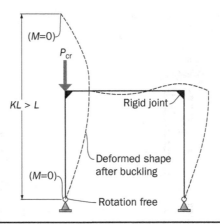

$(M=0)$

P_{cr}

$KL > L$

Rigid joint

Deformed shape
after buckling

$(M=0)$

Rotation free

Figure 4.16   Effective length of a column in a rigid (moment) frame.

the deformed shape changes at the point where the lateral movement is prevented, and forms an S-shape due to the lateral restrain in the plane of buckling. Thus, the effective length in y-direction, KL_y, is taken as the larger of the two buckling lengths, KL_y1 and KL_y2. Comparing the two possibilities, one can conclude that the column would buckle about x-x axis if r_x/r_y ratio is smaller than KL_x/KL_y, column would buckle about y-y axis if otherwise.

Effective length of a compression member may increase or decrease depending on its end conditions. We will consider the two extreme cases given in Fig. 4.15b for two identical columns in unbraced frames: Case (a) Fixed at base and pinned at joint, and Case (b) Fixed at both base and joint. As given in Fig. 4.15b, theoretical effective length factors for case (a) and case (b) would be 2.0 and 1.0, respectively. In other words, buckling length of a column increases as the end rotations become more substantial. Since a rigid joint is not fully fixed (no rotation) or perfectly pinned (free to rotate), the resistance against the end rotation provided by the members connected to a joint, such as girders, would be somewhere between fixed and pinned conditions. For determining K factor of a column in a moment frame, in which the beams and columns are rigidly connected, using alignment charts are the most common method. These charts are developed to account for the relative member stiffness connected to a rigid joint and accordingly their impact on the end rotations and buckling length. However, one should notice certain assumptions on which the alignment charts are based. For instance, columns are assumed to be prismatic and elastic, and all columns in a story buckle simultaneously and so on. Therefore, effective length factors can be adjusted by introducing a stiffness reduction factor, τ_b, that accounts for the inelasticity.

The alignment charts for sway and nonsway frames are shown in Fig. 4.18. As seen in Fig. 4.18b, the chart applicable to nonsway frames (i.e., the frames whose lateral stability is provided by braces, shear walls, or equivalent means), K factors for nonsway frames vary between 0.5 and 1.0. However, K factors for the members in braced frames are often conservatively taken as 1.0 in design. K factors for the unbraced (moment) frames range from 1.0 to infinity and can be determined using the chart in Fig. 4.18a. To use these

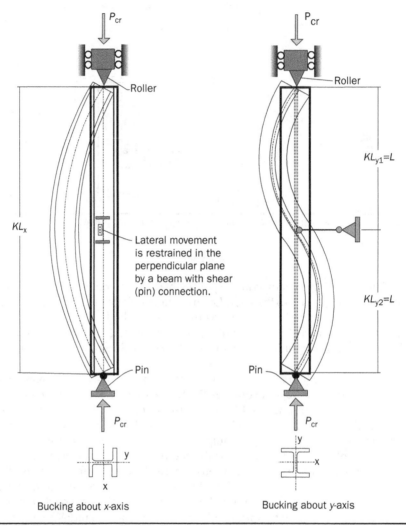

FIGURE 4.17 Buckled shapes of a wide flange.

charts, G_A and G_B values, which represent the relative rigidity of a column connected to joints A and B, respectively, need to be determined using the following formula:

$$G = \frac{\sum(E_c I_c / L_c)}{\sum(E_g I_g / L_g)} = \frac{\sum(I_c / L_c)}{\sum(I_g / L_g)} \tag{4.5}$$

where the subscripts c and g refer to columns and girders that are connected to a joint. As can be interpreted from the alignment charts, effective length factor, K and G values are directly proportional to each other. Furthermore, as expressed in Eq. (4.5), G values are larger when columns are stiffer than girders. In other words, with the same column size, if a stiffer girder is connected to a joint, the buckling strength of the column would

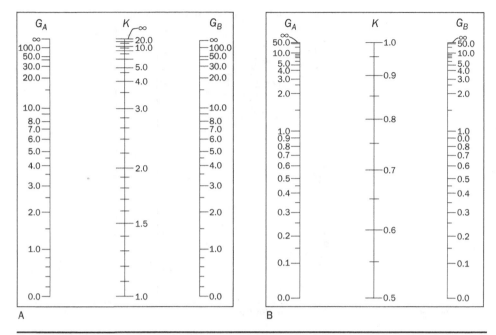

FIGURE 4.18 Alignment charts. (*a*) Sidesway uninhibited. (*b*) Sidesway inhibited. (Adapted from AISC Specification Comm. 7.2.)

increase owing to the fact that the stiffer girder minimizes the rotation at the column end more than its less stiffer counterpart. Note that G is taken as 10 and 1.0 for pin and fixed support, respectively, in lieu of the theoretical values.

Example 4.1

Determine the effective length factor, K, for the first and second story columns in the moment frame shown in Fig. 4.19. Moment frame is braced at each story level in the perpendicular direction.

$$\underline{W12 \times 79}: I_x = 662 \text{ in.}^4; \underline{W12 \times 30}: I_x = 238 \text{ in.}^4; \underline{W16 \times 40}: I_x = 518 \text{ in.}^4$$

Determine G_A, G_B, and G_C.

$$G_A = 1.0 \text{ (Fixed support)}$$

$$G_B = \frac{(662 \,/\, 12) + (662 \,/\, 15)}{(518 \,/\, 25)} = 4.80$$

$$G_C = \frac{(662 \,/\, 12)}{238 \,/\, 25} = 5.80$$

Enter the alignment chart for sway frames with G_A, G_B, and G_C values.

$$\text{Member AB} \rightarrow \begin{vmatrix} G_A = 1.0 \\ G_B = 4.80 \end{vmatrix} \rightarrow K_{x,\text{AB}} \approx 1.68 \qquad \text{(Fig. 4.20a)}$$

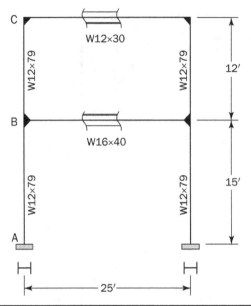

FIGURE **4.19** A two-story one-bay moment frame.

$$\text{Member BC} \rightarrow \begin{vmatrix} G_B = 4.80 \\ G_C = 5.80 \end{vmatrix} \rightarrow K_{x,BC} \approx 2.27 \qquad \text{(Fig. 4.20}b\text{)}$$

4.2.5 Concluding Remarks

Buckling of various compression members has been classified based on their buckling and postbuckling behavior. In view of the simulation-based discussions on column behavior, "efficiency" of a column section in terms of high load-bearing capacity-to-weight ratio is closely related to its cross-sectional properties and interrelation between them. Omitting the impact of uncertainties, such efficiency can be associated with the following characteristics (Ziemian, 2010):

1. Intrinsic buckling tendency of a cross-section (e.g., a cruciform shape build up from thin plates or a circular tubing);

2. The magnitude of the governing radius of gyration;

3. The ratio of radii of gyration about the principal axes (e.g., the ratio of r_x/r_y of a W-shape);

4. Slenderness (i.e., width-to-thickness ratio) of the compression elements composing the cross-section.

Hence, considering the applicable limit states and the parameters that affect the governing limit state, a column can be designed more efficiently. The following section aims to set forth the design requirements stipulated in the most recent *AISC Specification* as well as to demonstrate the procedure to achieve a safe and economical design together.

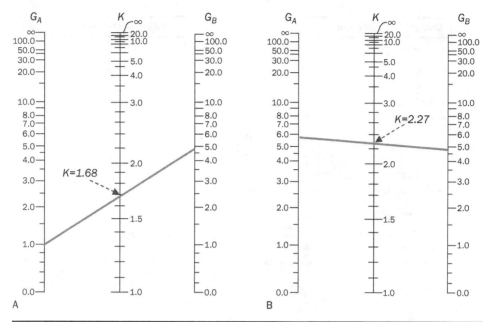

Figure 4.20 Effective length factors for the frame given in Fig. 4.18. (a) Member AB. (b) Member BC.

4.3 AISC Design Requirements

4.3.1 General Remarks

The design strength for compression members are given as $\phi_c P_n$ and P_n/Ω_c for LRFD and ASD approaches, respectively, where P_n is the nominal compressive strength, $\phi_c = 0.90$ (LRFD) and $\Omega_c = 1.67$ (ASD). *The Specification* defines the nominal compressive strength, P_n, as the lowest compressive strength obtained from the applicable limit states, which are flexural, torsional, and flexural-torsional buckling.

Design requirements for compression members are divided into subsections to study the major limit states separately. Therefore, the requirements for the members vulnerable to global and local buckling limit states will be presented as compression members with and without slender elements, respectively. In addition, design requirements for built-up compression members and their connectors are also summarized without limit state distinction.

4.3.2 Compressive Strength of Nonslender Members

Compressive strength of a compression member is associated with the failure mode. As explained in the previous sections, the limit states applicable to columns are basically global buckling and local buckling failure modes. This section will mainly cover design for global buckling limit states with an emphasis on flexural buckling in particular, which is the most common type that usually controls the design. Classification of compression elements for local buckling is provided in *Section B4, AISC Specification* based on the boundary conditions (stiffened or unstiffened) and width-to-thickness ratio of

Compression element	Limiting width-to-thickness ratio, λ_r
 I-Shape Channel Tee Shape	$\lambda_r = 0.56\sqrt{\dfrac{E}{F_y}}$
 Unequal-leg Angle Equal-leg Angle Double Angle	$\lambda_r = 0.45\sqrt{\dfrac{E}{F_y}}$
 I-Shape Channel Built-up I-Shape	$\lambda_r = 1.49\sqrt{\dfrac{E}{F_y}}$
 Built-up I-Shape	$\lambda_r = 0.64\sqrt{\dfrac{k_c E}{F_y}}$ $0.35 \leq k_c = 4/\sqrt{h/t_w} \leq 0.76$
 Tee Section	$\lambda_r = 0.75\sqrt{\dfrac{E}{F_y}}$
 Rectangular Hollow Structural Section (HSS)	$\lambda_r = 1.40\sqrt{\dfrac{E}{F_y}}$
 Circular Hollow Structural Section (HSS)	$\lambda_r = 0.11\dfrac{E}{F_y}$

Source: Adapted from Table B4.1a, AISC 360-16.

TABLE 4.1   Limiting Width-to-Thickness Ratios for Compression Elements Subject to Axial Compression

the element. Compression elements are classified as slender and nonslender. The limiting width-to-thickness ratios for common shapes are shown in Table 4.1. If a compression element's width-to-thickness ratio, such as b/t or D/t, is less than the limiting width-to-thickness ratio, λ_r, that element is considered as nonslender element and thus local buckling limit state does not apply.

4.3.2.1   Flexural Buckling

The nominal compressive strength, P_n, for flexural buckling of members without slender elements is determined based on the following formula in terms of average critical stress, F_{cr}, over the gross area, A_g:

$$P_n = F_{cr} A_g \tag{4.6}$$

As indicated in Fig. 4.21, AISC column design curve refers to two flexural buckling types as inelastic and elastic buckling and thus consists of two segments defined by the following equations:

$$F_{cr} = \left(0.658^{F_y/F_e}\right) F_y \text{ when} \frac{L_c}{r} \le 4.71 \sqrt{\frac{E}{F_y}} \left(\text{or } \frac{F_y}{F_e} \le 2.25 \right) \tag{4.7}$$

$$F_{cr} = 0.877 F_e \text{ when} \frac{L_c}{r} > 4.71 \sqrt{\frac{E}{F_y}} \left(\text{or } \frac{F_y}{F_e} > 2.25 \right) \tag{4.8}$$

where Fe is the elastic buckling stress (Euler's equation). Note that Eqs. (4.7) and (4.8) apply to inelastic and elastic buckling, respectively. Elastic buckling stress is determined according to Eq. (4.9), where L_c and r represent effective length (KL) and radius of gyration, respectively.

$$F_e = \frac{\pi^2 E}{\left(L_c / r\right)^2} \tag{4.9}$$

AISC Specifications based the column design formulas on the multiple columns strength curves developed by SSRC. In order to account for the initial imperfections, an extensive collection of experimental results from different research groups (Tall, 1964; Bjorhovde et al., 1972) was utilized for the development of these curves. A summary of the development of these curves can be found in AISC Comm. E1 and Ziemian (2010). Figure 4.21 shows the experimental data on the strength of real steel columns representatively along with Euler's curve and the current AISC column design curve in terms of the critical stress and slenderness ratio (L_c/r stands for KL/r). Each dot represents a tested column specimen and the AISC column design curve (presented with a solid line in Fig. 4.21) is the envelope of the test data obtained by curve fitting. By comparing Euler's theoretic column strength of an ideal column (dashed line in Fig. 4.21) with the tested real columns, the following observations can be made:

1. Euler's buckling load, based on ideal column, is the upper bound of column strength.

2. The strength of a "short" column (with small L_c/r) and slender column (with large L_c/r) are controlled by yielding (F_{cr} is close to F_y) and elastic buckling (close to Euler's theoretical estimation), respectively.

3. Euler's and the column design curves incline to converge as overall slenderness ratio increases, where columns tend to buckle elastically. Still, Euler's formula gives a larger estimation due to the ideal column assumptions other than elastic material.

4. Strength of a real column exhibits a large range of variations due to deviations from an "ideal" column case. Note that this variation originates mainly from the uncertainties in the initial imperfections.

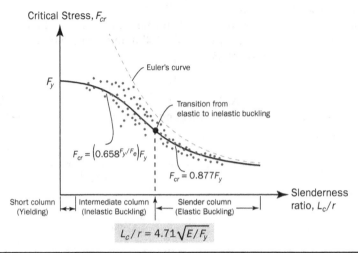

FIGURE 4.21 AISC column design curve.

AISC column strength curve consists of three regions. Referring to Fig. 4.21 and Eqs. (4.7) through (4.9), the following can be discussed:

- In "short column" region, slenderness ratio is so small that the critical stress approaches to yielding stress column $F_{cr} \cong F_y$. Therefore, column undergoes compressive yielding rather than buckling.

- In "intermediate column" region, where slenderness ratio varies roughly between 20 and 130, inelastic buckling occurs and the columns with slenderness ratios within this region are sensitive to both slenderness ratio (L_c/r) and material strength (F_y).

- In "elastic column" region, compressive strength of column ($P_n = F_{cr}A_g$) is fully dependent on L_c/r, and independent of material strength (F_y) because of elastic buckling. In this case, high strength of material (high F_y) does not increase the column strength, P_n. Note that the reduction factor of 0.877 in Eq. (4.8) accounts for initial imperfections that exist in any steel column.

Basic examples of column strength calculation can be seen in Examples 4.2 and 4.3.

Example 4.2

Determine the nominal compressive strength of the column considering the following two cases depicted in Fig. 4.22. The 30-ft-long column is made of W14 × 74 with $F_y = 70$ ksi.

(a) In Case A, the column is pinned from both ends.

(b) In Case B, with the same boundary conditions, the column is restrained against the lateral movement in y-direction.

Solution:

<u>W14 × 74:</u> $A_g = 21.8 \ in.^2;\ r_x = 6.04 \ in.;\ r_y = 2.48 \ in.;\ b/2t_f = 6.41;\ h/t_w = 25.4;\ L = 30 \ ft;$
$F_y = 70 \ ksi$

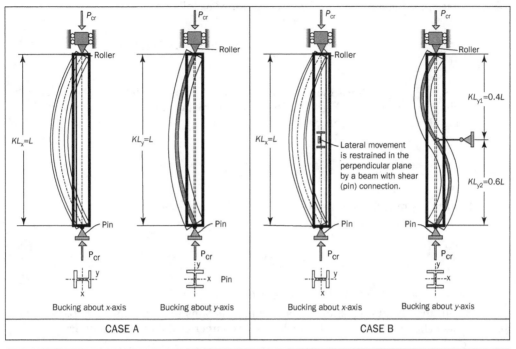

FIGURE 4.22 Example 4.2.

Check width-to-thickness ratio according to Table B4.1a, AISC 360.

$$\lambda_{r,f} = 0.56\sqrt{\frac{E}{F_y}} = 0.56\sqrt{\frac{29,000}{70}} = 11.39 > b \, / \, 2t_f = 6.41$$

$$\lambda_{r,w} = 1.49\sqrt{\frac{E}{F_y}} = 1.49\sqrt{\frac{29,000}{70}} = 30.3 > h/t_w = 25.4$$

Thus, nonslender compression element. Local buckling limit state does not apply. *Determine the nominal compressive strength for Case A.*

$$K_x = K_y = 1.0$$

$$\left. \begin{array}{l} \left(\dfrac{L_c}{r}\right)_x = \left(\dfrac{KL}{r}\right)_x = \dfrac{1.0 \times (30 \times 12)}{6.04} = 59.6 \\[2mm] \left(\dfrac{L_c}{r}\right)_y = \left(\dfrac{KL}{r}\right)_y = \dfrac{1.0 \times (30 \times 12)}{2.48} = 145.2 \end{array} \right\} \left(\dfrac{L_c}{r}\right)_x < \left(\dfrac{L_c}{r}\right)_y$$

Thus, buckling occurs about *y-y* axis. The governing slenderness ratio is the larger of $(L_c/r)_x$ and $(L_c/r)_y$.

$$\frac{L_c}{r} = \left[\left(\frac{L_c}{r}\right)_x ; \left(\frac{L_c}{r}\right)_y\right]_{\max} = 145.2$$

To determine the buckling type, L_c/r is examined as follows:

$$\frac{L_c}{r} = 145.2 > 4.71\sqrt{\frac{E}{F_y}} = 96 \text{ (For } F_y = 70 \text{ ksi steel)}$$

where $4.71\sqrt{E/F_y}$ defines the slenderness ratio at transition from elastic to inelastic buckling, as illustrated in Fig. 4.21. Therefore, elastic buckling occurs about y-y axis.

Elastic critical stress can be computed according to *Eq. E3-4, AISC Specification* as follows:

$$F_e = \frac{\pi^2 E}{(L_c/r)^2} = \frac{\pi^2 \times 29,000}{(145.2)^2} = 13.6 \text{ ksi}$$

The critical stress, F_{cr}, and the nominal strength, P_n, become

$$F_{cr} = 0.877 F_e = 0.877 \times 13.6 = 11.9 \text{ ksi}$$

$$P_n = F_{cr} A_g = 11.9 \times 21.8 \cong 260 \text{ kips}$$

Determine the nominal compressive strength for Case B.

The effective length for strong-axis buckling remains the same for Case B. However, KL_y will change in Case B. KL_y is taken as the larger of the two buckling lengths, KL_{y1} and KL_{y2}.

$$KL_x = L = 30 \text{ ft.}$$
$$KL_y = \{KL_{y1}; KL_{y2}\}_{max} = 0.6L = 18 \text{ ft.}$$

$$\left(\frac{L_c}{r}\right)_x = \frac{(30 \times 12)}{6.04} = 59.6 < \left(\frac{L_c}{r}\right)_y = \frac{(18 \times 12)}{2.48} = 87 \text{ (Buckling about y-y axis)}$$

$$\frac{L_c}{r} = \left[\left(\frac{L_c}{r}\right)_x ; \left(\frac{L_c}{r}\right)_y\right]_{max} = 87$$

$$\frac{L_c}{r} = 87 < 4.71\sqrt{\frac{E}{F_y}} = 96$$

Therefore, inelastic buckling occurs about y-y axis.
The elastic critical stress is as follows:

$$F_e = \frac{\pi^2 E}{(L_c/r)^2} = \frac{\pi^2 \times 29,000}{(87)^2} = 37.8 \text{ ksi}$$

The critical stress, F_{cr}, and the nominal strength, P_n, are

$$F_{cr} = \left(0.658^{F_y/F_e}\right)F_y = \left(0.658^{70/37.8}\right) \times 70 = 32.2 \text{ ksi}$$

$$P_n = F_{cr} A_g = 32.2 \times 21.8 \cong 703 \text{ kips}$$

Example 4.3
Determine the nominal compressive strength of column AB in Example 4.1 (Fig. 4.19).

Solution:
W12 × 79: $A_g = 23.2$ in.^{2}; $r_x = 5.34$ in.; $r_y = 3.05$ in.; $b/2t_f = 6.41$; $h/t_w = 25.4$; $F_y = 50$ ksi
Check width-to-thickness ratio according to Table B4.1a, AISC 360.

$$\lambda_{r,f} = 0.56\sqrt{\frac{E}{F_y}} = 0.56\sqrt{\frac{29,000}{50}} = 13.5 > b/2t_f = 8.22$$

$$\lambda_{r,w} = 1.49\sqrt{\frac{E}{F_y}} = 1.49\sqrt{\frac{29,000}{50}} = 35.9 > h/t_w = 20.7$$

W12 × 79 is a nonslender compression member (i.e., local buckling limit state does not apply) and the unbraced torsional and flexural buckling lengths are the same. Thus, the limit state of flexural buckling applies (*Section E3, AISC 360*).

Determine the nominal compressive strength.

$K_x = 1.68$; $K_y = 1.0$ (Braced in the perpendicular direction)

$$\left(\frac{L_c}{r}\right)_x = \frac{1.68 \times (15 \times 12)}{5.34} = 56.6 < \left(\frac{L_c}{r}\right)_y = \frac{1.0 \times (15 \times 12)}{3.05} = 59 \text{(Buckling about } y\text{-}y \text{ axis)}$$

Since $\dfrac{L_c}{r} = 59 < 4.71\sqrt{\dfrac{29,000}{50}} = 113$, the column falls into Region II (intermediate column) of the column strength curve given in Fig. 4.21.

Thus, inelastic flexural buckling occurs about y-y axis.

$$F_e = \frac{\pi^2 E}{\left(L_c/r\right)^2} = \frac{\pi^2 \times 29,000}{(59)^2} = 82.2 \text{ ksi}$$

$$F_{cr} = \left(0.658^{F_y/F_e}\right)F_y = \left(0.658^{50/82.2}\right) \times 50 = 38.8 \text{ ksi}$$

$$P_n = F_{cr}A_g = 38.8 \times 23.2 = 899 \text{ kips}$$

4.3.2.2 Torsional and Flexural-Torsional Buckling
Design engineers are usually inclined to use symmetrical shapes, such as W-shapes or HSS, as columns. When such shapes are utilized, the critical buckling load is often (always for HSS) determined by the flexural buckling stress, given that the buckling lengths for torsion and flexure are close. However, in certain situations, use of singly symmetric or even unsymmetrical shapes might be necessary. Therefore, *AISC Specification* requires investigation of the critical stresses due to torsional and flexural-torsional buckling as per *Section E4* when applicable. *The Specification* defines three

cases for torsional and flexural-torsional buckling of nonslender columns, which are listed in what follows:

(a) Doubly symmetric compression members twisting about the shear center. Note that the shear center and the center of gravity coincide for symmetrical shapes. Therefore, this item defines torsional buckling. The elastic torsional buckling stress is determined as follows:

$$F_e = \left(\frac{\pi^2 E C_w}{L_{cz}^2} + GJ\right)\frac{1}{I_x + I_y} \tag{4.10}$$

(b) For singly symmetric compression members, which are prone to flexural-torsional buckling, the elastic flexural-torsional buckling stress is

$$F_e = \left(\frac{F_{ey} + F_{ez}}{2H}\right)\left[1 - \sqrt{1 - \frac{4F_{ey}F_{ez}H}{\left(F_{ey} + F_{ez}\right)^2}}\right] \tag{4.11}$$

Note that the above equation is applicable when the axis of symmetry is y-axis. F_{ey} term should be replaced with F_{ex} when a member is symmetrical about x-axis.

(c) For unsymmetric compression members, which are prone to flexural-torsional buckling, the elastic flexural-torsional buckling stress, $F_{e'}$ is the lowest root of Eq. (4.12).

$$\left(F_e - F_{ex}\right)\left(F_e - F_{ey}\right)\left(F_e - F_{ez}\right) - F_e^2\left(F_e - F_{ey}\right)\left(\frac{x_0}{\bar{r}_0}\right)^2 - F_e^2\left(F_e - F_{ex}\right)\left(\frac{y_0}{\bar{r}_0}\right)^2 = 0 \tag{4.12}$$

The terms used in the above equations are defined as follows:

L_{cz} = effective length for torsional buckling
C_w = warping constant
G = shear modulus of elasticity of steel
J = torsional constant
H = flexural constant
$\bar{r}_0$ = polar radius of gyration about the shear center
x_0 and y_0 = coordinates of the shear center with respect to the centroid

The stress components and section properties required to calculate the terms given in Eq. (4.12) can be computed using Eqs. (4.13) through (4.17).

$$F_{ex} = \frac{\pi^2 E}{\left(L_{cx}/r_x\right)^2} \tag{4.13}$$

$$F_{ey} = \frac{\pi^2 E}{\left(L_{cy}/r_y\right)^2} \tag{4.14}$$

$$F_{ez} = \left(\frac{\pi^2 E C_w}{L_{cz}^2} + GJ\right)\frac{1}{A_g \bar{r}_0^2} \tag{4.15}$$

$$H = 1 - \frac{x_0^2 + y_0^2}{\overline{r}_0^2} \tag{4.16}$$

$$\overline{r}_0 = \sqrt{x_0^2 + y_0^2 + \frac{I_x + I_y}{A_g}} \tag{4.17}$$

Example 4.4

Compute the nominal compressive strength of a C10 × 25 made of A36 steel ($F_y = 36$ ksi). The effective lengths with respect to x-, y-, and z-axes are 15 ft., 5 ft., and 15 ft., respectively (Fig. 4.23).

Solution:

Design of a single channel is controlled either by flexural buckling about the axis of unsymmetry (y-axis in this case) or by flexural-torsional buckling about the axis of symmetry (x-axis) and the longitudinal axis (z-axis). Thus, the applicable limit states are

1. Flexural buckling about y-axis (the axis of unsymmetry)
2. Flexural-torsional buckling about x- and z-axes (axis of symmetry and longitudinal axis, respectively)

Since the section is nonslender, the smaller of the stresses due to the two potential limit states will govern.

<u>C10 × 25:</u> $A_g = 7.34$ in.^{2}; $r_x = 3.52$ in.; $r_y = 0.675$ in.; $C_w = 68.3$ in^6; $H = 0.912$; $J = 0.69$ in^4; $\overline{r}_0 = 3.76$ in; $G = 11,200$ ksi; $F_y = 36$ ksi; $L_{cx} = L_{cz} = 15$ ft.; $L_{cy} = 5$ ft.

Determine the elastic flexural buckling stress about y-axis.

$$F_e = \frac{\pi^2 E}{\left(L_{cy}/r_y\right)^2} = \frac{\pi^2 \times 29,000}{\left(\dfrac{5 \times 12}{0.675}\right)^2} = 36.2 \text{ ksi}$$

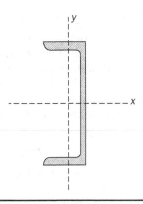

FIGURE 4.23 Example 4.4.

Determine the elastic flexural-torsional buckling stress about x- and z-axes.

$$F_{ex} = \frac{\pi^2 E}{(L_{cx}/r_x)^2} = \frac{\pi^2 \times 29,000}{\left(\dfrac{15 \times 12}{3.52}\right)^2} = 109.5 \text{ ksi}$$

$$F_{ez} = \left(\frac{\pi^2 E C_w}{L_{cz}^2} + GJ\right)\frac{1}{A_g \bar{r}_0^2} = \left(\frac{\pi^2 \times 29,000 \times 68.3}{(15 \times 12)^2} + 11,200 \times 0.69\right) \times \frac{1}{7.34 \times 3.76^2} = 80.3 \text{ ksi}$$

$$F_e = \left(\frac{F_{ex} + F_{ez}}{2H}\right)\left[1 - \sqrt{1 - \frac{4F_{ex}F_{ez}H}{(F_{ex} + F_{ez})^2}}\right] = \left(\frac{109.5 + 80.3}{2 \times 0.912}\right)\left[1 - \sqrt{1 - \frac{4 \times 109.5 \times 80.3 \times 0.912}{(109.5 + 80.3)^2}}\right]$$

$$F_e = 69.6 \text{ ksi}$$

The elastic buckling stress about y-axis is lower than that of flexural buckling stress. Thus, flexural buckling about y-axis governs.

$$F_e = \{36.2 \text{ ksi}; 69.6 \text{ ksi}\}_{min} = 36.2 \text{ ksi}$$

The critical stress, F_{cr}, can be computed as follows:

$$F_{cr} = \left(0.658^{F_y/F_e}\right)F_y \qquad \text{when} \frac{F_y}{F_e} \leq 2.25$$

$$F_{cr} = 0.877 F_e \qquad \text{when} \frac{F_y}{F_e} > 2.25$$

Since $\dfrac{F_y}{F_e} = \dfrac{36 \text{ ksi}}{36.2 \text{ ksi}} \cong 1.0 < 2.25$, inelastic flexural buckling about y-y axis occurs.

$$F_{cr} = \left(0.658^{F_y/F_e}\right)F_y = \left(0.658^{36/36.2}\right) \times 36 = 23.6 \text{ ksi}$$

$$P_n = F_{cr}A_g = 23.6 \times 7.34 = 173 \text{ kips}$$

4.3.3 Compressive Strength of Slender Members

The Specification considers the impact of potential local instabilities (i.e., local buckling) by introducing reduction factors for the area of each slender compression element. When an element's width-to-thickness ratio exceeds the limits summarized in Table 4.1, it is considered slender and thus the reduction in the load-carrying capacity of the member due to local buckling of that element is accounted for by replacing the gross area with an effective area when computing the available strength. To determine whether or not a reduction in the gross area is required, the following conditions given in *Eqs. E7-2 and E7-3, the Specification* need to be checked first.

$$\text{When } \lambda \leq \lambda_r \sqrt{F_y / F_{cr}} \rightarrow \text{No reduction is required} (b_e = b) \tag{4.18}$$

$$\text{When } \lambda > \lambda_r \sqrt{F_y / F_{cr}} \rightarrow b_e = b\left(1 - c_1 \sqrt{\frac{F_{el}}{F_{cr}}}\right)\sqrt{\frac{F_{el}}{F_{cr}}} \tag{4.19}$$

where F_{el} is the elastic local buckling stress, b is the width of the element of which the area is to be reduced, b_e is the effective width, and $c1$ is the empirical correction factor. Based on the width-to-thickness ratio of the element, elastic local buckling stress is determined as follows:

$$F_{el} = \left(c_2 \frac{\lambda_r}{\lambda}\right)^2 F_y \tag{4.20}$$

where c_2 is the effective width adjustment factor dependent on c_1. Effective width factors, c_1 and c_2, are given in Table 4.2.

The compressive strength is determined by assuming that the average critical compressive stress is uniformly distributed over the effective area, as indicated in Eq. (4.21).

$$P_n = F_{cr} A_e \tag{4.21}$$

where A_e is the effective area.

Likewise, to account for the interaction between local and global buckling of round HSS and pipe, *the Specification* considers the following diameter-to-wall thickness ratio (D/t) limits.

$$\text{When } \frac{D}{t} \leq 0.11 \frac{E}{F_y} \rightarrow \text{No reduction is required}\left(A_e = A_g\right) \tag{4.22}$$

$$\text{When } 0.11 \frac{E}{F_y} < \frac{D}{t} < 0.45 \frac{E}{F_y} \rightarrow A_e = \left[\frac{0.038E}{F_y(D/t)} + \frac{2}{3}\right] A_g \tag{4.23}$$

It should be noted that employing a round HSS (or pipe) with a D/t ratio of greater than $0.45E/F_y$ is not recommended.

Example 4.5

Determine the nominal compressive strength of an 8-ft-long W14×26 (A992) column with pinned-pinned boundary conditions.

<u>W14 × 26:</u> $A_g = 7.69$ in.²; $r_x = 5.65$ in.; $r_y = 1.08$ in.; $b_f/2t_f = 5.98$; $h/t_w = 48.1$; $h = 12.26$ in.; $t_w = 0.255$ in.

<u>ASTM A992:</u> $F_y = 50$ ksi

Slender element description	c_1	c_2
Stiffened elements* (e.g., webs of channels or I-shapes)	0.18	1.31
Walls of rectangular HSS	0.20	1.38
All other elements	0.22	1.49

*Walls of rectangular HSS are omitted.

Source: Adapted from *Table E7.1, Spec.*

TABLE 4.2 Effective Width Imperfection Adjustment Factors

Solution:

Applicable limit states:

1. Flexural buckling about strong or minor axis
2. Torsional buckling about longitudinal axis
3. Local buckling of web and/or flanges

Note that torsional buckling limit state is very unlikely to govern for doubly symmetric shapes (e.g., W-shapes) when $L_{cx} = L_{cy} = L_{cz}$. Therefore, the elastic torsional buckling stress calculations are not shown here.

Determine the buckling lengths.

$L_{cx} = L_{cy} = 8$ ft. Thus, flexural buckling occurs about the minor axis (y-y axis).
Determine elastic flexural buckling stress.

$$L_{cy}/r_y = \frac{(8 \times 12)}{1.08} = 88.9$$

$$F_e = \frac{\pi^2 E}{\left(L_{cy}/r_y\right)^2} = \frac{\pi^2 \times 29,000}{88.9^2} = 36.2 \text{ ksi}$$

Since $\dfrac{F_y}{F_e} = \dfrac{50 \text{ ksi}}{36.2 \text{ ksi}} = 1.35 < 2.25$ (or $L_{cy}/r_y = 88.9 < 4.71\sqrt{E/F_y} = 113$), inelastic

buckling occurs.

$$F_{cr} = \left(0.658^{F_y/F_e}\right)F_y = \left(0.658^{50/36.2}\right) \times 50 = 28.1 \text{ ksi}$$

Check slender/nonslender criteria according to Table B4.1a, AISC 360-16.

$$\lambda_{r,f} = 0.56\sqrt{\frac{E}{F_y}} = 0.56\sqrt{\frac{29,000}{50}} = 13.5 > \lambda_f = b_f/2t_f = 5.91 \rightarrow \text{Nonslender flange}$$

$$\lambda_{r,w} = 1.49\sqrt{\frac{E}{F_y}} = 1.49\sqrt{\frac{29,000}{50}} = 35.9 < \lambda_w = h/t_w = 48.1 \rightarrow \text{Slender web}$$

Hence, effective area of the member might be reduced as per *Section E7, Specification* to account for the potential capacity reduction due to local buckling of the web.

$$\lambda_w = 48.1 > \lambda_{r,w}\sqrt{F_y/F_{cr}} = 35.9 \times \sqrt{50/28.1} = 47.9$$

Therefore, instead of gross area, effective area of the slender web should be used when computing the compressive strength. The reduction according to *Eq. E7-3* is as follows:

$$F_{el} = \left(c_2\frac{\lambda_r}{\lambda}\right)^2 F_y = \left(1.31 \times \frac{35.9}{48.1}\right)^2 \times 50 = 47.8 \text{ ksi}$$

$$h_e = h\left(1 - c_1\sqrt{\frac{F_{el}}{F_{cr}}}\right)\sqrt{\frac{F_{el}}{F_{cr}}} = 12.26 \times \left(1 - 0.18 \times \sqrt{\frac{47.8}{28.1}}\right) \times \sqrt{\frac{47.8}{28.1}} = 12.24 \text{ in.}$$

Note that the effective width imperfection factors, c_1 and c_2, are taken from Table 4.2 as 0.18 and 1.31, respectively.

The reduction in the web area is less than 1% ($h_e \cong h$).

$$A_e = A_g - A_{\text{reduction}} = A_g - (h - h_e)t_w = 7.69 - (12.26 - 12.24) \times 0.255 = 7.68 \text{ in.}^2$$

Determine the nominal strength.

$$P_n = F_{cr}A_e = 28.1 \times 7.68 = 215.9 \text{ kips}$$

Example 4.6

Determine the nominal compressive strength of the WT6 × 8 (A572 Gr. 50) shown in Fig. 4.24 based on the AISC design requirements. The lateral movement in y-direction prevented by an additional restraint provided at the flange of the WT (close to the shear center).

Solution:

WT6 × 8: $A_g = 2.36$ in.^{2}; $r_x = 1.92$ in.; $r_y = 0.773$ in.; $b_f/2t_f = 7.53$; $d/t_w = 27.3$; $\bar{y} = 1.74$ in.; $C_w = 0.0678$ in.^{6}; $J = 0.0511$ in.^{4}; $I_x = 8.70$ in.^{4}; $I_y = 1.41$ in.^{4}

ASTM A572 Gr. 50: $F_y = 50$ ksi

Applicable limit states:

1. Flexural buckling about x-x axis
2. Flexural-torsional buckling about y-y and z-z axes
3. Local buckling of flange and/or stem

Determine the buckling lengths.

$$L_{cx} = 12 \text{ ft.}; L_{cy} = 6 \text{ ft.}; L_{cz} = 12 \text{ ft.}$$

Determine elastic flexural buckling stress about x-axis.

$$L_{cx}/r_x = \frac{(12 \times 12)}{1.92} = 75 \rightarrow F_{ex} = \frac{\pi^2 E}{(L_{cx}/r_x)^2} = \frac{\pi^2 \times 29,000}{75^2} = 50.9 \text{ ksi}$$

Determine elastic flexural buckling stress about y-axis.

$$L_{cy}/r_y = \frac{(6 \times 12)}{0.773} = 93.1 \rightarrow F_{ey} = \frac{\pi^2 E}{(L_{cy}/r_y)^2} = \frac{\pi^2 \times 29,000}{93.1^2} = 33 \text{ ksi}$$

Determine elastic flexural-torsional buckling stress about y- and z-axes.

$$y_0 = \bar{y} - t_f/2 = 1.74 - 0.265/2 = 1.608 \text{ in.}$$

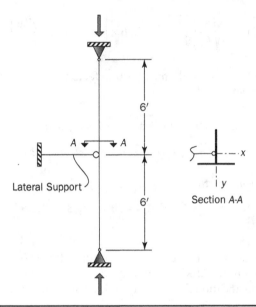

Figure 4.24 Example 4.6.

$$\bar{r}_0^2 = x_0^2 + y_0^2 + \frac{I_x + I_y}{A_g} = 0 + 1.608^2 + \frac{8.70 + 1.41}{2.36} = 6.87 \text{ in.}^2$$

$$H = 1 - \frac{x_0^2 + y_0^2}{\bar{r}_0^2} = 1 - \frac{1.608^2}{6.87^2} = 0.945$$

$$F_{ez} = \left(\frac{\pi^2 E C_w}{L_{cz}^2} + GJ\right)\frac{1}{A_g \bar{r}_0^2} = \left(\frac{\pi^2 \times 29,000 \times 0.0678}{(12 \times 12)^2} + 11,200 \times 0.0511\right) \times \frac{1}{2.36 \times 6.87} = 35.4 \text{ ksi}$$

$$F_e = \left(\frac{F_{ey} + F_{ez}}{2H}\right)\left[1 - \sqrt{1 - \frac{4 F_{ey} F_{ez} H}{\left(F_{ey} + F_{ez}\right)^2}}\right] = \left(\frac{33 + 35.4}{2 \times 0.945}\right)\left[1 - \sqrt{1 - \frac{4 \times 33 \times 35.4 \times 0.945}{(33 + 35.4)^2}}\right] = 27.6 \text{ ksi}$$

The lowest critical stress is obtained from the flexural-torsional buckling limit state. Thus, elastic torsional or flexural-torsional buckling stress, F_e=27.6 ksi will be used to compute F_{cr}.

$$\frac{F_y}{F_e} = \frac{50 \text{ ksi}}{27.6 \text{ ksi}} = 1.81 < 2.25 \rightarrow F_{cr} = \left(0.658^{F_y/F_e}\right)F_y = \left(0.658^{50/27.6}\right) \times 50 = 23.4 \text{ ksi}$$

Check slender/nonslender criteria according to Table B4.1a, AISC 360-16.

$$\lambda_{r,f} = 0.56\sqrt{\frac{E}{F_y}} = 0.56\sqrt{\frac{29,000}{50}} = 13.5 > \lambda_f = b_f/2t_f = 7.53 \rightarrow \text{Nonslender flange}$$

$$\lambda_{r,s} = 0.75\sqrt{\frac{E}{F_y}} = 0.75\sqrt{\frac{29,000}{50}} = 18.1 < \lambda_s = d/t_w = 27.3 \rightarrow \text{Slender stem}$$

Therefore, *Section E7, Specification* applies.

$$\lambda_s = \frac{d}{t_w} = 27.3 > \lambda_r\sqrt{F_y / F_{cr}} = 18.1\sqrt{50 / 23.4} = 26.5$$

Thus, a reduction in the stem area is required. Effective stem width is calculated as follows:
From Table 4.2, $c_1 = 0.22$ and $c_2 = 1.49$ (Case C)

$$F_{el} = \left(c_2\frac{\lambda_r}{\lambda}\right)^2 F_y = \left(1.49 \times \frac{18.1}{27.3}\right) \times 50 = 48.8 \text{ ksi}$$

$$d_e = d\left(1 - c_1\sqrt{\frac{F_{el}}{F_{cr}}}\right)\sqrt{\frac{F_{el}}{F_{cr}}} = 6 \times \left(1 - 0.22 \times \sqrt{\frac{48.8}{23.4}}\right) \times \sqrt{\frac{48.8}{23.4}} = 5.91 \text{ in.}$$

$$A_e = A_g - A_{\text{reduction}} = A_g - (d - d_e)t_w = 2.36 - (6 - 5.91) \times 0.22 = 2.34 \text{ in.}^2 (1\% \text{ reduction})$$

Determine the nominal strength.

$$P_n = F_{cr}A_e = 23.4 \times 2.34 = 54.8 \text{ kips}$$

4.3.4 Compressive Strength of Built-up Members

Built-up members can be fabricated by continuous welding or bolting plates, standard shapes with each other or combination of shapes and plates, or interconnecting two or more shapes by stitches, lacings, perforated plates, or battens. Per *Section E6, Specification*, the compressive strength of built-up members composed of two or more shapes connected by intermediate snug-tight bolts is computed by using the following modified slenderness ratio:

$$\left(\frac{L_c}{r}\right)_m = \sqrt{\left(\frac{L_c}{r}\right)_0^2 + \left(\frac{a}{r_i}\right)^2} \tag{4.24}$$

The modified slenderness ratio, $(L_c/r)_m$, is defined as follows when welded or pretensioned bolts are used:

$$\left(\frac{L_c}{r}\right)_m = \left(\frac{L_c}{r}\right)_0 \quad \text{when } \frac{a}{r_i} \leq 40 \tag{4.25}$$

$$\left(\frac{L_c}{r}\right)_m = \sqrt{\left(\frac{L_c}{r}\right)_0^2 + \left(\frac{K_i a}{r_i}\right)^2} \quad \text{when } \frac{a}{r_i} > 40 \tag{4.26}$$

where $(L_c/r)_m$ = modified slenderness ratio of built-up member
$\quad K_i = 0.50$ for angles back-to-back
$\quad\quad = 0.75$ for channels back-to-back
$\quad\quad = 0.86$ for all other cases

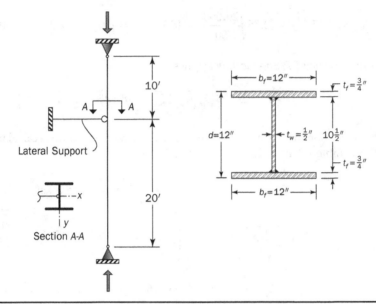

Figure 4.25   Example 4.7.

Figure 4.25   Example 4.7.

a = distance between connectors

r_i = minimum radius of gyration of individual component

$(L_c/r)_0$ = slenderness ratio of built-up member acting as a unit in the buckling direction being addressed

The connectors of built-up members are designed such that the member acts as a unit when buckling occurs. To avoid buckling of each component individually between the connectors, slenderness ratio of each interconnected part (a/r_i) should be less than 75% of the overall slenderness ratio. The other dimensional requirements for the connectors will be demonstrated through design examples.

Example 4.7

Determine the nominal strength of the built-up compression member fabricated by welding two $12 \times \frac{3}{4}''$ plates and a $10\frac{1}{2} \times \frac{1}{2}''$ plate continuously, as shown in Fig. 4.25. The plates are made of A36 steel.

Solution:

Applicable limit states:

  1. Flexural buckling about x-x or y-y axis

  2. Torsional buckling about z-z axis

  3. Local buckling of web and/or flanges

Section properties:

$$A_g = 2\left(b_f t_f\right) + h t_w = 2 \times (12 \times 3/4) + 10.5 \times 0.5 = 23.3 \text{ in.}^2 \left(\text{Area of fillet welds are ignored}\right)$$

$$I_x = 2\left[\frac{b_f t_f^3}{12} + b_f t_f \left(\frac{h+t_f}{2}\right)^2\right] + \frac{h^3 t_w}{12}$$

$$= 2\times\left[\frac{12\times 0.75^3}{12} + 12\times 0.75 \times \left(\frac{10.5+0.75}{2}\right)^2\right] + \frac{10.5^3 \times 0.5}{12}$$

$$= 619 \text{ in.}^4$$

$$I_y = 2\left(\frac{b_f^3 t_f}{12}\right) + \frac{h t_w^3}{12} = 2\times\left(\frac{12^3 \times 0.75}{12}\right)$$

$$= 216 \text{ in.}^4 \, (\text{The second term is negligible})$$

$$r_x = \sqrt{\frac{I_x}{A_g}} = \sqrt{\frac{619}{23.3}} = 5.16 \text{ in.}$$

$$r_y = \sqrt{\frac{I_y}{A_g}} = \sqrt{\frac{216}{23.3}} = 3.05 \text{ in.}$$

Since torsional buckling is one of the possible limit states, the torsional properties of the cross-section are needed for the available strength calculations.

$$J = \sum \frac{bt^3}{3} = \frac{2\times\left(12\times 0.75^3\right) + \left(10.5\times 0.5^3\right)}{3} = 3.81 \text{ in.}^4$$

$$C_w = \frac{I_y h_0^2}{4} = \frac{(216)\times(10.5+3/4)^2}{4} = 6834 \text{ in.}^6$$

Slenderness check:

$$\lambda_{r,f} = 0.64\sqrt{\frac{k_c E}{F_y}} = 0.64\sqrt{\frac{0.76\times 29{,}000}{36}} = 15.8 > \lambda_f = b_f/2t_f = 8 \rightarrow \text{Nonslender flange}$$

where

$$k_c = \frac{4}{\sqrt{h/t_w}} = \frac{4}{\sqrt{10.5/0.5}} = 0.873 > 0.76 \rightarrow \text{Use } k_c = 0.76$$

$$\lambda_{r,w} = 1.49\sqrt{\frac{E}{F_y}} = 1.49\sqrt{\frac{29{,}000}{36}} = 42.3 > \lambda_w = h/t_w \cong 21 \rightarrow \text{Nonslender web}$$

Determine elastic flexural buckling stress.

$$\begin{aligned} L_{cx} &= 30 \text{ ft} \\ L_{cy} &= 20 \text{ ft} \end{aligned} \bigg| \quad \frac{L_{cx}}{r_x} = \frac{30\times 12}{5.16} = 69.8 < \frac{L_{cy}}{r_y} = \frac{20\times 12}{3.05} = 78.7$$

Therefore, the elastic buckling stress is $F_e = \dfrac{\pi^2 E}{\left(L_{cy}/r_y\right)^2}$.

$$F_e = \frac{\pi^2 E}{\left(L_{cy}/r_y\right)^2} = \frac{\pi^2 \times 29{,}000}{78.7^2} = 46.2 \text{ ksi}$$

Determine elastic torsional buckling stress.

$$F_e = \left(\frac{\pi^2 E C_w}{L_{cz}^2} + GJ\right)\frac{1}{I_x + I_y} = \left(\frac{\pi^2 \times 29,000 \times 6834}{(30 \times 12)^2} + 11,200 \times 3.81\right) \times \frac{1}{619 + 216} = 69.2 \text{ ksi}$$

Note that the lateral support provided between the supports does not prevent rotation about z-axis. The torsional buckling length (L_{cz}) is, therefore, taken as 30 ft. in the above calculation.

Determine the critical buckling stress, F_{cr}.

The smaller of the two elastic stresses will determine the governing limit state. The two potential limit states yield elastic critical stresses of 46.2 and 69.3 ksi, respectively. Thus, flexural buckling about y-y axis controls:

$$\frac{L_{cy}}{r_y} = 78.7 < 4.71\sqrt{\frac{E}{F_y}} = 4.71 \times \sqrt{\frac{29,000}{36}} = 134 \left(\text{or } \frac{F_y}{F_e} = \frac{36^{ksi}}{46.2^{ksi}} = 0.78 < 2.25\right)$$

Inelastic flexural buckling occurs about y-y axis. Thus, use *Eq. E3-2, Specification.*

$$F_{cr} = \left(0.658^{F_y/F_e}\right)F_y = \left(0.658^{36/46.2}\right) \times 36 = 26 \text{ ksi}$$

The nominal strength is $P_n = A_g F_{cr} = 23.3 \times 26 = 605$ kips

Example 4.8
A built-up compression member with pinned ends consisting of two MC8×20 (A36) with a ¾″ separation is shown in Fig. 4.26. The total length of the member is 10 ft. and channels are connected through their back-to-back webs with a welded stitch at the midlength. Compute the nominal strength.

Solution:
<u>MC8×20:</u> $A_g = 5.87$ in.²; $I_x = 54.4$ in.⁴; $I_y = 4.42$ in.⁴; $r_x = 3.04$ in.; $r_y = 0.867$ in.; $\bar{x} = 0.84$ in.
<u>A36:</u> $F_y = 36$ ksi
Section properties of 2MC8×20:

$$A_g = 2A_{g1} = 2 \times 5.87 = 11.74 \text{ in.}^2$$

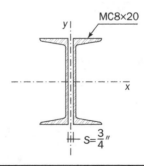

FIGURE 4.26 Example 4.8.

$$I_x = 2I_{x1} = 108.8 \text{ in.}^4$$

$$I_y = 2\left[I_{y1} + A_{g1}\left(\frac{s}{2} + \bar{x}\right)^2\right] = 2 \times \left[4.42 + 5.87 \times \left(\frac{3/4}{2} + 0.84\right)^2\right] = 26.2 \text{ in.}^4$$

$$r_x = r_{x1} = 3.04 \text{ in.}$$

$$r_y = \sqrt{\frac{I_y}{A_g}} = \sqrt{\frac{26.2}{11.74}} = 1.49 \text{ in.}$$

Check limiting width-to-thickness ratio according to Table B4.1a, AISC 360.

$$\lambda_{r,f} = 0.56\sqrt{\frac{E}{F_y}} = 0.56\sqrt{\frac{29,000}{36}} = 15.9 > b_f/t_f = 6.1 \rightarrow \text{Nonslender flange}$$

$$\lambda_{r,w} = 1.49\sqrt{\frac{E}{F_y}} = 1.49\sqrt{\frac{29,000}{36}} = 42.3 > h/t_w = 14.4 \rightarrow \text{Nonslender web}$$

Check the dimensional requirement for the connector spacing according to Section E6-2, Specification.

One intermediate connector is used along the length: $a = L/2 = 60$ in.

$$\frac{a}{r_i} = \frac{60}{0.867} = 69.2 < \frac{3}{4} \times 95.8 = 71.9$$

The slenderness ratio of an individual component does not exceed ¾ times the governing slenderness ratio of the built-up member. Therefore, the connector spacing is adequate.

Determine the failure mode.

Due to its symmetrical shape, torsional buckling is not applicable to built-up double channels when the buckling lengths about the three axes are identical. By comparing the slenderness ratios about x- and y-axes, the axis of flexural buckling can be determined. Since the two channels are interconnected with a welded intermediate connector, the slenderness ratio in y-direction should be modified as follows according to Section E6, *Specification*.

$$\frac{a}{r_i} = \frac{60}{0.867} = 69.2 > 40$$

Thus, the slenderness ratio will be modified using the following relationship between the overall slenderness ratio and the slenderness ratio of a single channel:

$$\left(\frac{L_c}{r}\right)_m = \sqrt{\left(\frac{L_c}{r}\right)_0^2 + \left(\frac{K_i a}{r_i}\right)^2} = \sqrt{\left(\frac{10 \times 12}{1.49}\right)_0^2 + \left(\frac{0.75 \times 60}{0.867}\right)^2} = 95.8 > \frac{L_{cx}}{r_x} = \frac{10 \times 12}{3.04} = 39.5$$

where $\left(\frac{L_c}{r}\right)_0^2 = \left(\frac{L_c}{r}\right)_y^2$ and $K_i = 0.75$. Thus, inelastic buckling about y-axis will take place.

$$F_e = \frac{\pi^2 E}{\left(L_{cy} / r_y\right)^2} = \frac{\pi^2 \times 29{,}000}{(95.8)^2} = 31.2 \text{ ksi}$$

$$F_{cr} = \left(0.658^{F_y/F_e}\right)F_y = \left(0.658^{36/31.2}\right) \times 36 = 22.2 \text{ ksi}$$

$$P_n = A_s F_{cr} = 11.74 \times 22.2 = 260 \text{ kips}$$

4.4 Design Examples

Design of a compression member can be done either by using the design tables of the *AISC Manual* (AISC, 2011) or by a trial-and-error process using the general design formulas. The latter process might seem to require too many iterations but the number of iterations can be substantially decreased by using engineering intuition. The step-by-step iterative design process is explained below:

1. Determine the required strength of the column ($P_{required}$).
2. Assume a slenderness ratio or F_{cr}.
3. Compute the required area using the required strength and assumed F_{cr}.
4. Select a trial section based on the required area.
5. Calculate the actual compressive strength of the trial section.
6. If the design strength is greater than the required strength, the design is adequate. If not, select another trial section for the next iteration. Note that if the required strength is somewhat larger than the design strength, the trial section can be used.

The two methods will be illustrated by the design examples provided in this section.

Example 4.9

Select the lightest W-shape (A992 steel, $F_y = 50$ ksi) for the 25-ft-long column shown in Fig. 4.27 using LRFD. The estimated concentrated loads on the column due to self-weight and snow are $P_D = 300$ kips and $P_S = 200$ kips, respectively.

Solution:

Determine the required strength.

$$P_r = P_u = 1.2P_D + 1.6P_S = 1.2 \times (300) + 1.6 \times (200) = 680 \text{ kips}$$

Determine the available strength.

Previously mentioned methods will be demonstrated through the following design strength calculations.

Method 1: Use *AISC Manual Table 4-1* (Applicable to $F_y = 50$ ksi only)

$$L_{cy} = L/2 = 12.5 \text{ ft}$$
$$L_{cx} = 2.1L = 2.1 \times 25 = 52.5 \text{ ft}$$

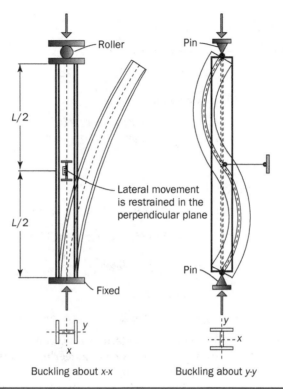

Figure 4.27 Example 4.9.

Given that r_x/r_y ratio for the most common wide-flange steel sections vary between 1.5 and 2.0 and L_{cx}/L_{cy} ratio is 4.2 ($r_x/r_y < K_x/K_y$), the column strength will most likely be governed by the strong-axis buckling. Therefore, when selecting a trial section, an r_x/r_y ratio of 1.75 is assumed. The table below summarizes the design strengths of the trial W-shapes that might satisfy the strength requirement ($P_r \leq \phi_c P_n$).

Enter *AISC Manual Table 4-1* with $\dfrac{L_{cx}}{\left(r_x / r_y \right)} = \dfrac{52.5}{1.75} = 30$ ft.

Trial sections	$\phi_c P_n$ for KL = 30 ft	r_x/r_y	A_g(in.²)	Weight per unit length
W14×109	729 kips	1.67	32.0	109 lb/ft
W14×99	658 kips	1.66	29.1	99 lb/ft
W12×136	695 kips	1.77	39.9	136 lb/ft

Even though W14×99 seems to be the lightest section, the available strength for the assumed r_x / r_y is smaller than the required strength. Since the actual $r_x / r_y = 1.66$ is lower than the assumed one, the available strength will be smaller than 658 kips, which is less than the required strength and therefore, the second lightest shape, which is W14×109 will be used as a trial section.

Trial I: Try W14×109 ($r_x / r_y = 1.67$)

Determine the actual design strength of the trial section.

$$\Rightarrow r_x / r_y = 1.76 < L_{cx} / L_{cy} = 4.2$$

Thus, the column will buckle about its *x*-axis.

$$KL = \text{larger of} \begin{cases} (KL)_y = 12.5 \text{ ft.} \\ \dfrac{(KL)_x}{r_x / r_y} = \dfrac{2.1 \times 25}{1.67} = 31.4 \text{ ft} \end{cases}$$

Enter *Table 4-1* with KL=31.4 ft.

By linear interpolation for $KL = 31.4$ ft $\rightarrow P_c = \phi_c P_n = 683$ kips

$$P_u = 680 \text{ kips} < \phi_c P_n = 683 \text{ kips}$$

Demand/Capacity ratio: $\dfrac{P_u}{\phi_c P_n} = \dfrac{680}{683} = 0.99$

Thus, W14×109 is safe and economical.

Method 2: Use formulas given by *the Specification* (A general method for any steel type)

To determine a trial section, either a slenderness ratio or a critical stress can be assumed. A critical stress value can be assumed using engineering intuition, which will be demonstrated in Example 4.10. In this example, however, a slenderness ratio is assumed based on the approximate radii of gyration in terms of section depth and flange width. Once a slenderness ratio is adopted for each section depth/width (i.e., W14 or W12), a critical stress, F_{cr}, can be found based on the assumed slenderness ratio and trial sections can be determined accordingly. The table below summarizes the assumed slenderness ratio and trial sections following the foregoing trial-and-error approach. A representative example for this approach is presented as follows:

$$\left. \begin{aligned} r_x &\approx \frac{d}{2.3} = \frac{14}{2.3} = 6.09 \text{ in.} \\ r_y &\approx \frac{b_f}{4.0} = \frac{14}{4.0} = 3.50 \text{ in.} \end{aligned} \right| \quad r_x/r_y = 1.74 < L_{cx}/L_{cy} = 4.2$$

$$L_c/r = \left(L_c/r\right)_x = \frac{\left(52.5^{\text{ft}} \times 12^{\text{in./ft}}\right)}{6.09} = 103 \rightarrow \phi_c F_{cr} = 20.7 \text{ ksi (Table 4-22 AISC Manual)}$$

$$A_{g,\text{required}} \geq \frac{P_u}{\phi_c F_{cr}} = \frac{680}{20.7} = 32.9 \text{ in.}^2$$

Trial Section#1 for W14 shapes: W14×120 ($A_g = 35.3$ in.², weight per unit length = 120 lb/ft)

Trial Section#2 for W14 shapes: W14×109 ($A_g = 32.0$ in.², weight per unit length = 109 lb/ft)

Trial Section#3 for W14 shapes: W14×99 ($A_g = 29.1$ in.², weight per unit length = 99 lb/ft)

Section	r_x/r_y	L_c/r	$\phi_c F_{cr}$ (ksi)	$A_{g,required}$ (in.)	Trial section	Weight per unit length (lb/ft)
W14	1.74	103	20.7	32.9	W14×120 (A_g=35.3 in.²)	120
					W14×109 (A_g=32.0 in.²)	109
					W14×99 (A_g=29.1 in.²)	99
W12	1.74	120	15.7	43.3	W12×152 (A_g=44.7 in.²)	152
					W12×136 (A_g=39.9 in.²)	136

Trial I: Try W14×109 (A_g=32.0 in.²; r_x=6.22 in.; r_y=3.73 in.; $b/2t_f$=8.49; h/t_w=21.7)

Check width-to-thickness ratio according to Table B4.1a, AISC 360.

$$\lambda_{r,f} = 0.56\sqrt{\frac{E}{F_y}} = 13.5 > b/2t_f = 8.49 \rightarrow \text{Nonslender flange}$$

$$\lambda_{r,w} = 1.49\sqrt{\frac{E}{F_y}} = 35.9 > h/t_w = 21.7 \rightarrow \text{Nonslender web}$$

Therefore, local buckling would not occur before global buckling.

Determine the available strength.

$$\left(\frac{L_c}{r}\right)_x = \frac{2.1\times(25\times12)}{6.22} = 101 > \left(\frac{L_c}{r}\right)_y = \frac{(12.5\times12)}{3.73} = 40 \text{ and } \frac{L_c}{r} = 79 < 4.71\sqrt{\frac{29,000}{50}} = 113$$

Thus, inelastic flexural buckling occurs about *x-x* axis.

$$F_e = \frac{\pi^2 E}{\left(L_c/r\right)^2} = \frac{\pi^2 \times 29,000}{(101)^2} = 28 \text{ ksi}$$

$$F_{cr} = \left(0.658^{F_y/F_e}\right)F_y = \left(0.658^{50/28}\right)\times 50 = 23.7 \text{ ksi}$$

$$P_n = F_{cr}A_g = 23.7 \times 32 = 759 \text{ kips}$$

$$P_c = \phi_c P_n = 0.9 \times 759 = 683 \text{ kips}$$

$$P_u = 680 \text{ kips} < \phi_c P_n = 683 \text{ kips}$$

Thus, use W14×109.

Example 4.10

As shown in Fig. 4.28, a three-story steel structure consists of braced frames along Lines 1 and 3, and moment frames along Lines A and C. The 3-½-inch-thick concrete is used

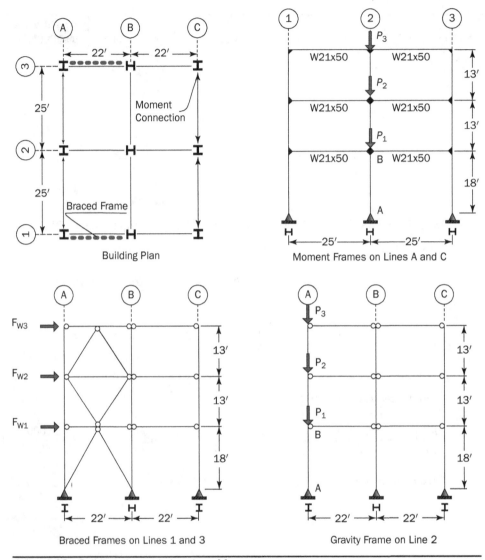

FIGURE 4.28 Building plan and elevation of frames.

for floors and roof. The orientation of columns (strong and weak axes) is shown in the figures. W-shapes with A992 steel ($F_y = 50$ ksi) are used for all columns. It has been estimated that the loads acting at beam-to-column joints $P_1 = 250$ kips, $P_2 = 250$ kips, and $P_3 = 160$ kips, respectively. All vertical loads are due to dead (70%) and live load (30%). The first and second story columns are continuous (i.e., the same W12 shape is used). Select the lightest W12 shape for the first story interior column, Column AB, using LRFD.

Solution:

Determine the required strength.

Axial force due to service loads $\rightarrow P_{AB} = P_1 + P_2 + P_3 = 250 + 250 + 160 = 660$ kips

$$P_r = P_u = 1.2P_D + 1.6P_L = 1.2 \times (0.7 \times 660) + 1.6 \times (0.3 \times 660) = 871 \text{ kips}$$

Determine the available strength.

Method 1: Use *AISC Manual Table 4-1*

$K_y = 1.0$ Sidesway is inhibited by the braced frames.

Since L_{cx} is not known until the column size is determined, use L_{cy} when selecting a trial section.

Enter *AISC Manual Table 4-1* with $P_u = 871$ kips and $KL = 18$ ft.

By investigating the trial W12 sections given in the below table, it can be seen that the available strength of W12×96 is the closest to the required strength.

Trial section	$\phi_c P_n$ for KL=18 ft.	r_x/r_y	A_g(in.²)	I_x(in.⁴)
W12×106	987 kips	1.76	31.2	933
W12×96	888 kips	1.76	28.2	833
W12×87	802 kips	1.75	25.6	740

Trial I: Try W12×96 ($r_x/r_y = 1.76$, $I_x = 833$ in.⁴)

Determine the actual design strength of the trial section.

According to *Appendix 7, Specification,* a reduction factor, τ_b can be introduced to account for the inelasticity effect in the column. In spite of that, τ_b is conservatively taken as 1.0 in this example. Note that the actual $\tau_b = 0.944$, which would result in smaller G_b and K_x values.

$$\left.\begin{array}{l} G_a = 10 \text{ (Pin)} \\ G_b = \dfrac{\sum I_c/L_c}{\sum I_g/L_g} = \dfrac{833/18 + 833/13}{2 \times 984/25} = 1.40 \end{array}\right\} \quad K_x \cong 1.98$$

(From the alignment chart for sway frames)

$$\Rightarrow r_x/r_y = 1.76 < K_x/K_y = 1.98$$

Thus, the column will buckle about its *x*-axis.

$$KL = \text{larger of} \begin{cases} (KL)_y = 18 \text{ ft.} \\ \dfrac{(KL)_x}{r_x / r_y} = \dfrac{1.98 \times 18}{1.76} = 20.25 \text{ ft.} \end{cases}$$

Enter *Table 4-1* with $KL = 20.25$ ft. for W12×96.

By linear interpolation for $KL = 20.25$ ft. $\rightarrow P_c = \phi_c P_n = 808$ kips

$$P_u = 871 \text{ kips} > \phi_c P_n = 808 \text{ kips}$$

Therefore, W12×96 is not adequate.

Trial II: Try W12×106 ($r_x/r_y = 1.76$, $I_x = 933$ in.^{4})
Determine the actual design strength of the trial section.

$$\left.\begin{array}{l} G_a = 10 \text{ (Pin)} \\ G_b = 1.57 \end{array}\right\} \quad K_x \cong 2.0 \Rightarrow r_x/r_y = 1.76 < K_x/K_y = 2.0$$

Thus, the column will buckle about its *x*-axis.

$$KL = \text{larger of} \begin{cases} (KL)_y = 18 \text{ ft.} \\ \dfrac{(KL)_x}{r_x/r_y} = \dfrac{2.0 \times 18}{1.76} = 20.45 \text{ ft.} \end{cases}$$

Enter *Table 4-1* for W12×106 with $KL = 20.45$ ft.

$$KL = 20.45 \text{ ft} \rightarrow P_c = \phi_c P_n = 890 \text{ kips}$$

$$P_u = 871 \text{ kips} < \phi_c P_n = 890 \text{ kips}$$

Therefore, use W12×106.

Method 2: Use formulas given by *the Specification*

As demonstrated in Example 4.9, approximate radii of gyration can be assumed to reduce the number of iterations. In this example, however, a critical stress value is directly assumed using engineering intuition. Note that assuming a slenderness ratio between 60 and 80 or a critical stress on the order of 0.60 to 0.70F_y usually gives a good approximation for the columns in an actual moment frame.
Assume $F_{cr} \approx 0.65F_y$

$$\phi_c P_n \geq P_u \quad (\phi_c = 0.9)$$
$$P_n = F_{cr}A_g = 0.65F_yA_g \rightarrow A_g \geq \frac{P_u}{\phi F_{cr}} = \frac{871}{0.9 \times (0.65 \times 50)} = 29.7 \text{ in.}^2$$

Trial I: Try W12×106 ($A_g = 31.2$ in.^{2}; $r_x = 5.47$ in.; $r_y = 3.11$ in.; $I_x = 933$ in.^{4}; $b/2t_f = 6.17$; $h/t_w = 15.9$)

Check width-to-thickness ratio according to Table B4.1a, AISC 360.

$$\lambda_{r,f} = 0.56\sqrt{\frac{E}{F_y}} = 13.5 > b/2t_f = 6.17$$

$$\lambda_{r,w} = 1.49\sqrt{\frac{E}{F_y}} = 35.9 > h/t_w = 15.9$$

Section is nonslender. Therefore, local buckling would not occur before global buckling.

$$K_x = 2.0; \quad K_y = 1.0$$

$$\left(\frac{L_c}{r}\right)_x = \frac{2.0 \times (18 \times 12)}{5.47} = 79 > \left(\frac{L_c}{r}\right)_y = \frac{1.0 \times (18 \times 12)}{3.11} = 69.5 \text{ (Buckling about } x\text{-}x \text{ axis)}$$

$$\frac{L_c}{r} = 79 < 4.71\sqrt{\frac{29,000}{50}} = 113\,(\text{Inelastic buckling})$$

Thus, inelastic flexural buckling occurs about x-x axis.

$$F_e = \frac{\pi^2 E}{\left(L_c/r\right)^2} = \frac{\pi^2 \times 29,000}{\left(79\right)^2} = 45.9 \text{ ksi}$$

$$F_{cr} = \left(0.658^{F_y/F_e}\right)F_y = \left(0.658^{50/45.9}\right) \times 50 = 31.7 \text{ ksi}$$

$$P_n = F_{cr}A_g = 31.7 \times 31.2 = 989 \text{ kips}$$

$$P_c = \phi_c P_n = 0.9 \times 989 = 890 \text{ kips}$$

$$P_u = 871 \text{ kips} < \phi_c P_n = 890 \text{ kips}$$

Likewise, $\phi_c P_n = 808$ kips can be obtained for W12×96 by following the procedure given in Trial I. Therefore, W12×106 is the most economical W12 section for column AB.

Example 4.11

Select the lightest square HSS (A500 Gr. B) for the second story braces of the braced frame given in Fig. 4.29 using LRFD. Concentrated wind loads acting on each story level are calculated as $F_{w1} = 86$ kips, $F_{w2} = 70$ kips, and $F_{w3} = 35$ kips for each braced frame along lines 1 and 3.

Solution:

Determine the required strength.

Governing LRFD combination: $1.2D + 1.0W + L$

Using the free-body diagram given in Fig. 4.29, tension and compression forces on the second-story bracings can be computed as follows:

$$C_2 = T_2 = \frac{F_{w3} + F_{w2}}{2\cos\alpha} = \frac{70 + 35}{2\cos 50^0} = 82 \text{ kips}$$

Determine the available strength.

Even though one of the braces is in tension, the same HSS section will be employed for the braces located at the same story. The reason being that wind load may change direction. Therefore, the brace sections will be selected based on the available strength in axial compression.

$K = 1.0$ for both in-plane and out-of-plane buckling.

Enter *AISC Manual Table 4-4* for Square HSS with $F_y = 46$ ksi with $P_u = 82$ kips and $KL = 17$ ft.

As indicated in the table below, there exist three square HSS sections with F_y=46 ksi that satisfy the strength requirement ($P_u \leq \phi_c P_n$). Among those, the most economical one, the one with the lowest unit weight, is selected. Thus, HSS5½×5½×$^3/_{16}$ is adequate.

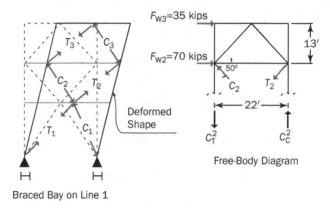

Braced Bay on Line 1

Figure 4.29 Deformed shape (left) and free-body diagram (FBD) (right) of the braced bay under wind load from left to right.

Section	$\phi_c P_n$ for $KL = 17$ ft.	Demand/Capacity Ratio $P_u / \phi_c P_n$	Weight per unit length (lb/ft)
HSS5×5×¼	84 kips	0.98	15.6
HSS5½×5½×³⁄₁₆	82.5 kips	0.99	**13.3**
HSS4½×4½×⅜	83 kips	0.99	19.8

Example 4.12

Select a double angle for the top chord of the truss given in Fig. 4.30, using LRFD approach. The service load, $P_{\text{service}} = P = 20$ kips consists of 60% dead load and 40% snow load.

Solution:

Determine the required strength.

$$P_u = 1.2P_D + 1.6P_S = 1.2 \times (0.6 \times P_{\text{service}}) + 1.6 \times (0.4 \times P_{\text{service}})$$
$$= 1.2 \times (0.6 \times 20) + 1.6 \times (0.4 \times 20) = 27.2 \text{ kips}$$

$P_{\text{max}} = 12.26P$ is obtained from the structural analysis. Thus, the required strength, P_r, can be calculated as $P_r = 12.26P_u = 12.26 \times 27.2 = 343$ kips .

Method 1: Use *AISC Manual Table 4-8, Table 4-9*, and *Table 4-10* for equal- and unequal-leg (LLBB and SLBB) double angles with $F_y = 36$ ksi.

The following table summarizes the available compressive strength of equal- and unequal-leg double angles based on $KL = 8$ ft. and the required strength ($P_r = 343$ kips). Note that *AISC Manual* provides the design strengths only for the double angles with ⅜ in. separation. It is, however, possible to use the design tables for different separation values by simply modifying the $(KL)_y$ to be used by the actual r_y of the section.

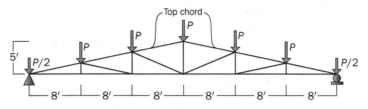

FIGURE 4.30 A roof truss of an industrial building.

Type (orientation)	Trial section	$\phi_c P_n$ (kips)*	Weight per unit length (lb/ft)
Equal leg	2L6×6× 9/16 - 3/8 in.	362	**43.8**
	2L5×5× 3/4 - 3/8 in.	366	47.2
Unequal leg (LLBB)	2L8×4× 5/8 - 3/8 in.	341	48.4
	2L6×4× 3/4 - 3/8 in.	363	47.2
Unequal leg (SLBB)	2L8×4× 3/4 - 3/8 in.	355	57.4
	2L6×4× 7/8 - 3/8 in.	347	54.4

*The available strengths are based on two welded intermediate connectors (a=32 in.).

By investigating the above table, it is apparent that 2L6×6×9/16 with 3/8 in. separation is the lightest shape among others that satisfy the strength requirement. Note that the available strength for 2L6×6×9/16 given in the above table corresponds to 8 ft. buckling length for the sake of simplicity. The design strength for the actual length (8.2 ft) is still greater than the required strength. It should also be noted that the requirements for the welded stitches are satisfied by using two connectors. The details of the connector requirements will be demonstrated in detail in the second method.

Thus, use 2L6×6×9/16-3/8 in. interconnected with two welded intermediate connectors within the 8.2-ft length.

Method 2: Use AISC column strength formulas

In order to find the lightest double angle that can be employed in the top chord, preliminary sections need to be determined based on the approximate radii of gyration. Then, using the governing slenderness ratio and its critical stress counterpart, a required gross area can be found by assuming that flexural buckling stress controls. It is noteworthy that the radii of gyration can be roughly estimated for built-up shapes without knowing the actual thickness (Waddell, 1916; Salmon and Johnson, 1980). The following example shows the details of how the trial section table is established.

For 2L6×6 with 3/8″ separation:

$$r_x \approx 0.31h = 0.31 \times 6 = 1.86 \text{ in.}$$
$$r_y \approx 0.215(2b + s) = 0.215 \times (2 \times 6 + 3/8) = 2.66 \text{ in.} \quad \Big| \quad r_x/r_y = 0.70$$

where h and b are the length of each leg.

$$L_c/r = (L_c/r)_x = \frac{\left(8.2^{ft} \times 12^{in./ft}\right)}{1.86} = 53 \rightarrow \phi_c F_{cr} = 28.0 \text{ ksi (Table 4-22 AISC Manual)}$$

$$A_{g,required} \geq \frac{P_r}{\phi_c F_{cr}} = \frac{343}{28} = 12.3 \text{ in.}^2$$

Trial Section#1: 2L6×6×9/16 (A_g = 12.9 in.², weight per unit length = 43.8 lb/ft)

The other equal- and unequal-leg double angles summarized in the table can be determined in a similar manner.

Section	Separation (in.)	r_x/r_y*	L_c/R	$\phi_c F_{cr}$ (ksi)	$A_{g,required}$ (in.²)	Trial section	Weight per unit length (lb/ft)
2L6×6	⅜ or ¾	0.70 (0.68)	53	28.0	12.3	2L6×6×⁹/₁₆ (A_g=12.9 in.²)	**43.8**
2L5×5	⅜ or ¾	0.70 (0.67)	64	26.1	13.1	2L5×5×³/₄ (A_g=14 in.²)	47.2
2L8×4 (LLBB)	⅜ or ¾	1.46 (1.39)	56	27.5	12.5	2L8×4×⁵/₈ (A_g=14.3 in.²)	48.4
2L7×4 (LLBB)	⅜ or ¾	1.27 (1.22)	56	27.5	12.5	2L7×4×⁵/₈ (A_g=13.0 in.²)	44.2
2L6×4 (LLBB)	⅜ or ¾	1.09 (1.05)	56	27.5	12.5	2L6×x4×³/₄ (A_g=13.9 in.²)	47.2

*The r_x/r_y ratios given in parenthesis represent cases with ¾" separation.

Trial 1: 2L6×6×9/16 separated by 3/8-in. welded intermediate connectors (Fig. 4.31)

Section properties of 2L6×6×⁹/₁₆—³/₈" (F_y=36 ksi):

A_g = 12.9 in.²; s = 3/8 in.; r_x = 1.85 in.; r_y = 2.64 in.; r_z = 1.18 in.; $\bar{r}_0$ = 3.52 in.; H = 0.838 in.; J = 1.41 in.⁴

Since double angles are symmetrical about y-axis, along with local buckling, the limit state of flexural buckling about x-axis and flexural-torsional buckling about y- and z-axes might occur.

Check limiting width-to-thickness ratio.

$$\lambda_r = 0.45\sqrt{\frac{E}{F_y}} = 0.45\sqrt{\frac{29,000}{36}} = 12.8 > b/t = \frac{6}{9/16} = 10.7$$

2L6×6×9/16 is a nonslender compression member. Thus, local buckling limit state does not apply.

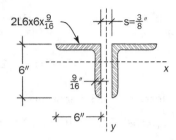

FIGURE 4.31 2L6×6×9⁄16 separated by ⅜-in. connectors.

Compute elastic flexural buckling stress about x-x axis (L_{cx}=8.2 ft.).

$$\left(\frac{L_c}{r}\right)_x = \frac{(8.2 \times 12)}{1.85} = 53.2 \;\rightarrow\; F_e = \frac{\pi^2 E}{(L_c/r)_x^2} = \frac{\pi^2 \times 29{,}000}{(53.2)^2} = 101.2 \text{ ksi}$$

Compute elastic flexural-torsional buckling stress about y-y and z-z axes (L_{cy} = 8.2 ft.).

To make sure that the compression member acts as a unit when the member buckles about the axis of symmetry, the slenderness ratio of an individual component should not exceed ¾ times the governing slenderness ratio of the built-up member. Try two welded intermediate stitches ($a = 32.8$ in.).

$$\frac{a}{r_i} = \frac{32.8}{1.18} = 27.8 < \frac{3}{4} \times 53.2 = 40$$

Thus, the stitch spacing is adequate.

As per *Section E6, Specification* the modified slenderness ratio, $(L_c/r)_m$ is defined depending on the slenderness ratio of a single angle between the connectors.

$$\frac{a}{r_i} = 27.8 < 40 \;\rightarrow\; \left(\frac{L_c}{r}\right)_m = \left(\frac{L_c}{r}\right)_0$$

$$\left(\frac{L_c}{r}\right)_0 = \left(\frac{L_c}{r}\right)_y = \frac{(8.2 \times 12)}{2.64} = 37.3 \;\rightarrow\; F_{ey} = \frac{\pi^2 E}{(L_c/r)_y^2} = \frac{\pi^2 \times 29{,}000}{(37.3)^2} = 206 \text{ ksi}$$

The term with C_w can be omitted from *Eq. E4-7, Spec.* when double angles are used. Thus, the elastic torsional stress can be written as

$$F_{ez} = \frac{GJ}{A_g \overline{r_0}^2} = \frac{11{,}200 \times 1.41}{12.9 \times 3.52^2} = 98.8 \text{ ksi}$$

$$F_e = \left(\frac{F_{ey} + F_{ez}}{2H}\right)\left[1 - \sqrt{1 - \frac{4 F_{ey} F_{ez} H}{\left(F_{ey} + F_{ez}\right)^2}}\right]$$

$$= \left(\frac{206 + 98.8}{2 \times 0.838}\right) \times \left[1 - \sqrt{1 - \frac{4 \times 206 \times 98.8 \times 0.838}{(206 + 98.8)^2}}\right] = 88.1 \text{ ksi}$$

The flexural-torsional buckling stress, $F_e = 88.1$ ksi controls. Therefore, the critical stress, F_{cr}, is determined as follows:

$$\frac{F_y}{F_e} = \frac{36}{88.1} = 0.41 \; < \; 2.25$$

Thus, inelastic flexural-torsional buckling occurs.

$$F_{cr} = \left(0.658^{F_y/F_e}\right)F_y = \left(0.658^{36/88.1}\right) \times 36 = 30.4 \text{ ksi}$$

$$P_n = F_{cr}A_g = 30.4 \times 12.9 = 391 \text{ kips}$$

$$\phi_c P_n = 0.9 \times 391 = 352 \text{ kips} > P_r = 343 \text{ kips}$$

Use 2L6×6× $\%_{16}$-$\%$ " with two welded intermediate connectors within the 8.2-ft length.

Example 4.13

Select a double channel made of A36 steel for the 30-ft-long column with pinned ends, shown in Fig. 4.32. The required LRFD strength, P_r, is 600 kips.

Solution:

To find an adequate channel considering strength and economy, first, approximate radii of gyration need to be determined. The radii of gyration for the given built-up column can be estimated as $r_x \approx 0.36d$ and $r_y \approx 0.45b$ (Waddell, 1916; Salmon and Johnson, 1980). By assuming that $b = d = 15$ in.,

$$r_x \approx 0.36d = 0.36 \times 15 = 5.40 \text{ in.}$$
$$r_y \approx 0.45b = 0.45 \times 15 = 6.75 \text{ in.}$$

$$L_c/r = \left(L_c/r\right)_x = \frac{\left(30^{ft} \times 12^{in./ft}\right)}{5.40} = 67 \rightarrow \phi_c F_{cr} = 25.6 \text{ ksi (Table 4-22 AISC Manual)}$$

$$A_{g,\text{required}} \geq \frac{P_r}{\phi_c F_{cr}} = \frac{600}{25.6} = 23.4 \text{ in.}^2 x$$

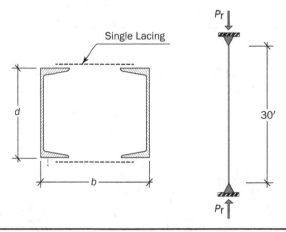

FIGURE 4.32 Example 4.13.

Trial Section#1: C15×40 (A_g = 11.8 in.²; I_x = 348 in.⁴; I_y = 9.17 in.⁴; r_x = 5.43 in.; r_y = 0.883 in.; $\bar{x}$ = 0.778 in.)

Section properties of 2C15×40:

$$I_x = 2 \times 348 = 696 \text{ in.}^4; \ r_x = 5.43 \text{ in.}$$

$$I_y = 2 \times \left[9.17 + 11.8 \times (7.5 - 0.778)^2\right] = 1084.7 \text{ in.}^4 \rightarrow r_y = \sqrt{\frac{I_y}{A}} = \sqrt{\frac{1084.7}{2 \times 11.8}} = 6.78 \text{ in.}$$

It can be noticed that the actual radii of gyration are very close to the initial estimation.

Check limiting width-to-thickness ratio.

$$\lambda_{r,f} = 0.56\sqrt{\frac{E}{F_y}} = 0.56\sqrt{\frac{29,000}{36}} = 15.9 > b/t_f = 5.42$$

$$\lambda_{r,w} = 1.49\sqrt{\frac{E}{F_y}} = 1.49\sqrt{\frac{29,000}{36}} = 42.3 > h/t_w = 23.3$$

Determine the design strength.

$$K_x = K_y = 1.0$$

$$\frac{L_c}{r} = \left(\frac{L_c}{r}\right)_x = \frac{1.0 \times (30 \times 12)}{5.43} = 66.3$$

$$F_e = \frac{\pi^2 E}{(L_c/r)^2} = \frac{\pi^2 \times 29,000}{(66.3)^2} = 65.1 \text{ ksi}$$

$$F_{cr} = \left(0.658^{F_y/F_e}\right)F_y = \left(0.658^{36/65.1}\right) \times 36 = 28.6 \text{ ksi}$$

$$\phi_c P_n = 0.9 \times 28.6 \times (2 \times 11.8) = 606 \text{ kips} > P_r = 600 \text{ kips}$$

Use 2C15×40.

Design of lacing bars and end tie plates.

Since the distance between the connectors (a = 1' 11/16″) is less than 15 in., single lacings made of flat bars can be used. Lacing bars should be designed based on a shearing strength equal to 2% of the available compressive strength of the member ($P_c = \phi_c P_n = 606$ kips).

$$V_{u,lacing} = 0.02 \times P_c = 0.02 \times 606 = 12.1 \text{ kips}$$

It is suggested that single-lacing bars are inclined at a 60-degree angle with respect to the axis of the member. The axial force acting on the plates each side of the member is

$$F_{lacing} = \pm\frac{V_{u,lacing}}{2\cos 30^0} = \pm\frac{12.1}{2 \times \cos 30^0} = \pm 7 \text{ kips}$$

Select a flat bar size so that the slenderness ratio of single-lacing bars is not less than 140.

$$\frac{L}{r_{plate}} \leq 140 \rightarrow r_{plate} \geq \frac{12 - 11/16''}{140} = 0.091 \text{ in. } \left(a = L_{lacing}\right)$$

$$r_{plate} = \sqrt{\frac{bt^3/12}{bt}} = 0.289t \geq 0.091 \text{ in. } \rightarrow t_{min} = 0.315 \text{ in.}$$

Use 3/8" flat bar.

The compressive strength of the lacing plate is determined as follows:

$$\frac{L}{r_{plate}} = \frac{12.688}{0.289 \times 3/8} = 117 \rightarrow F_e = \frac{\pi^2 E}{\left(L_c/r\right)^2} = \frac{\pi^2 \times 29,000}{\left(117\right)^2} = 20.9 \text{ ksi}$$

$$F_{cr} = 0.877 F_e = 0.877 \times 20.9 = 18.3 \text{ ksi}$$

$$A_{g,required} = \frac{F_{plate}}{\phi_c F_{cr}} = \frac{6.03}{0.9 \times 18.3} = 0.37 \text{ in.}^2 \rightarrow b_{plate} = 1 \text{ in.}$$

Use Pl. 1'-11/16"×1"×3/8" for lacing.

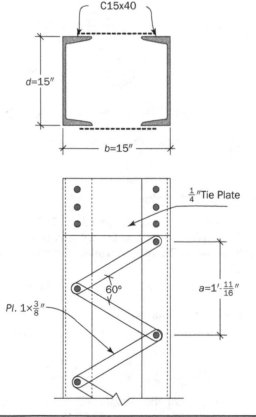

FIGURE 4.33   Design of the laced built-up column with end ties.

It is also required that the slenderness ratio of each individual channel between the fasteners should be less than 75% of the governing slenderness ratio of the built-up member.

$$\frac{a}{r_i} = \frac{a}{r_y} = \frac{12\frac{11}{16}}{0.883} = 14.4 < \frac{3}{4} \times 66.3 = 49.7$$

The thickness of the end tie plates should not be less than 0.02 times the distance between the fasteners.

Min. tie plate thickness $= 0.02 \times (15 - 2 \times 2) = 0.22$ in.

Use 1/4-in.-thick plate for end tie plates.

Therefore, the built-up column shown in Fig. 4.33 meets the requirements and recommendations stipulated in *the Specification*.

4.5 Problems

4.1 Select the lightest W-shape to support a factored axial compression of 900 kips based on LRFD. The buckling lengths in the perpendicular directions are 25 ft. and 10 ft., respectively. Use A992 steel ($F_y = 50$ ksi).

4.2 A 20-ft-long wide-flange column given in Fig. 4.34 is subject to $P_D = 500$ kips and $P_L = 240$ kips.

    a. Show the column orientation required to obtain maximum compressive strength possible.

    b. Select the lightest W12 ($F_y = 50$ ksi) for the column using LRFD.

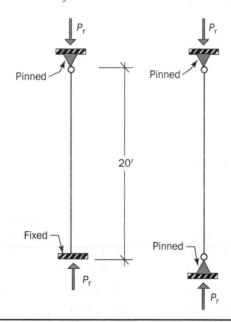

Figure 4.34 Problems 4.2 through 4.4.

4.3   Find the available strength of the column given in Fig. 4.34 using LRFD approach. The column is made of W14×120 ($F_y = 50$ ksi) and the boundary conditions are pin-pin and fix-pin in x-x and y-y directions, respectively.

4.4   Redesign the column given in Problem 4.2 by assuming that a lateral support is provided at the midlength in the plane with pin-pin boundary conditions.

  a. Show the column orientation required to obtain maximum compressive strength possible.

  b. Select the lightest W12 ($F_y = 50$ ksi) for the column using LRFD.

4.5   A frame given in Fig. 4.35 is composed of W-shapes (F_y=50 ksi). Lateral joint translation is prevented by the braced frames in the perpendicular direction. Determine the following:

  a. Show the column orientation required to obtain maximum compressive strength possible.

  b. Draw the deformed shape of the frame when column AB and CD buckles about x- and y-axes.

  c. Find the effective length factors for columns AB and CD.

  d. Find the maximum axial load in compression that can be supported by column AB and CD.

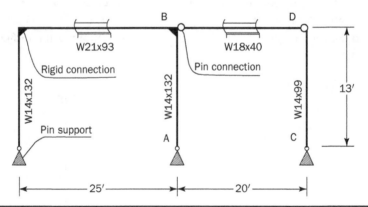

Figure 4.35   Problem 4.5.

4.6   A moment frame is loaded with the concentrated loads at its beam-to-column joints as shown in Fig. 4.36. Columns ab and bc are continuous (i.e., the same section is used). Wide-flange shapes are made of A992 steel (F_y=50 ksi) and braced fra   mes are used in the perpendicular direction. P=300 kips consists of 70% dead load and 30% live load. Select the most economical W-shape for column ab and bc based on LRFD.

4.7   Using the information given in Problem 4.6 select the most economical W-shape for column ef based on LRFD. Then, draw the deformed shape when column ef buckles about x-axis.

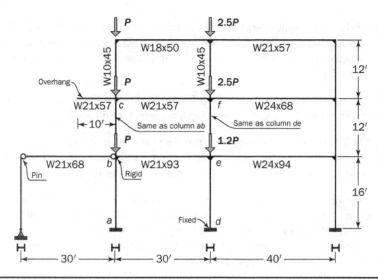

FIGURE 4.36   Problems 4.6 and 4.7.

4.8   A 12-ft-long truss member is shown in Fig. 4.37. Find the most economical WT section made of A992 steel that can support P_r. P_r consists of a dead load of P_D=250 kips and a snow load of P_S=100 kips. Use LRFD.

4.9   Find the maximum P_r that can be carried by the truss member shown in Fig. 4.37. WT8×50 with F_y=50 ksi is employed and L=10 ft. Use LRFD.

4.10   Find the maximum P_r that can be carried by the truss member shown in Fig. 4.37 by assuming that an additional lateral bracing in the out-of-plane direction is provided at the midlength. Note that lateral support is attached to the flange. WT8×50 with F_y=50 ksi is employed and L=10 ft. Use LRFD.

4.11   Find the maximum P_r that can be carried by the built-up column shown in Fig. 4.38. Plates are made of A36 steel. The column is pinned at both ends and the total length is 20 ft. Use LRFD.

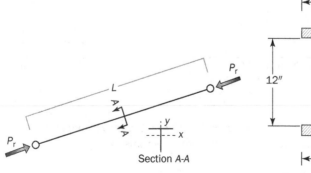

FIGURE 4.37   Problems 4.8 through 4.10.

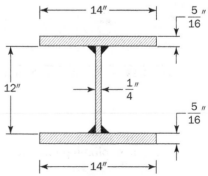

FIGURE 4.38   Problem 4.11.

4.12 Select the lightest square HSS for the first- and second-story braces to support the wind loads shown in Fig. 4.39 based on LRFD. The wind loads at each story level are estimated as F_w1=81 kips and F_w2=36 kips. Use A500 Gr. C steel (F_y=50 ksi). Neglect gravity load effect.

4.13 Select the lightest round HSS for the braces in Fig. 4.39 using the lateral loads given in Problem 4.12. Neglect gravity load effect. Use LRFD.

4.14 Select the same W-shape for the first- and second-story columns shown in Fig. 4.39 based on LRFD. The wind loads are given in Problem 4.12. P=100 kips consists of 60% dead and 40% live load. Use A992 steel (F_y=50 ksi).

4.15 Design a 40-ft-long laced built-up column that consists of four equal-leg angles, as shown in Fig. 4.40, using LRFD. The column with pin-pin boundary conditions is loaded with P_D=600 kips and P_L=300 kips. Use A36 steel.
Hint: Radii of gyration for the given built-up column can be roughly taken as r_x=0.42d and r_y=0.42b when selecting trial sections.

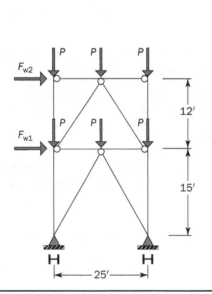

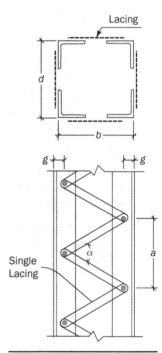

FIGURE 4.39 Problems 4.12 through 4.14.   **FIGURE 4.40** Problem 4.15.

4.16 Write a python code that computes the nominal compressive strength, P_n, of a wide flange column pinned at both ends in both directions.

Related to Building Project in Chapter 1:

4.17 Please design the following structural members in the building project:
   a. Columns in the first floor along Lines B and 3

   b. Columns in the first floor along Lines 2, 4, 5, and 7 in the unbraced bays

Bibliography

AISC, *Specification for Structural Steel Buildings*, ANSI/AISC Standard 360-16, American Institute of Steel Construction, Chicago, IL, 2016.

AISC, *Steel Construction Manual*, 14th ed., American Institute of Steel Construction, Chicago, IL, 2011.

ASTM A6, *Standard Specification for General Requirements for Rolled Structural Steel Bars, Plates, Shapes, and Sheet Piling*, 2017.

Bjorhovde, R., J. Brozzetti, G. A. Alpsten, and L. Tall, "Residual Stresses in Thick Welded Plates," *Welding Journal*, 51 (8): 392–405, 1972.

Engesser, F., "Die Knickfestigkeit gerader Stabe," *Zentralbl. Bauverwaltung*, 11: 483–486, 1891.

Euler, L., "De curvis elasticis: additamentum I to his Methodus inveniendi lineas curvas maximi minimive proprietate gaudentes, Lausanne and Geneva," 1744.

Salmon, C. G., and J. E. Johnson, *Steel Structures: Design and Behavior*, 2nd ed., Harper & Row, New York, 1980.

Shanley, F. R., "Inelastic Column Theory," *Journal of Aeronautical Science*, 14 (5): 261, 1947.

Smith, M., *ABAQUS/Standard User's Manual, Version 6.9*, Simulia, Providence, RI, 2009.

Tall, L., "Recent Developments in the Study of Column Behavior," The Journal of Institutions of Engineers, Australia, 36 (12): 319–333, 1964.

Waddell, J. A. L., *Bridge Engineering, Volume I and Volume II*, John Wiley & Sons, New York, 1916.

Ziemian, R. D. (ed.), *Guide to Stability Design Criteria for Metal Structures*, 6th ed., John Wiley & Sons, Inc., Hoboken, NJ, 2010.

CHAPTER 5

Design of Beams

5.1 Introduction

Beams are the structural members that are subjected primarily to bending moment and associated shear force. Thus, dominant internal forces at a typical beam cross-section are bending moment (M) and shear force (V) (Fig. 5.1). As such, beams can be found in most of the structures in different forms, depending on their use. For instance, in a building structure, some of the beam members (e.g., secondary beams, purlins, or lintels shown in Fig. 5.2a and b) solely support the gravity loads transferred from the floor or roofing system (i.e., dead, live, snow, etc.), whereas girders (main beams connected to columns at a joint) mainly support lateral loads together with a small fraction of gravity loads in lateral force–resisting systems, such as moment or braced frames (Fig. 5.2c). Contrary to reinforced concrete structures in which solid rectangular sections are often used, steel beams in steel structures are made of the thin flange and web plates (Fig. 5.1). Most commonly used beam cross-sections are hot-rolled or built-up wide-flange I-shapes (W-shapes) for their efficiency in terms of high strength-to-weight ratio. In cases where there is a long span and not enough lateral support or these plates are too slender, local and/or lateral-torsional buckling might occur before the element reaches its full plastic moment strength. In this case, hot-rolled or built-up wide-flange I-shapes will no longer be effective. The top and bottom flanges resist the applied bending moment (M), whereas the web is effective in resisting the shear force (V). Besides, I-shapes, single- or double-channel, hollow structural sections (HSS), Tee sections, and double-angles can be used as beam members based on the need (Fig. 5.3). In some cases, with the intention of further reducing the material cost, nonprismatic cross-sections, such as tapered (Fig. 5.2d), deep plate girders (i.e., built-up from welded slender plates) (Fig. 5.2e), cellular or castellated beams (Fig. 5.2e), open-web steel joists (Fig. 5.2f), or stringer beams (Fig. 5.2g) can also be utilized. Thus, the selection of a proper cross-section for a beam member is, in general, dependent on the economic, structural, and architectural constraints, type of loading, and intrinsic design considerations. This chapter will provide an overview of beam behavior and design of prismatic beam members according to the current AISC Specification (AISC 360). Girders are also horizontal structural members and also a member of the lateral force–resisting system (Fig. 5.2c); that is, they also transfer lateral forces in addition to gravity loads.

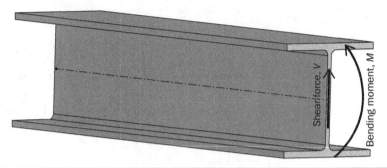

Figure 5.1   Elastic normal stress distribution of a rectangular beam.

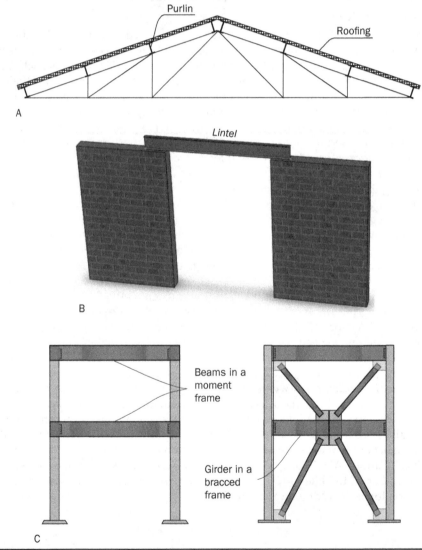

Figure 5.2   Typical beam members: (a) purlins, (b) lintel, (c) girders in framed structures, (d) tapered multispan bridge girder, (e) plate girders, (f) open-web steel joist, and (g) stringer.

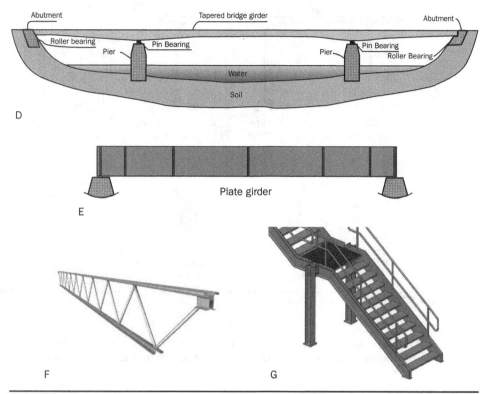

FIGURE 5.2   *(Continued)*

Steel beams usually are double-symmetric rolled I or built-up shapes. Unsymmetric sections, if used, to carry bending will create unsymmetric bending as well as torsion stresses reducing efficiency.

Section in reinforced concrete structures is solid rectangular sections. However, steel sections for beams are always open shape, consisting of thin plates (such as web and flanges) in the rolled shapes, or welded together in built-up sections. This type of shape is particularly efficient for bending about its strong axis. However, it is extremely vulnerable to torsion. Let's take wide-flange I-shaped (commonly called W-shape) section as an example (Fig. 5.3). An I-shaped section is more efficient for bending resistance because its large percentage of areas of steel is located far away from strong axis subject to bending moment. Such efficiency might be compromised if flanges and/or web are made too slender ("thin"), or the beam is over a long span without lateral supports, when local and/or lateral-torsional buckling might occur before available plastic moment strength is reached.

General assumptions made to establish beam design procedure are as follows:

i. Member has one or two axes of symmetry.

ii. Member is subjected to only bending about one axis of symmetry.

iii. Load is static.

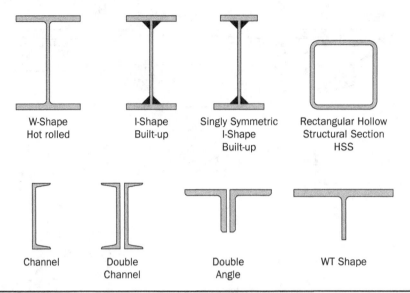

W-Shape	I-Shape	Singly Symmetric	Rectangular Hollow
Hot rolled	Built-up	I-Shape	Structural Section
		Built-up	HSS

Channel   Double Channel   Double Angle   WT Shape

FIGURE 5.3   Typical beam cross-sections.

Note that the behavior and design of steel flexural members are discussed in this chapter with the exception of plate girders, which will be discussed in detail in Chapter 9.

5.2   Behavior of Laterally Supported Beams

Beams are primarily subject to flexure along with shear deformation, while axial deformation is negligible. Accordingly, we design beam members considering the flexure- and shear-associated limit states. Shear-associated limit states will be covered separately in detail in Chapter 9 when the design of plate girders is introduced. This section intends mainly to delineate the characteristics of flexural limit states. We can classify these limit states as plastic hinge formation (yielding), lateral-torsional buckling (LTB), and local buckling (LB). In addition, the shear capacity of the members with nonslender webs and the limit states applicable when the member is subject to a single or double concentrated forces are also briefly summarized herein.

In order to understand the flexural behavior and related limit states, one should first examine the normal elastic stress (σ) distribution over a rectangular cross-section presented in Fig. 5.4. As given in Eq. (5.1), the linear stress distribution of a typical rectangular cross-section (Section A-A) is defined by the product of the bending moment-to-moment of inertia ratio (M/I) and the distance from the neutral axis (c). The beam element behaves elastically, developing tensile and compressive strains and stresses acting normal to the cross-section along the longitudinal axis. These stresses and strains are tensile in one-half of the beam and compressive in the other. As seen in Fig. 5.4, stress increases when moving toward the top and bottom fibers. This basic knowledge led to the utilization of the steel shapes other than solid rectangular sections for flexure.

$$\sigma = \frac{M}{I}c \tag{5.1}$$

where σ = normal stress, ksi

M = bending moment, kip-in.

I = moment of inertia, in.^{4}

c = largest distance of extreme fiber from the neutral axis, in.

As the transverse (gravity) loading gradually increased, the bending moment developing as a result in the cross-section will increase, which in turn increases the strains and stresses on the cross-section, and the extreme fiber will reach yield stress, and cross-section starts yielding. The corresponding bending moment is called the yield moment, M_y (Fig. 5.4):

$$M_y = F_y\left(\frac{I}{c}\right) = F_y S \tag{5.2}$$

where

M_y = yield moment, kip-in.

F_y = yield stress, ksi

S = elastic section modulus about the axis of bending, in.^{3}

If $M > M_y$, the stress at the top and bottom of the section remains F_y, and the inner fibers will gradually reach F_y as M keeps increasing until all cross-section reaches yield stress, F_y. Plastic bending moment (M_p) is the bending moment corresponding to the case where all fibers of the cross-section have reached yield stress (Fig. 5.4). Thus, M_p of the cross-section can be calculated by taking a moment about the plastic neutral axis:

$$M_p = \sum M_x = C\frac{h}{4} + T\frac{h}{4} = \left(\frac{bh^2}{4}\right)F_y \tag{5.3}$$

Letting $Z_x = \dfrac{bh^2}{4}$, M_p becomes

$$M_p = Z_x F_y \tag{5.4}$$

where

$C = \left(b\dfrac{h}{2}\right)F_y$, compression force, kips

$C = \left(b\dfrac{h}{2}\right)F_y$, tension force, kips

Z_x = plastic section modulus, in.^{3}

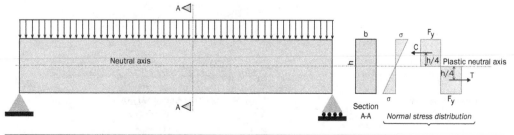

FIGURE 5.4   Elastic normal stress distribution over a rectangular cross-section.

Note that $\dfrac{M_p}{M_y} = \dfrac{Z_x}{S_x} = 1.5$. In other words, plastic moment capacity, M_p, is 1.5 times elastic moment capacity, M_y, in a rectangular section. Thus, the failure mode in a solid section will be plastic hinge formation when $M=M_p$ (Fig. 5.5a).

A plastic hinge differs from a real hinge in a structural system as follows (Fig. 5.5):

- A real hinge in a beam:

    i. Free rotation

    ii. $M = 0$

- Plastic hinge (P.H.) in a beam:

    i. Free rotation

    ii. $M=M_p$

To better understand why solid sections are not "proper" when designing members for flexure, Movie 5.1 shows the finite-element simulation of the simple representative beam given in Fig. 5.4 under uniformly distributed loads. The following can be interpreted by examining the side and cross-section views of the simple simulated beam given in Movie 5.1:

(1) Under the increasing transverse loading, the stress, σ, remains zero (indicated in green color) throughout the loading history along the horizontal axis of symmetry, which is also referred to as the neutral axis;

(2) The areas above and below the neutral axis are subject to compression (indicated in tones of blue) and tension (indicated in tones of red) stresses, respectively;

(3) The normal stress grows in magnitude as moving away from the axis about which the cross-section bends (i.e., neutral axis).

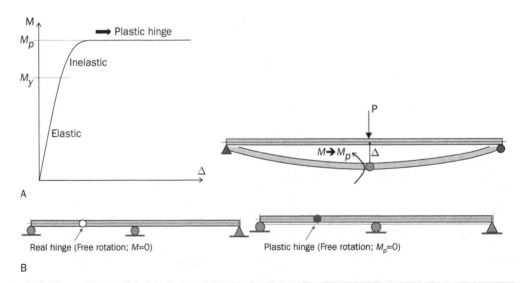

FIGURE 5.5   Plastic hinge and real hinge in a beam.

Therefore, higher efficiency can be achieved by removing the less stressed portions from the rectangular cross-section. In other words, a steel section becomes more economical as the percentage of the cross-sectional area located away from the neutral axis, by which the major part of a bending moment is resisted, increases. By this logic, instead of solid rectangular cross-sections, thin-walled I-shapes are preferred for their economy. Such efficiency might be damaged due to the local limit states if the section consists of excessively thin elements (i.e., web or flange), which will be discussed in the following sections. Furthermore, steel sections for beams are often open shapes, consisting of plates, such as web and flanges in the rolled shapes, or welded together in built-up sections. These shapes are particularly efficient for bending about their strong axis while they are vulnerable to torsion. These limit states will be discussed thoroughly in the following sections of this chapter.

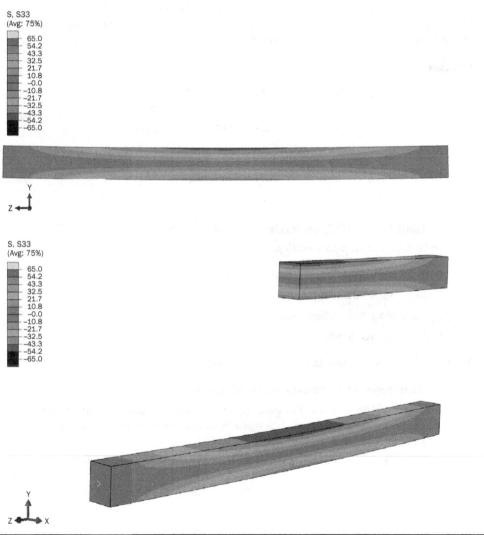

Movie 5.1 Flexural deformation of a rectangular solid section. (To view the entire movie, go to www.mhprofessional.com/design-of-steel-structures.)

Lateral support to a beam can be in any form, such as floor deck on the top of a floor beam, a joint normal to the beam, or a specially designed lateral bracing (Fig. 5.6).

Lateral support (or bracing) of the compression flange for steel beams is required to reach the beam's plastic bending moment capacity, M_p. In cases where there is not sufficient lateral support, the overall buckling of the compression flange will occur, causing a reduction in M_p. Even if there is sufficient lateral support of the compression flange, to reach M_p, local buckling in compression flange and/or web elements should also be prevented (Fig. 5.6). The bracing can be in any form, such as floor deck on the top of a floor beam, a joist normal to the beam, or specially designed bracing. Note that the bracing should be at or near the flange under compression to be effective.

Example 5.1

Compute the yield and plastic bending moments, M_y and M_p, of W12×82 section in Fig. 5.7 and determine the M_p/M_y ratio ($F_y = 36$ ksi). Cross-section properties can be found from *AISC Manual*, pages 1-24 and 1-25.

Solution:

$$M_y = F_y S_x = 36^{ksi} \times 123^{in.^3} = 4,428 \text{ kip-in.}$$
$$= 369 \text{ kip-ft}$$
$$M_p = F_y Z_x = 36^{ksi} \times 139^{in.^3} = 5,004 \text{ kip-in.}$$
$$= 417 \text{ kip-ft}$$
$$\frac{M_p}{M_y} = \frac{417^{kip-ft}}{369^{kip-ft}} = 1.13$$

5.2.1 Limit States (Failure Modes) in Pure Bending

A typical steel beam in pure bending can fail due to the following:

1. Reaching its bending moment capacity
2. Reaching its shear capacity
3. Inadequate connection capacity
4. Concentrated loads

There are three limit states (failure modes) in pure bending:

1. Plastic hinge formation due to bending moment
2. Local buckling of the flange and/or web under compression (due to axial compression forces created as a result of bending moment at the flanges and concentrated forces)
3. Lateral-torsional buckling (LTB) (will be treated in the next section)

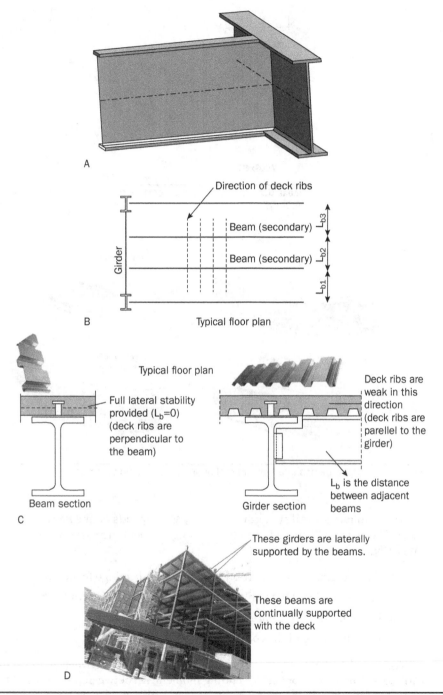

FIGURE 5.6 Laterally supported beams and girders in a typical floor plan.

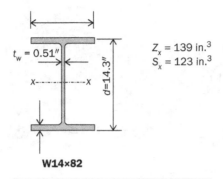

$Z_x = 139$ in.^{3}
$S_x = 123$ in.^{3}

$t_w = 0.51''$

$d = 14.3''$

W14×82

Figure 5.7   Cross-section properties of W12 ×82.

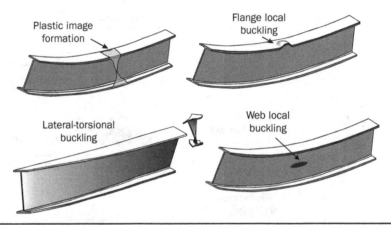

Plastic image formation

Flange local buckling

Lateral-torsional buckling

Web local buckling

Figure 5.8   Failure modes in a typical steel beam under bending moment.

The nominal flexural strength of a beam, M_n, depends on the failure mode that will occur. The beam is likely to fail as the bending moment increases due to the following cases: (Fig. 5.8)

1. Full yielding of cross-section leading to plastic hinge formation
2. Lateral-torsional buckling (LTB) of compression flange of the beam
3. Local buckling of compression flange
4. Local buckling of the compression region of the web

Assume that the number of lateral bracing points, L_b, between A and B in Fig. 5.9 varies. As P increases from zero, bending moment, M, between points A and B, and the beam end rotation will respond. Depending on λ_f, λ_w, and L_b, we can observe one of the following flexural failure modes (or called limit states) ($\lambda_f =$ width-to-thickness ratio for flange, $\lambda_w =$ width-to-thickness ratio for web) (Tables 5.1 and 5.2):

1) Plastic behavior:

    i. Plastic hinge is formed, and plastic (maximum) moment strength is reached, $M_n = M_p$, where M_n is the nominal moment capacity or flexural strength of the beam.

    ii. A significant deformation is expected at the point of failure, $\mu > 1.0$ (Point 2 in Fig. 5.9), where μ is rotational ductility. In some cases $\mu > 4.0$, the beam shows a very ductile behavior (Point 1 in Fig. 5.9), $\mu = \theta_u$.

    iii. The beam has to have a compact section with closely spaced lateral supports to have a plastic behavior.

2) Inelastic behavior:

    i Plastic (maximum) moment strength is not reached before the failure occurs, $M_y < M_n < M_p$ (Point 3 in Fig. 5.9).

    ii. No significant deformation is observed at the point of failure.

    iii. The observed failure modes include local flange buckling, web buckling, lateral-torsional buckling, or the combination of them.

    iv. The beam has either a noncompact section (causing flanges or web to buckle) or not enough number of lateral supports over a long beam span length (causing lateral-torsional buckling) to have an inelastic behavior.

3) Elastic buckling limit state:

    i. $M_n < M_p$ (Point 4 in Fig. 5.9). This means that all steel remains elastic when some type(s) of buckling occur(s). The available design strength is low.

    ii. Little deformation is shown at the point of failure (brittle failure pattern).

    iii. The observed failure modes include local flange buckling, local web buckling, lateral-torsional buckling, or the combination of them.

    iv. The beam has either a slender section (causing elastic flanges or web buckling) or no lateral supports over a long beam span length (causing elastic lateral-torsional buckling).

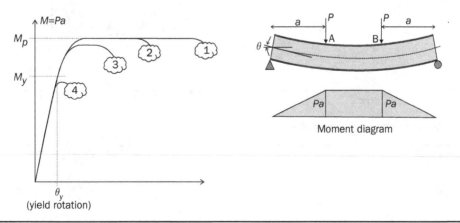

Figure 5.9 Moment–rotation relationship of beams with different design parameters (☁ = failure point).

5.2.2 Plastic Hinge Formation

As discussed before, bending moments cause normal-tension/compression stresses to develop in the beam's cross-section. First, the yield moment is gradually reached, then the entire cross-section will start yielding in tension and/or compression. As demonstrated in Movie 5.2, a plastic hinge will form when the entire cross-section becomes plastic (Fig. 5.10 and Table 5.3). As different from a real hinge where free rotation is allowed and $M = 0$, the bending moment will be equal to M_p, and free rotation will also be allowed (Fig. 5.10).

5.2.3 Local Buckling

The failure mode buckling, due to its nature, needs compression to occur (Fig. 5.11). During the bending, if the compression flange or part of the web subject to compression is too thin, the plate may actually fail by buckling, before the full plastic moment is reached. It should also be noted that local buckling cannot be prevented in a steel member, but can be delayed until the desired inelastic deformation has been reached. Local buckling has a substantial effect on both inelastic deformation capacity (or ductility) and strength. AISC 360 provides width/thickness ratios, λ, for common steel structural shapes that should be satisfied to prevent local buckling from happening (Table 5.3).

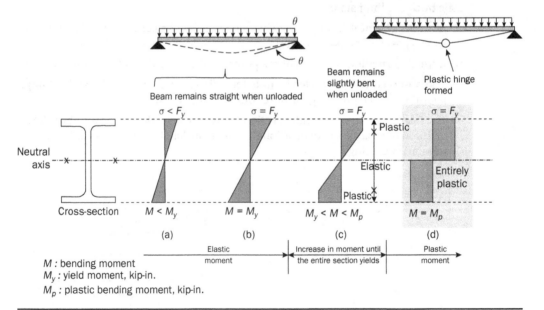

FIGURE 5.10 Normal stress distribution at different stages of loading of an I-shaped section.

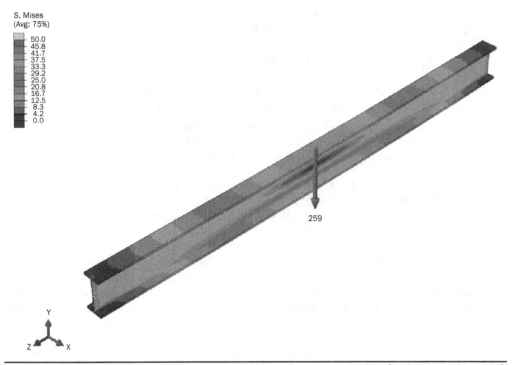

Movie 5.2 Plastic hinge formation. (To view the entire movie, go to www.mhprofessional.com/design-of-steel-structures.)

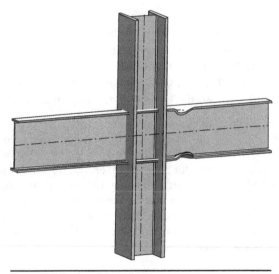

Figure 5.11 Local buckling.

The above cases for different W-shapes with different λ_f, λ_w values are explained in Fig. 5.9. From Fig. 5.9, the following can be observed:

1. A plastic hinge is formed, and plastic (maximum) moment strength is reached, M_p (Cases 1 and 2).

2. Local buckling of flange/beam or LTB or both occurred, and plastic moment strength, M_p, is not reached before the failure occurs, $M_y < M_n < M_p$ (Case 3).

3. Local buckling of flange/beam or LTB or both occurred, and elastic moment strength, M_y, is not reached before the failure occurs, $M_n < M_y$ (Case 4). The available design strength will be quite low.

A compact section is defined as the one that is capable of developing a fully plastic stress distribution before buckling. For a compression member to be classified as compact, its flanges must be continuously connected to its web or webs, and the width/thickness ratio of its compression elements ($\lambda = \lambda_f$ or λ_w) should not be greater than the limiting ratios, λ_p. A noncompact section is defined as the one for which the yield stress can be reached in some but not all of its compression elements before buckling occurs. A noncompact section is not capable of reaching a fully plastic stress distribution. The width/thickness ratio of noncompact sections (λ_f, λ_w) is greater than λ_p but not greater than λ_r. The width/thickness ratio of slender sections (λ_f, λ_w) is greater than λ_r.

$$\lambda_f \leq \lambda_p \text{ and } \lambda_w \leq \lambda_p, \text{ compactness requirement} \qquad (5.5a)$$

$$\lambda_p < \lambda_f \leq \lambda_r \text{ or } \lambda_p < \lambda_w \leq \lambda_r, \text{ noncompactness requirement} \qquad (5.5b)$$

$$\lambda_f > \lambda_r \text{ or } \lambda_w > \lambda_r, \text{ slenderness requirement} \qquad (5.5c)$$

where
λ = width-to-thickness ratio for the element
λ_f = width-to-thickness ratio for flange
λ_w = width-to-thickness ratio for web
λ_p = limiting width-to-thickness ratio for compact element
λ_r = limiting width-to-thickness ratio for noncompact element

The limiting width-to-thickness ratios are given in accordance with the classification of the members being unstiffened or stiffened. Stiffened elements are restrained at both ends, whereas unstiffened elements are only restrained at one hand. For example, when checking local buckling for a W-shape, each flange is considered to be made of two plates, and both plates are supported by the web in the middle. That is why the flange plates are named as unstiffened elements. This allows each flange to be treated as if the flanges of the shape are supported by the web at one hand and are free on the other hand (Fig. 5.12). On the other hand, the web of a W-shape is constrained by both flanges. Hence it is a stiffened element (Fig. 5.12).

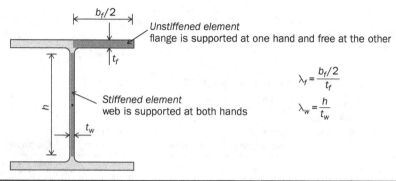

$$\lambda_f = \frac{b_f/2}{t_f}$$

$$\lambda_w = \frac{h}{t_w}$$

FIGURE 5.12 Limiting width/thickness ratios for local buckling.

Section	Failure mode	Bending moment capacity M_n
Highly/Moderately ductile and compact	Plastic hinge formation	$M_n = M_p$
Noncompact	Inelastic buckling of flange and/or web	$M_y < M_n < M_p$
Slender-element	Elastic buckling of flange and/or web	$M_n < M_y$

Note: Rolled I-shaped W-sections are manufactured so that $\lambda_w \leq \lambda_p$ for $F_y \leq 50$ ksi. In other words, the web local buckling does not occur before M_p is reached in these sections with $F_y \leq 50$ ksi.

TABLE 5.1 Flexural (Bending Moment) Strength of Various Section Types

Classification of compression element	Limiting width-to-thickness ratio (Table B4.1b)	Failure mode	Nominal flexural capacity
Compact	$\lambda \leq \lambda_p$	Plastic hinge formation	$M_n = M_p$
Noncompact	$\lambda_p < \lambda \leq \lambda_r$	Inelastic local buckling	$M_y < M_n < M_p$
Slender	$\lambda_r \leq \lambda$	Elastic local buckling	$M_n < M_y$

TABLE 5.2 Local Buckling Limit State-Flexural Strength Relationship

	Description of element	Width-to-thickness ratio	Limiting width-to-thickness ratios: compression elements			Example
			For members subject to flexure		**For members subject to axial compression**	
			λ_p (compact/ noncompact)	λ_r (noncompact/ slender)	λ_r (nonslender/ slender)	
unstiffened elements	Flanges of rolled I-shaped section, channels, and tees (*unstiffened element*)	b/t	$0.38\sqrt{\dfrac{E}{F_y}}$	$1.0\sqrt{\dfrac{E}{F_y}}$	$0.56\sqrt{\dfrac{E}{F_y}}$	
	Legs of single angles	b/t	$0.54\sqrt{\dfrac{E}{F_y}}$	$0.91\sqrt{\dfrac{E}{F_y}}$	$0.45\sqrt{\dfrac{E}{F_y}}$	
stiffened elements	Flanges of rectangular HSS	b/t	$1.12\sqrt{\dfrac{E}{F_y}}$	$1.40\sqrt{\dfrac{E}{F_y}}$	$1.40\sqrt{\dfrac{E}{F_y}}$	
	Webs of I-shaped sections and channels	h/t_w	$3.76\sqrt{\dfrac{E}{F_y}}$	$5.70\sqrt{\dfrac{E}{F_y}}$	$1.49\sqrt{\dfrac{E}{F_y}}$	
	Webs of rectangular HSS	h/t	$2.42\sqrt{\dfrac{E}{F_y}}$	$5.70\sqrt{\dfrac{E}{F_y}}$	$1.40\sqrt{\dfrac{E}{F_y}}$	
	Round HSS	D/t	$0.07\dfrac{E}{F_y}$	$0.31\dfrac{E}{F_y}$	$0.11\dfrac{E}{F_y}$	

TABLE 5.3 Width-to-Thickness Ratios for Members Subject to Flexure and Axial Compression (AISC 360)

5.3   Lateral-Torsional Buckling

In a steel framing system, two alternative lateral supports are available: (a) continuous lateral support, (b) lateral support at certain intervals (Fig. 5.13). A typical steel beam without any lateral support can rotate and move laterally mainly because of the buckling of compression flange, which becomes unstable and bends sideways (Fig. 5.14). Let's think of an I-shaped section as if it is made of three plates (two flange plates and one web plate) (Fig. 5.14). The rectangular flange of the I-shaped beam has its weak axis in the vertical plane. However, since the flange plate is connected to the web, its weak axis becomes out-of-the-plane direction. Thus, the rectangular flange under increasing compression loads will buckle and rotate about the out-of-plane direction and also experience lateral displacement. This behavior is called the *lateral-torsional buckling*. Thus, lateral-torsional buckling is a combination of torsional rotation and lateral bending due to the buckling of the compression flange. Wide-flange sections (W-shapes) have a higher resistance to lateral-torsional buckling than S-shapes because they have wider flanges (Fig. 5.14). To prevent lateral-torsional buckling from occurring, lateral support or bracing should be provided in the compression region, i.e., between the neutral axis and the compression flange (Fig. 5.15). If the beam is part of a structural framing system transferring gravity loads as well as lateral loads (girder), both top and bottom flanges of the beam might be subject to compression due to the cyclic nature of the lateral loads (earthquake, wind). However, if the beam is only transferring gravity loads (e.g., secondary beams), since both ends are simply supported, only the top flange will be subject to compression, and lateral bracing needs only be provided between the neutral axis and top flange. It should be noted that the lateral support (or bracing) should be at or near the flange under compression to be effective. Floor deck on the top of a floor beam, a joist normal to the beam and supported by the beam, or specially designed bracing can be used as lateral support to a beam or girder. The smaller L_b

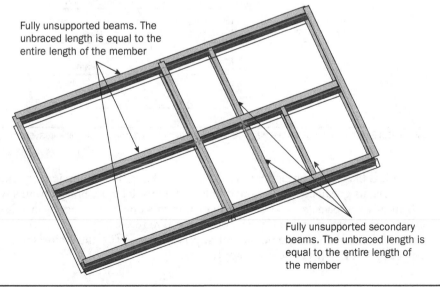

Fully unsupported beams. The unbraced length is equal to the entire length of the member

Fully unsupported secondary beams. The unbraced length is equal to the entire length of the member

FIGURE 5.13   Laterally unsupported beams.

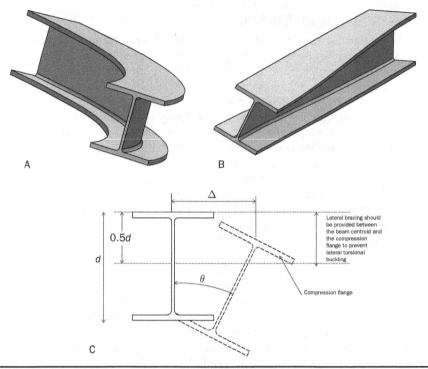

FIGURE 5.14 Lateral-torsional buckling: (*a*) simply supported beam, (*b*) cantilever beam, (*c*) lateral displacement and rotation of the cross-section.

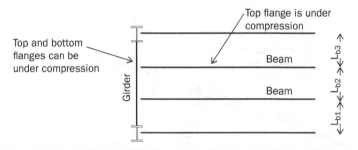

FIGURE 5.15 Plan view of a laterally supported girder in a typical steel framing system.

is, the less likely LTB occurs. Similar to local buckling, lateral-torsional buckling usually develops gradually during the cyclic loading for laterally braced beams with even seismically compact cross-sections. Lateral and torsional deformations increase in each loading cycle, and moment capacity of the beam reduces.

When there is continuous lateral support or $L_b \leq L_p$, the limit state of lateral-torsional buckling (LTB) does not apply.

$$M_n = M_p = F_y Z_x \tag{5.6}$$

Lateral-torsional buckling (LTB) is affected by the lateral bracing condition of the beam, as well as beam's section properties. The variation of the flexural strength with the increasing unbraced length is as follows:

When $L_p < L_b \leq L_r$

$$M_n = C_b\left[M_p - \left(M_p - 0.7F_yS_x\right)\left(\frac{L_b - L_p}{L_r - L_p}\right)\right] \leq M_p = F_yZ_x \tag{5.7}$$

where

L_b = laterally unbraced length (in.)
L_p = limiting laterally unbraced length to prevent LTB (in.)
L_r = limiting laterally unbraced length to prevent elastic LTB (in.)

When $L_b > L_r$

$$M_n = F_{cr}S_x \leq M_p \tag{5.8}$$

$$L_p = 1.76r_y\sqrt{\frac{E}{F_y}} \tag{5.9}$$

$$L_r = 1.95r_{ts}\frac{E}{0.7F_y}\sqrt{\frac{Jc}{S_xh_o} + \sqrt{\left(\frac{Jc}{S_xh_o}\right)^2 + 6.76\left(\frac{0.7F_y}{E}\right)^2}} \tag{5.10}$$

where

$c = 1$, for doubly symmetric I-shapes

$ = \dfrac{h_o}{2}\sqrt{\dfrac{I_y}{C_w}}$, for channels

$r_y = \sqrt{\dfrac{I_y}{A_g}}$, radius of gyration about y-axis, in.

$r_{ts}^2 = \dfrac{\sqrt{I_yC_w}}{S_x}$

J = torsional constant, in.[4]
S_x = elastic section modulus about the x-axis, in.[3]
h_o = distance between flange centroids, in.

$$F_{cr} = \frac{C_b\pi^2E}{\left(\dfrac{L_b}{r_{ts}}\right)^2}\sqrt{1 + 0.078\frac{Jc}{S_xh_o}\left(\frac{L_b}{r_{ts}}\right)^2} \tag{5.11}$$

C_b is a lateral-torsional buckling modification factor and permitted by AISC 360 to be conservatively taken as 1.0 for all cases in practice. C_b value considers the beneficial effect of moment gradient on bending moment strength (Fig. 5.17). If the moment gradient is uniform between the lateral bracing points, C_b will be equal to unity, according to Eq. (5.12) (Fig. 5.16). Table 5.5 provides C_b values for simply supported beams for different loading conditions.

$$C_b = \frac{12.5M_{max}}{2.5M_{max} + 3M_A + 4M_B + 3M_C} \qquad (5.12)$$

where

M_{max} = absolute value of the maximum moment in the unbraced segment, kip-in.

M_A = absolute value of moment at the quarter-point of the unbraced segment, kip-in.

M_B = absolute value of moment at the centerline of the unbraced segment, kip-in.

M_C = absolute value of moment at three-quarter point of the unbraced segment, kip-in.

It is apparent that the smaller the laterally unbraced length, L_b, is, the less likely LTB occurs. Table 5.4 summarizes the relation between laterally unbraced length, L_b, and failure mode, together with moment capacity. Lateral-torsional buckling (LTB) is affected by the lateral bracing condition of the beam, as well as the beam's radius of gyration (r_y).

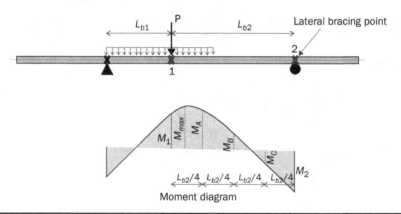

Moment diagram

Figure 5.16   Lateral-torsional buckling modification factor, C_b.

Laterally unbraced length	Failure mode	Nominal flexural capacity
$L_b \leq L_p$	Plastic hinge formation	$M_n = M_p$
$L_p < L_b \leq L_r$	Inelastic LTB	$M_y < M_n < M_p$
$L_r \leq L_b$	Elastic LTB	$M_n < M_y$

Table 5.4   Lateral-Torsional Buckling Limit State–Flexural Strength Relationship (When $C_b = 1.0$)

Loading	Lateral bracing	C_b values*			
Concentrated load at midpoint	None			1.32	
	At load point		1.67		1.67
Concentrated load at third points	None			1.14	
	At load points		1.67	1.00	1.67
Concentrated load at quarter points	None			1.14	
	At load points		1.67	1.11	1.11	1.67
Uniformly distributed load	None			1.14	
	At midpoint		1.30		1.30
	At third points		1.45	1.01	1.45
	At quarter points		1.52	1.06	1.06	1.52
	At fifth points		1.56	1.12	1.00	1.12	1.56

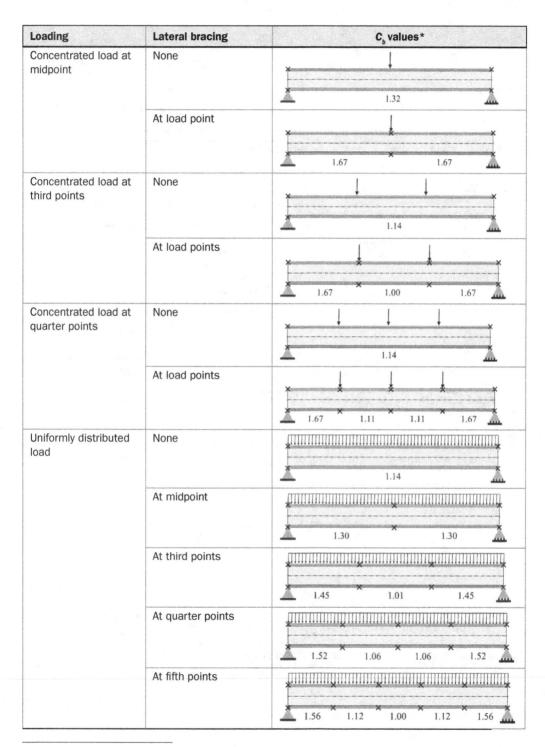

*Lateral bracing must always be provided at points of support

TABLE 5.5 C_b Values for Simply Supported Beams

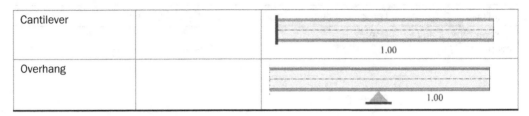

Cantilever		
Overhang		

TABLE 5.5 C_b Values for Simply Supported Beams (*Continued*)

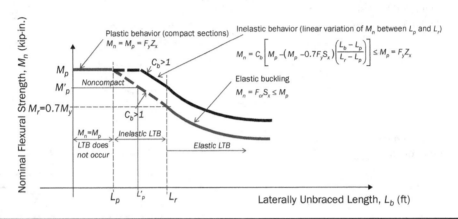

FIGURE 5.17 Nominal flexural strength with respect to laterally unbraced length.

The nominal strength of sections with compact webs and noncompact or slender flanges according to the limit state of flange local buckling (Fig. 5.18):

a. For sections with noncompact flanges:

$$M_n = M'_p = M_p - \left(M_p - 0.7F_yS_x\right)\left(\frac{\lambda - \lambda_{pf}}{\lambda_{rf} - \lambda_{pf}}\right) \tag{5.13}$$

$$L'_p = L_p + \left(L_r - L_p\right)\frac{\left(M_p - M'_p\right)}{\left(M_p - M_r\right)} \tag{5.14}$$

b. For section with slender flanges:

$$M_n = \frac{0.9Ek_cS_x}{\lambda^2} \tag{5.15}$$

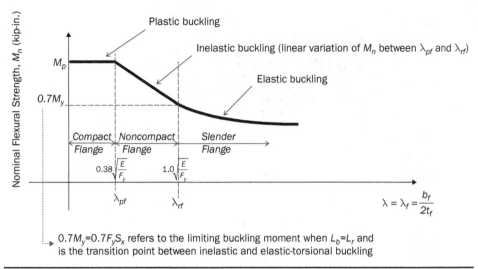

FIGURE 5.18 Nominal flexural strength with respect to the noncompact or slender flanges.

where

$$0.35 \le k_c = \frac{4}{\sqrt{\dfrac{h}{t_w}}} \le 0.76$$

h = clear distance between the flanges less the fillet or corner radius, in.

$$\lambda = \frac{b_f}{2t_f}$$

$\lambda_{pf} = \lambda_p$, limiting slenderness for a compact flange

$\lambda_{rf} = \lambda_r$, limiting slenderness for a noncompact flange

Example 5.2

Determine the design bending strength, $\phi_b M_n$, for W24×76 and W21×48 flexural members for the following cases ($E = 29{,}000$ ksi, $F_y = 50$ ksi):

a. Continuous lateral support

b. $L_b = 12$ ft, $C_b = 1.0$

Solution:

Section properties from *AISC Manual*:

	W24×76	**W21×48**
b_f (in.)	8.99	8.14
t_f (in.)	0.68	0.43
$b_f/(2t_f)$	6.61	9.47
h/t_w	49	53.6
t_w (in.)	0.44	0.35
S_x (in.³)	176	93
Z_x (in.³)	200	107
J (in.⁴)	2.68	0.803
C_w (in.⁶)	11,100	3,950
r_{ts} (in.)	2.33	2.05
h_o (in.)	23.2	20.2
r_y (in.)	1.92	1.66

Classification of the sections is provided in Table 5.6.

Section	$\lambda_f = \dfrac{b_f}{2t_f}$	$\lambda_{pf} = 0.38\sqrt{\dfrac{E}{F_y}}$	$\lambda_{rf} = 1.0\sqrt{\dfrac{E}{F_y}}$	$\lambda_w = \dfrac{h}{t_w}$	$\lambda_{pw} = 3.76\sqrt{\dfrac{E}{F_y}}$	
W24×76	6.61			49		Compact
W21×48	9.47	9.2	24.1	53.6	90.6	Noncompact flange

TABLE 5.6 Classification of the Sections

    a. Continuous lateral support

For W24×76 (compact):

$$M_n = M_p = F_y Z_x$$
$$= 50^{\text{ksi}} \times 200^{\text{in.}^3}$$
$$= 10{,}000 \text{ kip-in.}$$
$$= 833.3 \text{ kip-ft}$$

For W21×48 [noncompact shapes (noncompact flanges)]:

$$M_n = M_p' = M_p - \left(M_p - 0.7F_y S_x\right)\left(\frac{\lambda - \lambda_{pf}}{\lambda_{rf} - \lambda_{pf}}\right)$$

$$M_p = F_y Z_x$$
$$= 50^{\text{ksi}} 107^{\text{in.}^3}$$
$$= 5{,}350 \text{ kip-in.}$$
$$= 445.8 \text{ kip-ft}$$

$$M_n = 5{,}350^{\text{kip-in.}} - \left(5{,}350^{\text{kip-in.}} - 0.7 \times 50^{\text{ksi}} \times 93^{\text{in.}^3}\right)\left(\frac{9.47 - 9.2}{24.1 - 9.2}\right)$$

$$= 5{,}312 \text{ kip-in.}$$
$$= 442.7 \text{ kip-ft}$$

b. $L_b = 12$ ft, $C_b = 1.0$

For W24×76 (compact):

$$L_p = 1.76 r_y \sqrt{\frac{E}{F_y}}$$

$$= 1.76 \times 1.92^{\text{in.}} \sqrt{\frac{29{,}000^{\text{ksi}}}{50^{\text{ksi}}}}$$

$$= 81.4 \text{ in.}$$
$$= 6.78 \text{ ft}$$

Since $L_b = 12$ ft $> L_b = 6.78$ ft:

$$M_n = C_b \left[M_p - \left(M_p - 0.7 F_y S_x\right)\left(\frac{L_b - L_p}{L_r - L_p}\right)\right] \le M_p = F_y Z_x$$

$$L_r = 1.95 r_{ts} \frac{E}{0.7 F_y} \sqrt{\frac{Jc}{S_x h_o} + \sqrt{\left(\frac{Jc}{S_x h_o}\right)^2 + 6.76\left(\frac{0.7 F_y}{E}\right)^2}}$$

$$= 1.95 \times 2.33^{\text{in.}} \frac{29{,}000^{\text{ksi}}}{0.7 \times 50^{\text{ksi}}} \sqrt{\frac{2.68^{\text{in.}^4} \times 1.0}{176^{\text{in.}^3} \times 23.2^{\text{in.}}} + \sqrt{\left(\frac{2.68^{\text{in.}^4} \times 1.0}{176^{\text{in.}^3} \times 23.2^{\text{in.}}}\right)^2 + 6.76\left(\frac{0.7 \times 50^{\text{ksi}}}{29{,}000^{\text{ksi}}}\right)^2}}$$

$$= 234 \text{ in.}$$
$$= 19.5 \text{ ft}$$

$$M_n = 1.0 \left[10{,}000^{\text{kip-in.}} - \left(10{,}000^{\text{kip-in.}} - 0.7 \times 50^{\text{ksi}} \times 176^{\text{in.}^3}\right)\left(\frac{12^{\text{ft}} - 6.78^{\text{ft}}}{19.5^{\text{ft}} - 6.78}\right)\right] \le M_p = 10{,}000^{\text{kip-in.}}$$

$$= 8{,}424 \text{ kip-in.}$$
$$= 704 \text{ kip-ft}$$

For W21×48 [noncompact shapes (noncompact flanges)]:

$$L_p = 1.76 r_y \sqrt{\frac{E}{F_y}}$$

$$= 1.76 \times 1.66^{\text{in.}} \sqrt{\frac{29{,}000^{\text{ksi}}}{50^{\text{ksi}}}}$$

$$= 70.4 \text{ in.}$$
$$= 6.09 \text{ ft}$$

$$L_r = 1.95 r_{ts} \frac{E}{0.7F_y} \sqrt{\frac{Jc}{S_x h_o} + \sqrt{\left(\frac{Jc}{S_x h_o}\right)^2 + 6.76\left(\frac{0.7F_y}{E}\right)^2}}$$

$$= 1.95 \times 2.05^{in.} \frac{29,000^{ksi}}{0.7 \times 50^{ksi}} \sqrt{\frac{0.803^{in.^4} \times 1.0}{93^{in.^3} \times 20.2^{in.}} + \sqrt{\left(\frac{0.803^{in.^4} \times 1.0}{93^{in.^3} \times 20.2^{in.}}\right)^2 + 6.76\left(\frac{0.7 \times 50^{ksi}}{29,000^{ksi}}\right)^2}}$$

$$= 199 \text{ in.}$$
$$= 16.5 \text{ ft.}$$

Note: when $F_y = 50$ ksi steel is used, Table 3-2 in *AISC Manual* lists all W-shapes, and can be used directly. For W21 × 48, $F_y = 50$ ksi, page 3-25 in *AISC Manual* shows that $L_p = 6.09$ ft and $L_r = 16.5$ ft, the same as the calculated ones above. However, for $F_y \neq 50$ ksi, the above equations have to be used for L_p and L_r.

$$L_p' = L_p + \left(L_r - L_p\right)\frac{\left(M_p - M_p'\right)}{\left(M_p - M_r\right)}$$

$$M_r = 0.7F_y S_x$$

$$= 0.7 \times 50^{ksi} \times 93^{in.^3}$$
$$= 3,255 \text{ kip-in.}$$
$$= 271.3 \text{ kip-ft}$$

$$L_p' = 6.09^{ft} + \left(16.5^{ft} - 6.09^{ft}\right)\frac{\left(445.8^{kip-ft} - 442.7^{kip-ft}\right)}{\left(445.8^{kip-ft} - 271.3^{kip-ft}\right)}$$

$$= 6.27 \text{ ft}$$

Since $L_b = 12$ ft $> L_p' = 6.27$ ft:
Replace M_p with M_p' in Eq. (5.13):

$$M_n = C_b\left[M_p' - \left(M_p' - 0.7F_y S_x\right)\left(\frac{L_b - L_p'}{L_r - L_p'}\right)\right] \leq M_p'$$

$$M_n = 1.0\left[5,350^{kip-in.} - \left(5,350^{kip-in.} - 3,255^{kip-in.}\right)\left(\frac{12^{ft} - 6.27^{ft}}{16.5^{ft} - 6.27^{ft}}\right)\right] \leq M_p' = 5,350^{kip-in.}$$

$$= 4,177 \text{ kip-in.}$$
$$= 348 \text{ kip-ft}$$

The design bending strengths, $\phi_b M_n$ (Table 5.7).

Section	ϕ_b	Continuous lateral support		$L_b = 12$ ft, $C_b = 1.0$	
		M_n (kip-ft)	$\phi_b M_n$ (kip-ft)	M_n (kip-ft)	$\phi_b M_n$ (kip-ft)
W24×76	0.9	833.3	750	704	633.6
W21×48		442.7	398.4	348	313.2

TABLE 5.7 Design Bending Strengths for Both Cases

Example 5.3

Compute C_b for the simply supported beam with overhangs subjected to concentrated loads at the ends of the overhangs at both sides and midspan as shown in Fig. 5.19.

Solution:

C_b and L_b for unbraced segments ($M_r = M_u$):

Segment A: $L_b = 10'$, $C_b = 1.67$, $M_r = M$
Segment B: $L_b = 15'$, $C_b = 2.27$, $M_r = M$
Segment C: $L_b = 15'$, $C_b = 1.0$, $M_r = M$
Segment D: $L_b = 15'$, $C_b = 2.27$, $M_r = M$
Segment E: $L_b = 10'$, $C_b = 1.0$, $M_r = M$

Segment	L_b	C_b	M_r
A	10'	1.67	M
B	15'	2.27	M
C	**15'**	**1.0**	M
D	15'	2.27	M
E	10'	1.0	M

A section with the largest L_b and smallest C_b is the most critical for bending strength.

5.4   Shear Strength

Design of a beam is governed by flexural strength, shear strength, or deflection. Flexural strength dominates the design of medium length beams, whereas deflection is the main design criteria for long-span beams. However, shear becomes dominant for short-span beams ($L < 9'$), thin webs, or if there is a concentrated force near ends. A well-known shear stress equation for symmetrical sections is

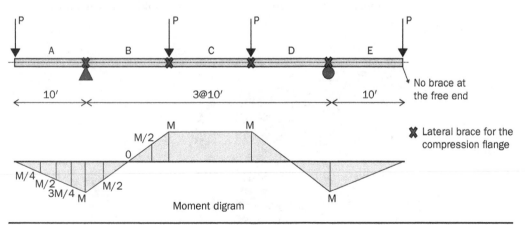

FIGURE 5.19   Simply supported beam with overhangs and its moment diagram under the given loading condition.

$$\tau = \frac{VQ}{Ib} \tag{5.16}$$

where

V = shear force, kips

Q = statical moment about the x-axis of the cross-sectional area between the extreme fiber and the particular location at which the shear stress is to be determined, in.[3]

I = moment of inertia, in.[4]

b = width of the cross-section, in.

Shear stress for a rectangular section is maximum in the middle and zero at the ends (Fig. 5.20). For an I-shaped section, shear stress is approximated by assuming it is uniformly distributed through the web area of the cross-section (Fig. 5.21).

5.4.1 Nominal Shear Strength (Vn)

Nominal shear strength is the gross area of the web multiplied by the shear yield stress, τ_y, and neglecting the effect of any fastener holes:

$$V_n = \tau_y A_w C_{v1} \tag{5.17}$$

where

τ_y = shear yield stress of the web steel, ksi

A_w = area of the web ($=dt_w$), in.[2]

C_{v1} = web shear strength coefficient (shear buckling reduction coefficient)

Using the Von Mises theory $\tau_y = 0.6F_y$, the nominal shear strength equation becomes

$$V_n = (0.6F_y A_w)C_{v1} = (0.6F_y dt_w)C_{v1} \tag{5.18}$$

AISC specification gives different expressions for different h/t_w ratios (web slenderness limit for local web buckling) where shear failures would be plastic, inelastic, or elastic (Fig. 5.22).

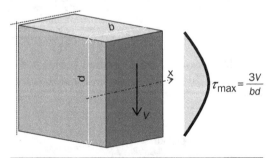

Figure 5.20   Shear stress distribution on a rectangular section.

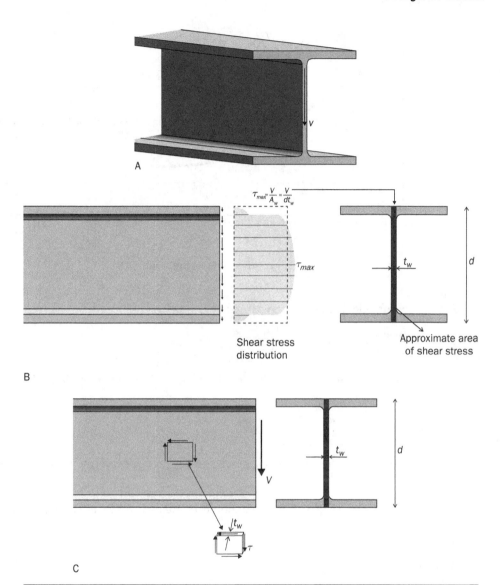

FIGURE 5.21   Shear stress: (*a*) shear force in an I-shaped section, (*b*) shear stress distribution.

The web in an I-shaped section takes all shear, *V*, and will have the following failure modes (or limit states) based on h/t_w value (Fig. 5.22):

a. For $\dfrac{h}{t_w} \leq 2.24 \sqrt{\dfrac{E}{F_{yw}}}$ (Rolled W shapes)

and

$\dfrac{h}{t_w} \leq 2.45 \sqrt{\dfrac{E}{F_{yw}}}$ (for other doubly symmetric, singly symmetric, open-section shapes, and channels)

The web will yield under shear, and

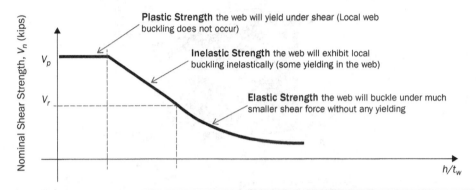

Figure 5.22   Nominal shear strength as a function of web slenderness (h/t_w).

$$\tau = \tau_y = 0.6F_{yw} \text{ shear yield stress (ksi)} \qquad (5.19)$$

where

F_{yw} = minimum yield stress of the web, ksi

b. For $2.45\sqrt{\dfrac{E}{F_{yw}}} < \dfrac{h}{t_w} \le 3.07\sqrt{\dfrac{E}{F_{yw}}}$

The web will show the local buckling inelastically (some yielding in the web) (inelastic buckling).

c. For $3.07\sqrt{\dfrac{E}{F_{yw}}} < \dfrac{h}{t_w} \le 260$

The web will buckle under much smaller shear force without any yielding (elastic buckling).

Note that for $\dfrac{h}{t_w} > 260$, the beam should be treated as "Plate Girder," which is covered in detail in Chapter 9.

Strength requirement for shear:

$$V_u \le \phi_v V_n \qquad (5.20)$$

where

V_u = nominal shear strength, kips
V_u = maximum required factored shear, kips
$\phi_v = 1.0$ for $\dfrac{h}{t_w} \le 2.24\sqrt{\dfrac{E}{F_y}}$ (web slenderness limit for local web buckling for

rolled W-sections)
$= 0.90$ for else

5.5 Serviceability

Serviceability limitations were first used in buildings for brittle finishes such as plaster in the early 1900s. The current engineering practice uses the same limits, even though plaster has been replaced by other materials, but the limits work just fine.

In building the type of structures, whether it is a steel or R/C building, serviceability requirements provide human comfort and are classified into two categories:

1. Deflection

2. Floor Vibration

Maximum deflection for a simply supported beam subjected to uniformly distributed loading is given in Fig. 5.23. To satisfy deflection limit and optimum design, beams are generally cambered upward to counteract the dead loads. The cambered beam becomes flat prior to the application of live or other loads if there is any. It should be noted that serviceability similar to drift limitation has nothing to do with the design method selected (LRFD or ASD) because serviceability needs to be satisfied at service load levels. It has more impact on nonstructural elements than on structural elements.

Depending on the span length, loading, and beam size, the design might be governed by serviceability instead of strength. In most practical applications, deflections due to service live load or dead + live loads is limited to $L/360$ and $L/240$ of the span length, respectively (for shored construction) (Table 5.8).

$$\Delta_{max} \leq \frac{L}{360}, \text{ live load deflection limit} \tag{5.21a}$$

$$\Delta_{max} \leq \frac{L}{240}, \text{ dead + live load deflection limit (including snow load, if there is any)} \tag{5.21b}$$

where

Δ_{max} = maximum deflection due to service live load or dead + live loads (not factored)

L = span length, in. (for cantilever beams, L is the twice the span of the cantilever)

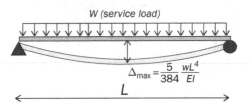

FIGURE 5.23 Maximum deflection for a simply supported beam subjected to uniformly distributed loading.

Member	Live load	Dead + live load	Snow or wind
Floor members	L/360	L/240	–
Roof members not supporting ceiling	L/180	L/120	L/180
Roof members supporting nonplaster ceiling	L/240	L/180	L/240
Roof members supporting plaster ceiling	L/360	L/240	L/360

TABLE 5.8 Deflection Limits (ASCE 7)

5.6 Concentrated Forces

Due to high vertical stresses directly over support or under a concentrated load, the beam web is likely to buckle or fail as a result of these stresses. The possible failure modes under concentrated forces are

    a. Flange local bending

    b. Web local yielding

    c. Web local crippling

    d. Web sidesway buckling

    e. Web compression buckling

    f. Web panel zone shear

Web local yielding and web crippling are mostly possible in compression-related cases.

A steel member should have sufficient *flange local bending, web yielding, web crippling, and sidesway buckling* strength, R_n, if it is subject to concentrated load perpendicular to one flange and symmetric to the web. In cases where both flanges are subject to concentrated forces, the member should have sufficient *web yielding, web crippling*, and *column web buckling* strength, R_n (Fig. 5.24).

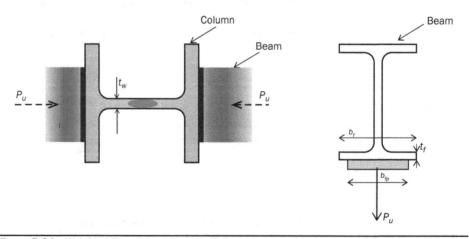

FIGURE 5.24 Web buckling and web bearing failures.

5.6.1 Flange Local Bending

Flange local bending applies only for tensile forces. There are two possible cases that flange local bending needs to be checked:

 i. A concentrated force might be acting on the bottom flange of a beam element (Fig. 5.25).

 ii. When an end moment from a beam is applied to the flanges of a column, the flange force due to the moment is transferred to the column through the flanges and to the web (Fig. 5.26). This force can cause localized bending of the flange, if it is tension.

To prevent such behavior, stiffener plates can be added to the flange and the web.

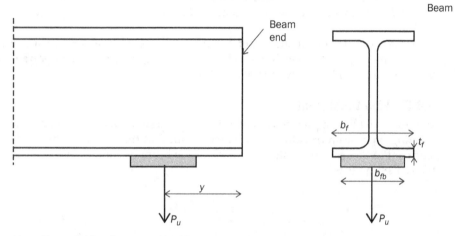

Note: if $b_{fp} < 0.15b_f$, flange local buckling need not be checked.

FIGURE 5.25 Flange local bending for beams subject to a concentrated force.

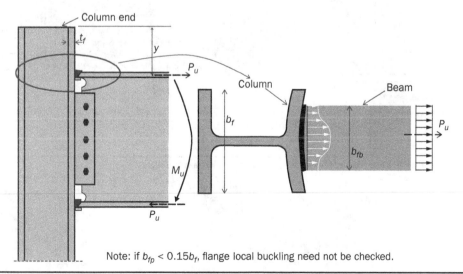

Note: if $b_{fp} < 0.15b_f$, flange local buckling need not be checked.

FIGURE 5.26 Flange local bending for columns.

Nominal flange local bending strength, R_n,

$$R_n = 6.25t_f^2 F_{yf} \text{ if } y \geq 10t_f \tag{5.22}$$

And the design strength for flange local bending, ϕR_n, is

$$P_u \leq \phi R_n \tag{5.23}$$

where

$\phi = 0.9$
F_{yf} = specified minimum yield stress of the flange, ksi
t_f = thickness of the loaded flange, in.
P_u = ultimate reaction or concentrated force, kips

If Eq. (5.23) is not satisfied, then a pair of transverse stiffeners extending at least one-half the depth of the web has to be provided adjacent to the concentrated tensile force centrally applied across the flange (Fig. 5.27).

5.6.2 Web Local Yielding

Web local yielding applies to concentrated compression forces. The concentrated force P acting on a beam or column is assumed critical at the toe of the fillet. The load is assumed to be distributed along the web of the beam or column at a slope of 2.5 to 1 (Figs. 5.28 and 5.29).

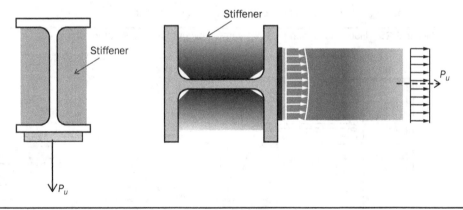

FIGURE 5.27 Transverse stiffeners for flange local bending.

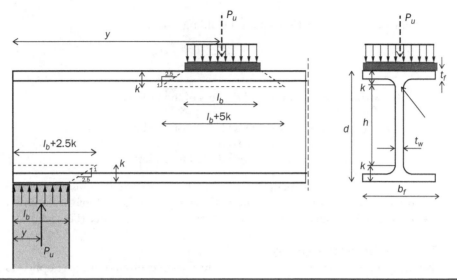

FIGURE 5.28 Web local yielding for beams.

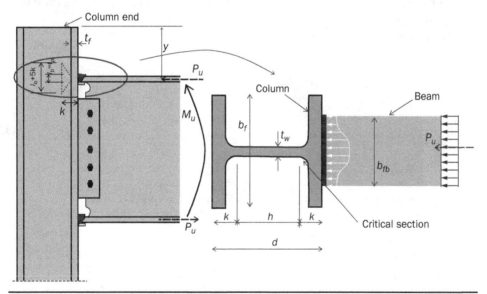

FIGURE 5.29 Web local yielding for columns.

Nominal web local yielding strength, R_n,

$$R_n = (5k + N)F_{yw}t_w \text{, for } y > d \tag{5.24a}$$

$$R_n = (2.5k + N)F_{yw}t_w \text{, for } y \leq d \tag{5.24b}$$

And the design strength for web local yielding, ϕR_n, is

$$P_u \leq \phi R_n \tag{5.25}$$

where

$\phi = 1.0$

F_{yw} = specified minimum yield stress of the web, ksi

t_w = web thickness, in.

N = length of bearing (not less than k for end beam reaction), in.

k = distance from outer face of the flange to the web toe of the fillet, in.

P_u = ultimate reaction or concentrated force, kips

If the strength requirement is not satisfied, either a pair of transverse stiffeners or a doubler plate, extending at least one-half the depth of the web of the column or beam has to be provided adjacent to a concentrated compressive force (Fig. 5.30).

5.6.3 Web Local Crippling

Web local crippling is defined as the case where a concentrated force is applied to beams, beam bearing at supports, and the reaction of beam flanges at connections to columns causes a localized yielding from high compressive stress followed by inelastic buckling in the web region adjacent to the toe of a fillet in the vicinity of the concentrated force (Figs. 5.31 and 5.32).

Nominal web local crippling strength, R_n,

$$R_n = 0.80 t_w^2 \left[1 + 3\left(\frac{l_b}{d}\right)\left(\frac{t_w}{t_f}\right)^{1.5} \right] \sqrt{\frac{EF_{yw}t_f}{t_w}} \text{ , for } y > d \tag{5.26a}$$

$$R_n = 0.40 t_w^2 \left[1 + 3\left(\frac{N}{d}\right)\left(\frac{t_w}{t_f}\right)^{1.5} \right] \sqrt{\frac{EF_{yw}t_f}{t_w}} \text{ , for } y \leq d \text{ and } N/d \leq 0.2 \tag{5.26b}$$

$$R_n = 0.40 t_w^2 \left[1 + 3\left(\frac{4N}{d} - 0.2\right)\left(\frac{t_w}{t_f}\right)^{1.5} \right] \sqrt{\frac{EF_{yw}t_f}{t_w}} \text{ , for } y \leq d \text{ and } N/d > 0.2 \tag{5.26c}$$

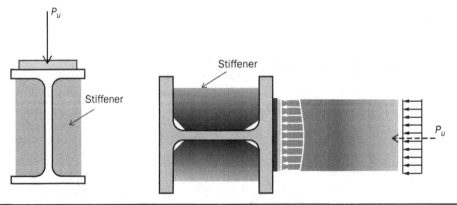

FIGURE 5.30 Transverse stiffeners for web local yielding.

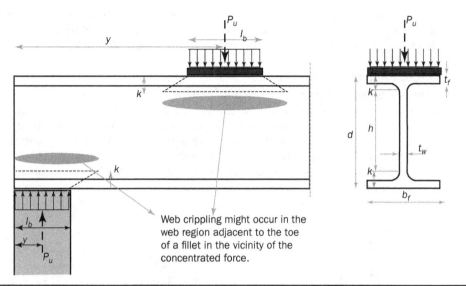

FIGURE 5.31   Web local crippling for beams.

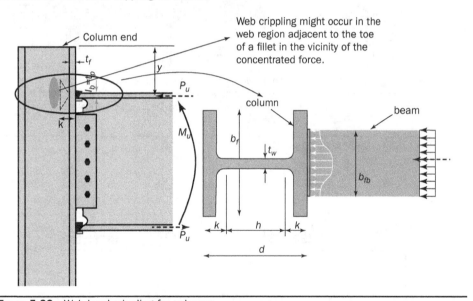

FIGURE 5.32   Web local crippling for columns.

And the design strength for web local yielding, ϕR_n, is

$$P_u \le \phi R_n \qquad (5.27)$$

where

$\phi = 0.75$
$t_f =$ flange thickness, in.
$d =$ overall depth of the member, in.
$l_b =$ length of bearing, in.
$P_u =$ ultimate reaction or concentrated force, kips

If the strength requirement is not satisfied, either a pair of transverse stiffeners or a doubler plate, extending at least one-half the depth of the web of the column or beam, has to be provided adjacent to a concentrated compressive force (Fig. 5.33).

5.6.4 Web Sidesway Buckling

If the compressive forces are applied to laterally braced compression flanges, and the relative lateral movement between the loaded compression flange and the tension flange is not restrained at the point of application of the concentrated force, the web will be put in compression, and the tension flange may buckle. The concentrated compressive force could be either an internal loading or an end reaction. This limit state only applies to compressive forces in bearing connection and does not apply to moment connections (Fig. 5.34).

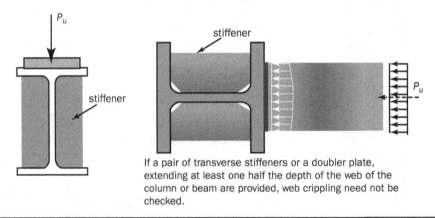

If a pair of transverse stiffeners or a doubler plate, extending at least one half the depth of the web of the column or beam are provided, web crippling need not be checked.

Figure 5.33 Transverse stiffeners for web local crippling.

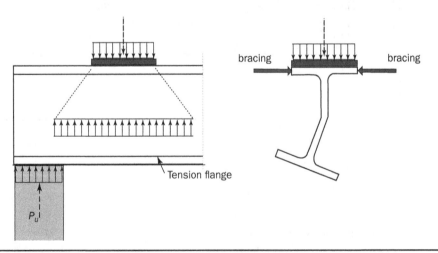

Figure 5.34 Web sidesway buckling.

It is possible to prevent web sidesway buckling by using lateral bracing or stiffeners properly designed at the load point. AISC 360 commentary suggests that local bracing at both flanges be designed for 1% of the concentrated force applied at that point. If stiffeners are used, they must extend from the load point through at least one-half the beam or girder depth. And also, the pair of stiffeners should be designed to carry the full load. Flange rotation must be prevented if stiffeners or doubler plates are to be effective.

Nominal web sidesway buckling, R_n:

If the compression flange is restrained against rotation:

$$R_n = \frac{C_r t_w^3 t_f}{h^2}\left[1 + 0.4\left(\frac{h / t_w}{L_b / b_f}\right)^3\right], \text{ for } (h / t_w) / (L_b / b_f) \leq 2.3 \tag{5.28}$$

For $(h / t_w) / (L_b / b_f) > 2.3$, web sidesway buckling does not occur.

If the compression flange is not restrained against rotation (Fig. 5.35):

$$R_n = \frac{C_r t_w^3 t_f}{h^2}\left[0.4\left(\frac{h / t_w}{L_b / b_f}\right)^3\right], \text{ for } (h / t_w) / (L_b / b_f) \leq 1.7 \tag{5.29}$$

For $(h / t_w) / (L_b / b_f) > 1.7$, web sidesway buckling does not occur.

And the design strength for web local yielding, ϕR_n, is

$$P_u \leq \phi R_n \tag{5.30}$$

where
$\phi = 0.85$
$C_r = 960,000$ ksi for $M_u < M_y$ at the location of the force
$\quad\quad 480,000$ ksi for $M_u \geq M_y$ at the location of the force
$L_b =$ largest laterally unbraced length along either flange at the point of load, in.
$M_u =$ required flexural strength, kip-in.
$P_u =$ ultimate reaction or concentrated force, kips

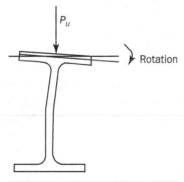

FIGURE 5.35 Compression flange not restrained against rotation.

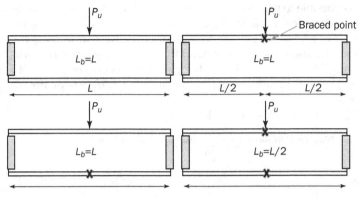

L_b=Largest laterally unbraced length along flange at the point of load

Figure 5.36   Laterally unbraced lengths for web sidesway buckling.

If the strength requirement is not satisfied, lateral bracing has to be provided at the tension flange (Fig. 5.36) or a pair of transverse stiffeners or a doubler plate has to be provided adjacent to a concentrated compressive force.

5.6.5   Web Compression Buckling

Compression buckling of the web may occur if a pair of double-concentrated forces are applied at both flanges of a member at the same location (i.e., moment connections at both flanges of a column) (Fig. 5.37).

Nominal web compression buckling strength, R_n,

$$R_n = \frac{24t_w^3\sqrt{EF_{yw}}}{h}, \text{ for } y \geq d/2 \qquad (5.31a)$$

$$R_n = \frac{12t_w^3\sqrt{EF_{yw}}}{h}, \text{ for } y < d/2 \qquad (5.31b)$$

And the design strength for web local yielding, ϕR_n, is

$$P_u \leq \phi R_n \qquad (5.32)$$

where
$\phi = 0.90$
P_u = ultimate reaction or concentrated force, kips

If the strength requirement is not satisfied, a single transverse stiffener, or a pair of transverse stiffeners or a doubler plate, extending the full depth of the web, has to be provided adjacent to concentrated compressive forces at both flanges.

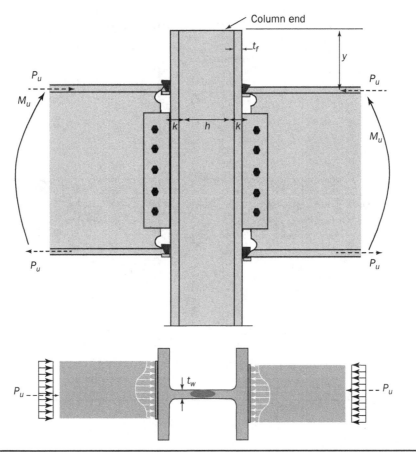

FIGURE 5.37 Web compression buckling for columns.

5.6.6 Web Panel Zone Shear

Panel zone is defined as the region in which concentrated forces on a column flange cause large shear forces across the column web. The column web shear stresses are generally high within the boundaries of the rigid connection of two or more members whose webs lie in a common plane.

The shear in the panel zone (or the required shear), ΣF_u, is the sum of the shear in the web and the shear due to flange force, P_{uf} or P_{af}. In such cases, the calculated total force, ΣF_u, should not exceed the column web design strength ϕR_n or such webs should be reinforced (Fig. 5.38).

The nominal strength of the web panel zone for the limit state of shear yielding, R_v, is determined as

    i. When the effect of panel zone deformation on frame stability is not considered in the analysis:

For $P_u \leq 0.4P_y$

$$R_v = 0.60F_y d_c t_w \tag{5.33a}$$

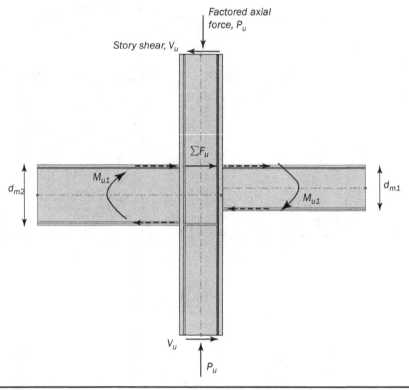

FIGURE 5.38 Panel zone shear.

For $P_u > 0.4P_y$

$$R_v = 0.60F_y d_c t_w \left(1.4 - \frac{P_u}{P_y}\right) \tag{5.33b}$$

ii. When frame stability, including plastic panel zone deformation, is considered in the analysis:

For $P_u \le 0.75P_y$

$$R_v = 0.60F_y d_c t_w \left(1 + \frac{3b_{cf} t_{cf}^2}{d_b d_c t_w}\right) \tag{5.34a}$$

For $P_u > 0.75P_y$

$$R_v = 0.60F_y d_c t_w \left(1 + \frac{3b_{cf} t_{cf}^2}{d_b d_c t_w}\right)\left(1.9 - \frac{1.2P_u}{P_y}\right) \tag{5.34b}$$

$$\sum F_u = \frac{M_{u1}}{d_{m1}} + \frac{M_{u2}}{d_{m2}} - V_u \tag{5.35}$$

And the design strength for web panel zone shear in a moment connection, ϕR_v, is

$$\phi R_v \leq \sum F_u \tag{5.36}$$

where

$\phi = 0.90$

M_{u1} = sum of the factored moments due to factored lateral loads and the moments due to gravity loads, kip-in.

M_{u2} = difference between the moment due to factored lateral loads and the moments due to gravity loads, kip-in.

V_u = factored story shear due to lateral load, kips

d_{m1}, d_{m2} = distance between the flange forces in the moment connection ($\approx 0.95 \times$ the beam depth), in.

b_{cf} = width of column flange, in.

d_c = column depth, in.

d_b = beam depth, in.

t_{cf} = thickness of the column flange, in.

t_w = column web thickness, in.

If the strength requirement is not satisfied, a doubler plate or a pair of diagonal stiffeners has to be provided.

5.7   Design of Bearing Stiffeners

Stiffeners for concentrated forces in beams are designed as short columns to reinforce the web of a beam along the length or the web of a beam at an end reaction. For columns, stiffener plates are required when the applied forces are greater than the design strength for each failure mode (flange local bending, web local yielding, web crippling, compression buckling). The stiffeners should extend the full depth of the column when there are applied forces on both sides of the column. For columns, stiffener plates are welded to both the web and the flange. The weld to the flange is designed for the difference between the required strength and the design strength of the controlling limit state (flange local bending, web local yielding, web crippling, compression buckling) (Fig. 5.39).

The bearing strength, R_n, is determined for the limit state of bearing as

$$R_n = 1.8 F_y A_{pb} \tag{5.37}$$

And the design strength for web local yielding, ϕR_n, is

$$P_u \leq \phi_{pb} R_n \tag{5.38}$$

where

$\phi_{pb} = 0.75$

P_u = ultimate reaction or concentrated force, kips

A_{pb} = projected area in bearing, in.^{2}

The available strength of the stiffener is determined for the limit states of yielding and buckling for the connecting elements as

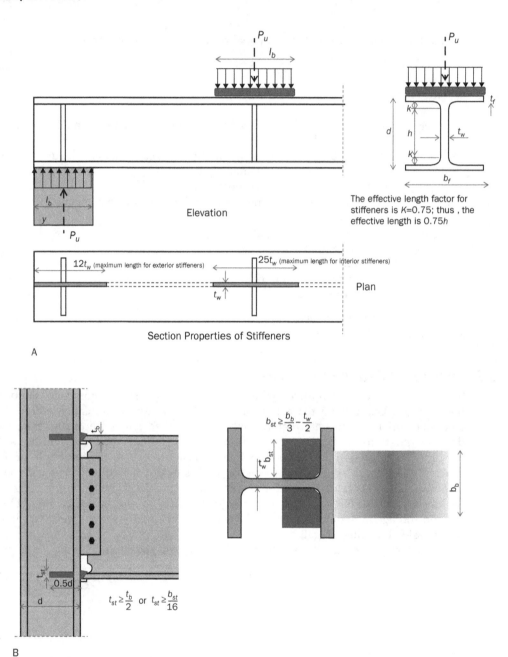

The effective length factor for stiffeners is $K=0.75$; thus, the effective length is $0.75h$

Elevation

$12t_w$ (maximum length for exterior stiffeners) $25t_w$ (maximum length for interior stiffeners)

Plan

Section Properties of Stiffeners

A

$b_{st} \geq \dfrac{b_b}{3} - \dfrac{t_w}{2}$

$t_{st} \geq \dfrac{t_b}{2}$ or $t_{st} \geq \dfrac{b_{st}}{16}$

B

Figure 5.39   Bearing stiffeners: (a) for beams, (b) for columns.

a. For $KL/r \leq 25$

$$P_n = F_y A_g \qquad (5.39)$$

b. For $KL/r > 25$

Determine the available strength based on elements in compression (Chapter 4).

And the compressive design strength of the stiffener, $\phi_c P_n$, is

$$P_u \leq \phi_c P_n \qquad (5.40)$$

where

$$\phi_c = 0.90$$

5.8   Design Summary

A summary of beam design is given below.

1. Design the beam based on bending:

$$M_u \leq \phi_b M_n \qquad (5.41a)$$

$$M_u \leq \phi_b F_y Z_x \qquad (5.41b)$$

$$Z_{x,\text{required}} \geq \frac{M_u}{\phi_b F_y} \qquad (5.41c)$$

where

   $\phi_b = 0.90$ for flexure
   $M_u = $ maximum required factored bending moment (obtained from structural analysis), kip-in.
   $M_n = $ nominal flexural strength, kip-in.

2. Select a section W ...×xxx and determine the selected section's L_p and L_r.
3. For $L_b \leq L_p$, the selected section is O.K.
4. For $L_p < L_b \leq L_r$,

$$\phi_b M_n = \phi_b C_b \left[M_p - \left(M_p - 0.7 F_y S_x \right) \left(\frac{L_b - L_p}{L_r - L_p} \right) \right] \leq \phi_b M_p = F_y Z_x \qquad (5.42)$$

5. If $\phi_b M_n \leq M_u$, the selected section is O.K.; if not, increase the section size.
6. For $L_b > L_r$, elastic buckling.
7. Check the section for shear.
8.

$$\phi_v V_n = \phi_v (0.6 F_y A_w) C_{v1} = (0.6 F_y d t_w) C_{v1} \geq V_u \qquad (5.43)$$

where
   $\phi_v = 1.0$ for $\dfrac{h}{t_w} \leq 2.24 \sqrt{\dfrac{E}{F_y}}$

   $= 0.90$ for else

9. Check the deflection under service load if required.
10. Check web failure under any concentrated forces.

Example 5.4

A 30' simply supported beam, **shown in Fig. 5.40,** is subjected to concentrated dead loads of P_D=20 kips and a series of live loads $P_L = 10$ kips with 7½' spacing. Assume that the beam is fully laterally supported by the floor system. Steel grade is A992 Grade 50 ($F_y = 50$ ksi).

a. Design the beam with $\Delta_{max} \leq L/360$ for the live load.

b. Design bearing stiffeners at the end and one-fourth points for a W24×76 beam.

1. Demand analysis.

Structural analysis is carried out for the dead and live loads. Moment and shear force diagrams are given in Fig. 5.23 with Fig. 5.41. Note that live loads are placed in a different way on the beam to get the maximum bending moment and shear force.

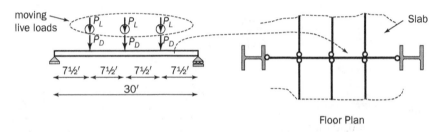

FIGURE 5.40 Simply supported beam and floor plan.

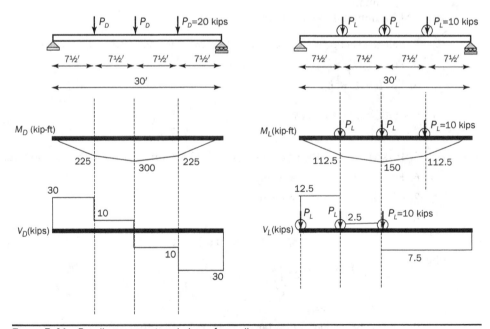

FIGURE 5.41 Bending moment and shear force diagrams.

$$M_u = 1.2M_D + 1.6M_L = 1.2(300^{kip\text{-}ft}) + 1.6(150^{kip\text{-}ft}) = 600 \text{ kip-ft}$$

$$V_u = 1.2V_D + 1.6V_L = 1.2(30^{kips}) + 1.6(12.5^{kip\text{-}ft}) = 56 \text{ kips}$$

2. Design the beam by bending moment.

$$\phi_b M_n = M_u$$

$$\phi_b F_y Z_x = M_u \quad \rightarrow \quad Z_x = \frac{M_u}{\phi_b F_y}$$

$$Z_x = \frac{600^{kip\text{-}ft} \times 12^{\frac{in.}{ft}}}{0.9 \times 50^{ksi}} = 160 \text{ in.}^3$$

Use W24×68 → Section properties from the *AISC Manual*:

$Z_x = 177$ in.^{3}
$I_x = 1{,}830$ in.^{4}
$d = 23.7$ in.
$t_w = 0.415$ in.
$t_f = 0.585$ in.
$b_f = 8.97$ in.
$k = 1.09$ in.

3. Check compactness (Table 5.9).

W24×68 is a compact section.

4. Check shear strength.

$$\frac{h}{t_w} = 52 \le 2.24\sqrt{\frac{E}{F_y}} = 2.24\sqrt{\frac{29{,}000^{ksi}}{50^{ksi}}} = 53.95 \rightarrow \begin{cases} \phi_v = 1.0 \\ C_{v1} = 1.0 \end{cases}$$

$$V_u \le \phi_v V_n$$

$$56^{kips} \le 1.0(0.6F_y A_w)C_{v1}$$
$$\le 1.0(0.6 \times 50^{ksi} \times 23.7^{in.} \times 0.415^{in.})1.0$$
$$\ll 295 \text{ kips}$$

Section	$\lambda_f = \dfrac{b_f}{2t_f}$	$\lambda_{pf} = 0.38\sqrt{\dfrac{E}{F_y}}$	$\lambda_w = \dfrac{h}{t_w}$	$\lambda_{pw} = 3.76\sqrt{\dfrac{E}{F_y}}$	Classification
W24×68	7.66	9.2	52	90.6	Compact

TABLE 5.9 Classification of the Section

5. Check deformation for service load using any approach to get deflection.

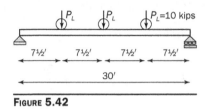

FIGURE 5.42

$$\Delta_{max} = 0.43 \text{ in.} \leq \frac{L}{360} = \frac{30^{ft} \times 12^{\frac{in.}{ft}}}{360} = 1.0 \text{ in.} \quad \text{O.K.}$$

6. Check web yielding under concentrated load (Fig. 5.43).

$$P_u \leq \phi R_n \qquad \phi = 1.0$$

$$\begin{aligned} 40 \text{ kips} &\leq \phi(l_b + 5k)F_{yw}t_w \\ &\leq 1.0(l_b + 5.45^{in.})50^{ksi}0.415^{in.} \\ &\leq 113^{kips} + 20.8l_b \quad \text{O.K. always} \end{aligned}$$

7. Check web crippling under concentrated load (Fig. 5.44).

$$P_u \leq \phi R_n \qquad \phi = 0.75$$

$$R_n = 0.80t_w^2\left[1 + 3\left(\frac{l_b}{d}\right)\left(\frac{t_w}{t_f}\right)^{1.5}\right]\sqrt{\frac{EF_{yw}t_f}{t_w}}$$

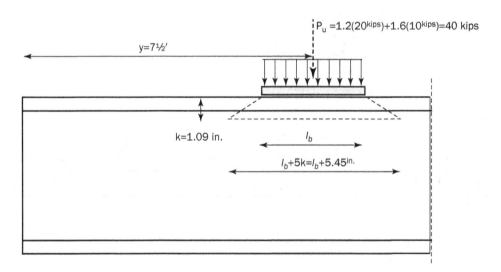

$P_u = 1.2(20^{kips}) + 1.6(10^{kips}) = 40 \text{ kips}$

$y = 7\frac{1}{2}'$

$k = 1.09 \text{ in.}$

l_b

$l_b + 5k = l_b + 5.45^{in.}$

FIGURE 5.43

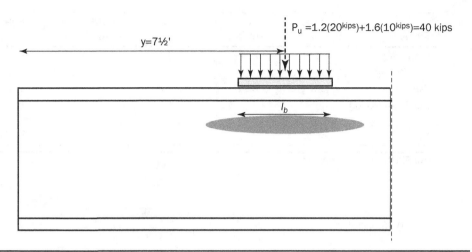

FIGURE 5.44

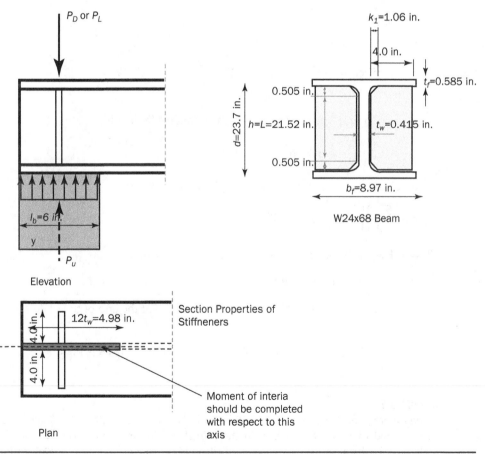

FIGURE 5.45

Shape	Area (in.^{2})	I (in.^{4})	d (in.)	Ad^2 (in.^{4})	$I+Ad^2$ (in.^{4})
Stiffener plate (0.4×4.0 in.)	1.6	4.27	2.21	7.815	12.09
Web (0.415×4.98 in.)	2.07	0.0297	0	0	0.0297 (may be ignored)
Σ	3.67				12.12

TABLE 5.10 Section Properties

$$= 0.80(0.415^{\text{in.}})^2\left[1 + 3\left(\frac{l_b}{23.7^{\text{in.}}}\right)\left(\frac{0.415^{\text{in.}}}{0.585^{\text{in.}}}\right)^{1.5}\right]\sqrt{\frac{29,000^{\text{ksi}}\,50^{\text{ksi}}\,0.585^{\text{in.}}}{0.415^{\text{in.}}}}$$

$40\text{ kips} \le 0.75R_n$ O.K. always

8. Final design.

Use W24×68 with F_y=50 ksi.

9. Design of bearing stiffeners (Fig. 5.45).

Try stiffeners with 0.4 in. thickness.

   a. Section properties of the area with stiffeners (Table 5.10):

$$r = \sqrt{\frac{I}{A}} = \sqrt{\frac{12.12}{3.67}} = 3.30\text{ in.}$$

   b. Check design compressive strength of the stiffener:

$$\frac{KL}{r} = \frac{0.75 \times 21.52^{\text{in.}}}{3.30} = 4.89 < 25$$

$$\phi_c P_n = \phi_c F_y A_g = 0.9 \times 50^{\text{ksi}} \times 3.67^{\text{in.}^2} = 165.2\text{ kips} > 56\text{ kips}$$

   c. Check bearing strength:

$A_{pb} = 2 \times 0.4^{\text{in.}} \times (4^{\text{in.}} - 1.06^{\text{in.}}) = 2.35\text{ in.}^2$ the area at the end of the plate excluding the fillets

$$\phi_{pb} R_n = \phi_{pb}1.8F_y A_{pb} = 0.75 \times 1.8 \times 50^{\text{ksi}} \times 2.35^{\text{in.}^2} = 158.63\text{ kips} > 56\text{ kips}$$

Example 5.5.

The beam between A and C in Fig. 5.46 has a W24×76, ASTM A992 (F_y=50 ksi) section. Concentrated live load P is acting in the middle of AB span and point C. Dead load w_D=0.5 kip/ft throughout the beam in addition to beam weight $L_b = 5$ ft. Determine the maximum service load P.

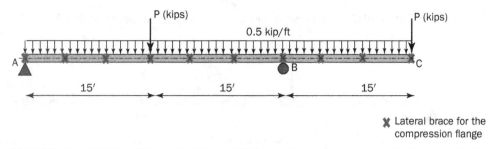

FIGURE 5.46

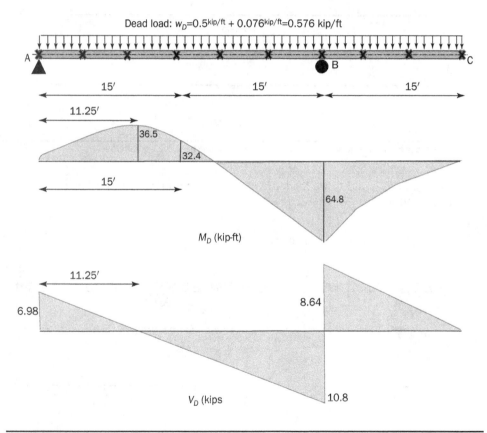

FIGURE 5.47

Solution:
1) Structural analysis to find M_u and V_u
Due to dead load (Fig. 5.47):
Due to live load:
Live loads should be placed on the beam so that maximum M_L and V_L will occur.

Case a: P at point C is not present for maximum M_L in the middle span (Fig. 5.48).

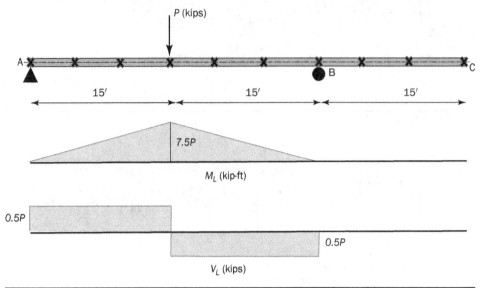

FIGURE 5.48   Moment and shear force diagrams due to live load for Case a.

Case b: Both A and C are loaded with P for maximum V_L, as well as M_L at support B (Fig. 5.49).

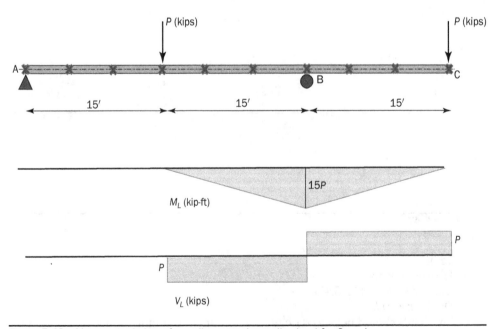

FIGURE 5.49   Moment and shear force diagrams due to live load for Case b.

From the analyses, the critical location is at support B.

$$M_L = 15\,P \qquad\qquad V_L = P$$
$$M_D = 64.8 \text{ kip-ft} \qquad V_D = 10.8 \text{ kips}$$
$$M_u = 1.2M_D + 1.6M_L = 77.8^{\text{kip-ft}} + 24P \text{ kip-ft}$$
$$V_u = 1.2V_D + 1.6V_L = 13.0^{\text{kips}} + 1.6P \text{ kips}$$

2) Bending strength requirement

    a. Check LTB and FLB

W24×76 (ASTM A992), $F_y = 50$ ksi
From Table 1-1, *AISC Manual* (Part 1):

$$\lambda_f = \frac{b_f}{2t_f} = 6.61\,; \ \lambda_w = \frac{h}{t_w} = 49$$

$$r_y = 1.92 \text{ in.}$$
$$Z_x = 200 \text{ in.}^3$$
$$d = 23.9 \text{ in.}$$
$$t_w = 0.44 \text{ in.}$$

$$L_p = 1.76r_y\sqrt{\frac{E}{F_y}} = 1.76 \times 1.92^{\text{in.}}\sqrt{\frac{29{,}000^{\text{ksi}}}{50^{\text{ksi}}}} = 81.4 \text{ in.} = 6.78 \text{ ft (or from } AISC\ Manual,$$

Part 3, Table 5-3: L_p=6.78 ft)

    $L_b = 5 \text{ ft} < L_p = 6.78 \text{ ft (no LTB)}$

    Check local buckling for flange: $\lambda_p = 0.38\sqrt{\dfrac{E}{F_y}} = 0.38\sqrt{\dfrac{29{,}000^{\text{ksi}}}{50^{\text{ksi}}}} = 9.15$

    $\lambda_f = 6.61 < \lambda_p = 9.15$ (compact section/no flange local buckling)

    W24×76 (ASTM A992, $F_y = 50$ ksi) is a compact section (as most 50-ksi-rolled W-shapes are).

    Note: Compactness of the section can also be checked using *AISC Manual*, Part 3, Table 3-2. In the table, most sections are compact ($\lambda_f < \lambda_p$), and a few noncompact sections are indicated by superscript "f" next to them. When a section with superscript "v" appears, the shape does not meet h/t_w limit, use $\phi_v = 0.9$ to replace $\phi_v = 1.0$.

    b. Design bending strength

$$M_n = F_y Z_x = 50^{\text{ksi}} \times 200^{\text{in.}^3} = 10{,}000 \text{ kip-in}$$
$$= 833 \text{ kip-ft}$$
$$\phi_b M_n = 0.9 \times 833^{\text{kip-ft}} = 750 \text{ kip-ft} \quad \text{(or from } ASIC\ Manual, \text{ Part 3, Table 5-3: } \phi_b M_n =$$
750 kip-ft)

    c. Maximum P

$$M_u \le \phi_b M_n$$

$$77.8^{\text{kip-ft}} + 24P \le 0.9 \times 833^{\text{kip-ft}}$$

$$P \le 28.0 \text{ kips}$$

3) Shear strength requirement

   a. Limit state in shear

$$\lambda_w = \frac{h}{t_w} = 49 < 2.45\sqrt{\frac{E}{F_y}} = 59 \text{ yielding of web in shear}$$

   Note: limit state in shear can also be checked using *AISC Manual*, Part 3, Table 3-2. When a section with superscript "v" appears, the shape does not meet h/t_w limit, use $\phi_v = 0.9$ to replace $\phi_v = 1.0$.

   b. Design shear strength

$$\phi_v V_n = \phi_v(0.6F_y)dt_w = 1.0 \times 0.6 \times 50^{ksi} \times 23.9^{in.} \times 0.44^{in.} = 315.5 \text{ kips (or from } AISC$$
Manual, Part 3, Table 5-3: $\phi_v V_n = 316$ kips)

   c. Maximum P based on shear strength

$$V_u \le \phi_v V_n$$

$$13^{kips} + 1.6P \le 315.5^{kips}$$

$$P \le 189.1 \text{ kips}$$

4) Maximum service load
$P = 28.0$ kips (from bending strength requirement)

Note: the steel beam design is often controlled by the bending instead of shear force, particularly when the beam has a span of 20 ft or longer.

Example 5.6
The simply supported beam in Fig. 5.50 has $L=40$ ft and unbraced length of $L_b=4$ ft (Case A) and $L_b=40$ ft/3 (lateral supports at ends and one-third points) (Case B). $w_D=1.5$ kip/ft, $w_L=2.0$ kip/ft. The serviceability requirement for live load only is $\Delta_{max} \le L / 360$. Select the lightest section with (a) $F_y=36$ ksi, (b) $F_y=50$ ksi for Case A and $F_y=50$ ksi for Case B.

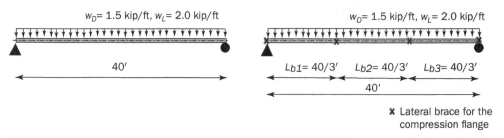

$w_D = 1.5$ kip/ft, $w_L = 2.0$ kip/ft

40'

$w_D = 1.5$ kip/ft, $w_L = 2.0$ kip/ft

$L_{b1} = 40/3'$ $L_{b2} = 40/3'$ $L_{b3} = 40/3'$

40'

✗ Lateral brace for the compression flange

Figure 5.50   Simply supported beam with/without lateral supports.

Solution:
Case A

  a. $F_y = 36$ ksi steel is used

Step 1: Compute M_u and V_u (Fig. 5.51)

$$w_u = 1.2w_D + 1.6w_L = 1.2 \times 1.5^{kip/ft} + 1.6 \times 2.0^{kip/ft}$$
$$= 5.0 \text{ kip/ft (no beam weight included yet)}$$

$$M_u = \frac{w_u L^2}{8} = \frac{5^{kip/ft} \times (40')^2}{8} = 1{,}000 \text{ kip-ft}$$

$$V_u = \frac{w_u L}{2} = \frac{5^{kip/ft} \times 40'}{2} = 100 \text{ kips}$$

Step 2: Select a trial shape by bending strength requirement

  Assume $\lambda \le \lambda_p$ (compact section/no flange local buckling)
  And $L_b \le L_p$ (braced beam/no LTB)
  Thus, no flange local buckling and LTB
  $$\phi_b M_n = \phi_b M_p = \phi_b F_y Z_x$$

  From $M_u \le \phi_b M_n$

$$Z_{x,required} \ge \frac{M_u}{\phi_b F_y} = \frac{1000^{kip\text{-}ft} \times (12^{in/ft})}{0.9 \times 36^{ksi}} = 370.4 \text{ in.}^3$$

  Table 3-2 of *AISC Manual* lists W-shapes with $F_y = 50$ ksi, normally used as beams in the order of decreasing plastic section modules, Z_x. They are grouped so that the shape at the top of the group (in bold type) is the lightest in the group with almost the same Z_x. For steels other than $F_y = 50$ ksi, we might use this table as a reference to get a trial W-shape, and then check it to ensure $M_u \le \phi_b M_n$ as follows:

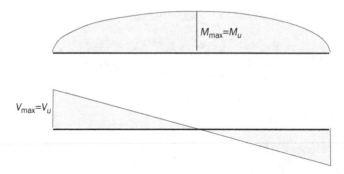

Figure 5.51   Moment and shear force diagrams.

i. Use the table (made for F_y=50 ksi) to find a trial section for F_y=36 ksi steel shape:
From Table 3-2, *AISC Manual*, on page 3-15:

Try W30×116, Z_x =378 in.^{3} > $Z_{x,required}$ = 370.4 in.^{3}

ii. Check W30×116, F_y=36.

Recalculate M_u, including the weight of W30×116, 116 lb/ft:

$$M_u = 1000 + 1.2 \left[\frac{0.116^{kip/ft} \times (40')^2}{8} \right] = 1{,}027.8 \text{ kip-ft}$$

$$\phi_b M_n = \phi_b M_p = \phi_b F_y Z_x = 0.9 \times 36^{ksi} \times 378^{in.^3} = 12{,}247.2 \text{ kip-in.}$$
$$= 1{,}021 \text{ kip-ft} \approx M_u \text{ (O.K. within 5\%)}$$

We know that W30×116, F_y=36 ksi is a compact section because W30×116, F_y=50 ksi is a compact section, and L_b increases as F_y decreases. Also, check LTB:

$$L_p = 1.76 r_y \sqrt{\frac{E}{F_y}} = 1.76 \times 2.19^{in.} \sqrt{\frac{29{,}000^{ksi}}{36^{ksi}}} = 109.4 \text{ in.} = 9.12 \text{ ft} \, (r_y=1.92 \text{ in. for W30×116})$$

$L_b = 4 \text{ ft} < L_p = 9.12 \text{ ft (no LTB)}$

Step 3: Check shear strength requirement
W30×116, d=30.0 in., t_w =0.565 in., I_x =4,930 in.^{4}

$$\lambda_w = \frac{h}{t_w} = 47.8 < 2.45 \sqrt{\frac{E}{F_y}} = 2.45 \sqrt{\frac{29{,}000^{ksi}}{36^{ksi}}} = 69.5 \text{ yielding of web in shear}$$

Design shear strength due to yielding under shear limit state.

$$\phi_v V_n = \phi_v (0.6 F_y) d t_w = 1.0 \times 0.6 \times 36^{ksi} \times 30^{in.} \times 0.565^{in.} = 366.1 \text{ kips}$$

$$V_u = 100 \text{ kips} \geq \phi_v V_n = 366.1 \text{ kips (O.K.)}$$

Step 4: Check deflection (for only live load)

$$\Delta_{max} = \frac{5 w_L L^4}{384 E I_x} = \frac{5 \times \left(\frac{2^{kip/ft}}{12^{in/ft}} \right) \times \left(40' \times 12^{in/ft} \right)^4}{384 \times 29{,}000^{ksi} \times 4{,}930^{in.^4}} = 0.81 \text{ in.}$$

$$\Delta_{max} = 0.81 \text{ in.} < \frac{L}{360} = \frac{40' \times 12^{in./ft}}{360} = 1.33 \text{ in.}$$

Use W30×116, F_y=36 ksi

b. F_y =50 ksi steel is used

Beam design tables are available for F_y =50 ksi.
Step 1: Compute M_u and V_u
The same as in (a).

$$M_u = \frac{w_u L^2}{8} = \frac{5^{kip/ft} \times (40')^2}{8} = 1{,}000 \text{ kip-ft}$$

$$V_u = \frac{w_u L}{2} = \frac{5^{kip/ft} \times 40'}{2} = 100 \text{ kips}$$

(the beam weight is not yet included)

Step 2: Select a trial shape by bending strength requirement using beam design Table 3-2 in *AISC Manual*, Part 3

Enter Table 3-2 with: $\phi_b M_n = \phi_b M_p = \phi_b M_{px} = 1{,}060$ kip-ft.
Try W30×90, $F_y = 50$ ksi, $L_b = 4$ ft $< L_p = 7.38$ ft (no LTB).
Include beam weight of W30×90, 90 lb/ft:

$$M_u = 1{,}000 + 1.2 \left[\frac{0.09^{kip/ft} \times (40')^2}{8} \right] = 1{,}021.6 \text{ kip-ft}$$

$$M_u = 1{,}021.6 \text{ kip-ft} \le \phi_b M_n = 1{,}060 \text{ kip-ft (O.K.)}$$

Step 3: Check shear strength requirement
W30×90:

$\phi_v V_n = 375$ kips (*AISC Manual*, Table 3-2)
$V_u = 100 \text{ kips} + 1.2 \times 0.09^{kip/ft} \times 40'/2 = 102.2 \text{ kips} \le \phi_v V_n = 375$ kips (O.K.)

Note: W30×90, $F_y = 50$ ksi, is one of the very few W-shapes with both superscripts "c" and "v," indicating that it is a slender section and does not meet the h/t_w limit for shear in Specification Section G2.1a. Table 3-2 includes the effects, such as FLB and $\phi_v = 0.9$ (instead of $\phi_v = 1.0$) of such a section on bending and shear strength.

Step 4: Check deflection (for only live load)
W30×90, $I_x = 3{,}610$ in.^{4}

$$\Delta_{max} = \frac{5 w_L L^4}{384 E I_x} = \frac{5 \times \left(\frac{2^{kip/ft}}{12^{in/ft}} \right) \times \left(40' \times 12^{in/ft} \right)^4}{384 \times 29{,}000^{ksi} \times 3{,}610^{in.^4}} = 1.10 \text{ in.}$$

$$\Delta_{max} = 1.10 \text{ in.} < \frac{L}{360} = \frac{40' \times 12^{in./ft}}{360} = 1.33 \text{ in.}$$

Use W30×90, $F_y = 50$ ksi.

Case B

$F_y = 50$ ksi steel is used.
$w_u = 1.2 w_D + 1.6 w_L$
   $= 1.2 \times 1.5^{kip/ft} + 1.6 \times 2.0^{kip/ft}$
   $= 5.0$ kip/ft

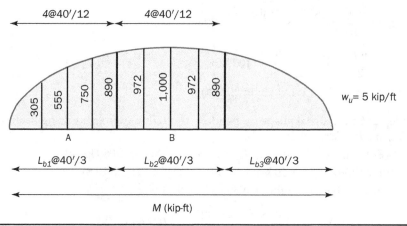

Figure 5.52

Step 1: compute M_u and V_u

Segment A (Fig. 5.52): $M_u = 890$ kip-ft, $V_u = 100$ kips, $L_b = L_{b1} = 13.3'$

$$C_b = \frac{12.5M_{max}}{2.5M_{max} + 3M_A + 4M_B + 3M_C}$$
$$= \frac{12.5(890)}{2.5(890) + 3(305) + 4(555) + 3(750)}$$
$$= 1.46 < 3$$

Segment B: $M_u = 1000$ kip-ft, $V_u = 33.3$ kips, $L_b = L_{b2} = 13.3'$ $(=L_{b1})$

$$C_b = \frac{12.5(1000)}{2.5(1000) + 3(972) + 4(1000) + 3(972)}$$
$$= 1.01$$

Segment B controls the design for bending strength, and Segment A controls the design for shear strength.

Step 2: Select a trial section

To get a first trial section, select a section assuming that it is a compact section and $L_b < L_p$ so that

$$M_u \leq \phi_b M_n = \phi_b M_p = \phi_b F_y Z_x \quad \text{when (as assumed for } L_b < L_p \text{ trial section)}$$

$$Z_{x,required} \geq \frac{M_u}{\phi_b F_y} = \frac{1000^{kip\text{-}ft} \times (12^{in/ft})}{0.9 \times 50^{ksi}} = 266.67 \text{ in.}^3$$

Based on either $Z_x \geq Z_{x,\text{required}}$ from Table 1-1 in Part 1 of *AISC Manual*, try W30×99, $Z_x = 312$ in.$^3 > Z_{x,\text{required}} = 266.67$ in.^{3}

$$M_p = F_y Z_x = 50^{\text{ksi}} \times 312^{\text{in.}^3} = 15{,}600 \text{ kip-in.}$$
$$= 1{,}300 \text{ kip-ft}$$

$$M_r = 0.7 M_y = 0.7 F_y S_x = 0.7 \times 50^{\text{ksi}} \times 269^{\text{in.}^3} = 9{,}415 \text{ kip-in.} \quad (S_x = 269 \text{ in.}^3 \text{ from}$$
Table 1-1 of *AISC Manual*)

$$= 785 \text{ kip-ft}$$

Or

$M_u \leq \phi_b M_n$ (Note that $M_{px} = M_p$ here) from Table 3-2 (Page 3-15 of the *AISC Manual*). Try W30×99: $\phi_b M_p = 1{,}170$ kip-ft $> M_u = 1{,}000$ kip-ft (ignoring self-weight).

Also from Table 3-2, *AISC Manual*,

$L_p = 7.42$ ft, $L_r = 21.4$ ft, $\phi_v V_n = 463$ kips

BF = 33.3 kips (Note: BF$= \phi_b = \dfrac{M_p - M_r}{L_r - L_p}$)

It appears that the beam may have potential to have an inelastic torsional buckling limit state since $L_p = 7.21$ ft $< L_b = 13.3$ ft $< L_r = 21.4$ ft. Therefore, M_n needs to be re-evaluated, and we need to recheck to make sure that $M_u \leq \phi_b M_n$:

$$M_n = C_b \left[M_p - \left(M_p - 0.7 F_y S_x \right) \left(\frac{L_b - L_p}{L_r - L_p} \right) \right] \leq M_p = F_y Z_x$$

$$= 1.01 \left[1{,}300^{\text{kip-ft}} - \left(1{,}300^{\text{kip-ft}} - 785^{\text{kip-ft}} \right) \left(\frac{13.3' - 7.42'}{21.4' - 7.42'} \right) \right] = 1{,}093 \text{ kip-ft} \leq 1{,}300 \text{ kip-ft}$$

$$M_n = 1{,}093 \text{ kip-ft}, \quad \phi_b M_n = 0.9 \times 1{,}093^{\text{kip-ft}} = 983 \text{ kip-ft}$$

Or, we can use Table 3-2, *AISC Manual* to get $\phi_b M_n$ (only when $F_y = 50$ ksi):

$$\phi_b M_n = \phi_b C_b \left[M_p - \left(M_p - 0.7 F_y S_x \right) \left(\frac{L_b - L_p}{L_r - L_p} \right) \right] \leq \phi_b M_p = F_y Z_x$$

$$= C_b \left[\phi_b M_p - \left(\phi_b \frac{M_p - M_r}{L_r - L_p} \right) \left(L_b - L_p \right) \right] \leq \phi_b M_p = F_y Z_x$$

$$= C_b \left[\phi_b M_p - \text{BF} \left(L_b - L_p \right) \right] \leq \phi_b M_p = F_y Z_x$$

$$\phi_b M_n = 1.01 \left[1{,}170^{\text{kip-ft}} - (33.3)(13.3' - 7.42') \right] = 983 \text{ kip-ft} \leq 1{,}170 \text{ kip-ft}$$

$$\phi_b M_n = 983 \text{ kip-ft}$$

Check if $M_u \leq \phi_b M_n$.

$M_u = 1{,}000$ kip-ft $\approx \phi_b M_n = 983$ kip-ft (it is acceptable if $\phi_b M_n$ is no more than 5% smaller than M_u).

Step 3: Check shear strength
$\phi_v V_n = 463$ kips (*AISC Manual*, Table 3-2)
$V_u = 100$ kips (ignoring self-weight)
$V_u = 100$ kips $\leq \phi_v V_n = 463$ kips (O.K.)

Step 4: Check deflection
W30×99, $I_x = 3{,}990$ in.[4]

$$\Delta_{max} = \frac{5 w_L L^4}{384 E I_x} = \frac{5 \times \left(\frac{2^{kip/ft}}{12^{in/ft}}\right) \times \left(40' \times 12^{in/ft}\right)^4}{384 \times 29{,}000^{ksi} \times 3{,}990^{in.^4}} = 1.0 \text{ in.}$$

$$\Delta_{max} = 1.0 \text{ in.} < \frac{L}{360} = \frac{40' \times 12^{in./ft}}{360} = 1.33 \text{ in.}$$

Use W30×99, $F_y = 50$ ksi.

Example 5.7.

As shown in Fig. 5.53, the uniform service dead load of 6 kip/ft (including the beam weight) is applied over a simply supported beam with a 20 ft-long overhang. The moving service live load, $P_L = 10$ kips might be located at any point between the supports or on the overhang. The beam is laterally supported continuously ($L_b < L_p$). Select a W36 or shallower section with ASTM A992 steel ($F_y = 50$ ksi). Only consider the flexural (bending) strength requirement.

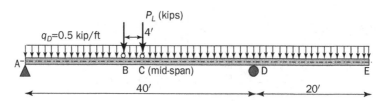

Figure 5.53 Simply supported beam subjected to uniformly distributed loading and moving live loads.

Solution:

1) Structural analysis (service loads)

Due to dead load (Fig. 5.54):

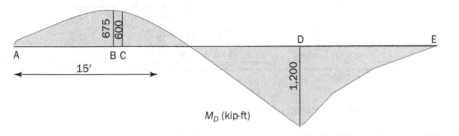

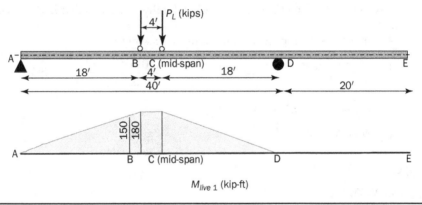

FIGURE 5.54   Moment diagram due to dead load.

Due to live load:
Loading Case 1 (Fig. 5.55):

FIGURE 5.55   Moment diagram due to live load for loading case 1.

Loading Case 2 (Fig. 5.56):

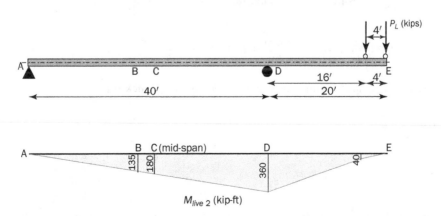

FIGURE 5.56   Moment diagram due to live load for loading case 2.

2) Determine factored loads for (Table 5.11)

| Section | Service load | | Factored load (kip-ft) |
	M_{Dead} (kip-ft)	Maximum M_{Live} (kip-ft)	
B	675	150 (Case 2)	1,050
C	600	180 (Case 1 & 2)	1,008
D*	1200	360 (Case 2)	2,016

TABLE 5.11. Factored Loads

Thus, D is the most critical section.

Factored loads: At point D
$M_D = 1,200$ kips-ft, $M_{L,max}=360$ kips-ft
$M_u=1.2M_D+1.6M_L=1.2(1,200^{\text{kip-ft}})+1.6(360^{\text{kip-ft}})=2,016$ kip-ft

3) Select a trial shape
Since we know that the beam is properly laterally braced so that $L_b<L_p$, no LTB would occur.

Assume $\lambda \le \lambda_p$ (compact section/no local buckling):

$$\begin{cases} \phi_b M_n = \phi_b M_p = \phi_b Z_X F_y \\ \phi_b M_n \ge M_u \end{cases} \Rightarrow Z_X \ge \frac{M_u}{\phi_b F_y} = \frac{2,016^{\text{kip-ft}} \times 12^{\text{in./ft}}}{0.9 \times 50^{\text{ksi}}} = 537.6 \text{ in.}^3$$

From Table 1-1 (*AISC Manual*):

Try W33×152, $Z_X = 559$ in.$^3 > Z_{X,req} = 537.6$ in.^{3}

4) Check W33×152

Check W33×152, $F_y = 50$ ksi: if $M_c \ge M_r$.

(1) Check to see if local buckling is possible:

$$\lambda_f = \frac{b_f}{2t_f} = 5.48 \ (\textit{AISC Manual}, \text{Table 1-1})$$

$$\lambda_p = 0.38\sqrt{\frac{E}{F_y}} = 0.38\sqrt{\frac{29,000^{\text{ksi}}}{50^{\text{ksi}}}} = 9.15$$

Thus, $\lambda_f < \lambda_p$, is a compact section (no local flange buckling).

(2) Check if $M_c \geq M_r$

$$M_n = M_p = Z_x F_y = \frac{559^{in.^3} \times 50^{ksi}}{12^{in./ft}} = 2,329 \text{ kip-ft}$$

$$M_c \geq M_r$$

$M_c = \phi_b M_n = 0.9 \times 2,329^{kip\text{-}ft} = 2,096 \text{ kip-ft}$
$M_c = 2,096 \text{ kip-ft} > M_r = 2,016 \text{ kip-ft}$
$0.95 < {M_c}/{M_r} = 1.04 < 1.20 \quad \text{O.K.}$

Use W33×152, $F_y = 50$ ksi.

Example 5.8.
As shown in Fig. 5.57, the beam is a W18×65 of ASTM A992 ($Fy = 50$ ksi). The live service loads, P and $3P$, are applied at both ends, in the middle span, respectively. The beam is laterally supported at its both ends and every 5 ft between them. Calculate the maximum service load P. Only consider one limit state: flexural (bending) strength requirement. Neglect the beam weight.

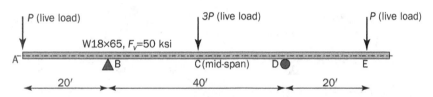

FIGURE 5.57 Simply supported beam with overhangs at both ends subjected to concentrated live loads at the ends and midspan.

Solution:
1) Structural analysis (service loads)

Due to live load:

Case 1: Apply $3P$ at the midspan (Point C) (Fig. 5.58)

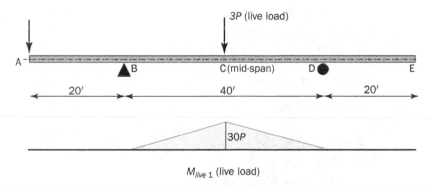

FIGURE 5.58 Moment diagram for case 1.

Case 2:   Apply *P* at Point A (or E) (Fig. 5.59)

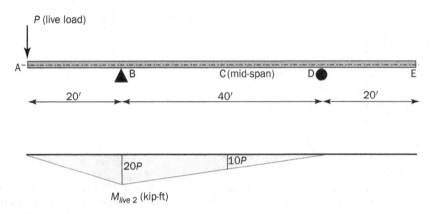

FIGURE 5.59   Moment diagram for case 2.

Case 3:   Apply P at Point A and E (Fig. 5.60)

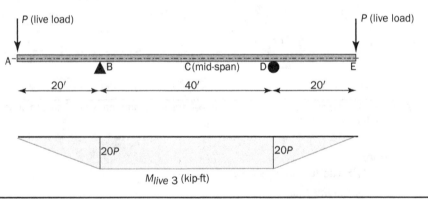

FIGURE 5.60   Moment diagram for case 3.

2) Determine factored loads
Section C is the most critical section!

$M_D = 0$ kip-ft, $M_L = 30P$ (Case 1)

$M_r = M_u = 1.2M_D + 1.6M_L = 1.2 \times 0 + 1.6 \times 30P = 48P$

3) Determine moment capacity of W18×65 considering LTB and local buckling

(1) Check local buckling:

$$\lambda_f = \frac{b_f}{2t_f} = 5.06 < \lambda_p = 0.38\sqrt{\frac{E}{F_y}} = 0.38\sqrt{\frac{29{,}000^{\text{ksi}}}{50^{\text{ksi}}}} = 9.15$$

Compact section (no local flange buckling).

(2) Check lateral-torsional buckling (LTB).

Since $F_y=50$ ksi, *AISC Manual*, Table 3-2 can be used to determine L_p and L_r values.

   AISC Manual, Table 3-2: **W18×65**→ $L_p=5.97$ ft, $L_r=18.8$ ft, $Z_x=133$ in.^{3}, $S_x=117$ in.^{3},
$F_y=50$ ksi
     $L_b=5.0$ ft $< L_p=5.97$ ft → No LTB ($M_n = M_p$).

$$M_n = M_p = Z_x F_y = \frac{133^{\text{in.}^3} \times 50^{\text{ksi}}}{12^{\text{in./ft}}} = 554 \text{ kip-ft}$$

4) Find P_{max}

$M_c \geq M_r$

$M_c = \phi_b M_n = 0.9 \times 554^{\text{kip-ft}} = 498.6$ kip-ft
$M_r = M_u = 48P$
$498.6 \geq 48P$
$10.4 \geq P \Rightarrow P_{max} = 10.4$ kips

Example 5.9
As shown in Fig. 5.61, the beam is a W18×71, $F_y=50$ ksi. The service dead load $Q = 40$ kips is applied in the middle span. The service live load $P = 40$ kips is applied at the right end and middle span. The beam is laterally supported continuously ($L_b < L_p$). Check if the beam is safe (i.e., if $M_u \leq \phi_b M_n$). Only consider the flexural (bending) strength requirement—neglect beam self-weight.

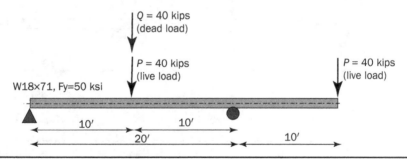

Figure 5.61   Simply supported beam subject to concentrated dead and live loads.

Solution:

1) Structural analysis (service loads)

Due to dead load (Fig. 5.62):

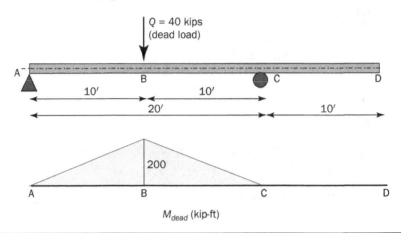

FIGURE 5.62 Moment diagram due to dead load.

Due to live load:

Case 1: Apply P at the midspan (Point B) (Fig. 5.63)

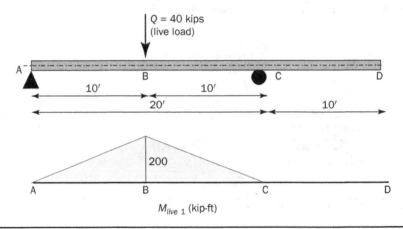

FIGURE 5.63 Moment diagram due to live load for case 1.

Case 2:   Apply P at Point D (Fig. 5.64)

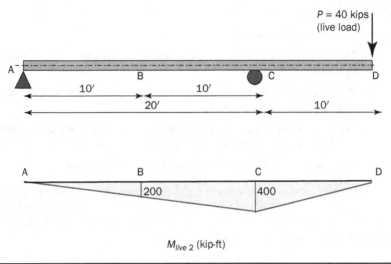

FIGURE 5.64   Moment diagram due to live load for case 2.

2) Determine factored loads
At Point B:
$\quad M_D = 200$ kips-ft, $M_L = 200$ kips-ft (Case 1)
$\quad M_U = 1.2M_D + 1.6M_L$
$\qquad = 1.2 \times 200^{\text{kip-ft}} + 1.6 \times 200^{\text{kip-ft}} = 560$ kip-ft

At Point C:
$\quad M_D = 0$ kip-ft, $M_L = 400$ kips-ft (Case 2)
$\quad M_U = 1.2M_D + 1.6M_L$
$\qquad = 1.2 \times 0 + 1.6 \times 400^{\text{kip-ft}} = 640$ kip-ft

Thus, Point C is the most critical section.

3) Check W18×71 ($F_y = 50$ ksi)
$\quad$ Since we know that the beam is properly laterally braced so that $L_b < L_p$, no LTB would occur.

(1) $\lambda_f = \dfrac{b_f}{2t_f}$

$\lambda_p = 0.38\sqrt{\dfrac{E}{F_y}} = 0.38\sqrt{\dfrac{29{,}000^{\text{ksi}}}{50^{\text{ksi}}}} = 9.15$

Thus, $\lambda_f < \lambda_p$ is a compact section (no local buckling).

(2) $\phi_b M_n = \phi_b M_p = \phi_b F_y Z_x = 0.9 \times 50^{\text{ksi}} \times 146^{\text{in.}^3}$
$\qquad = 6{,}570$ kip-in.
$\qquad = 547.5$ kip-ft $< M_u = 640$ kip-ft not safe.

5.9 Problems

Related to Laterally Supported Beams

5.1 Please answer the following questions.

    a. Define compact, noncompact, slender-element, and nonslender-element sections.

    b. What is the normal stress distribution in a compact section when the section reaches its bending capacity?

    c. Why is a noncompact section unable to develop its plastic bending moment strength, M_p, according to the limit state of yielding?

    d. Why do we use the elastic method in deflection control?

5.2 The simply supported beam in Fig. 5.65 is subjected to uniformly distributed dead and live loads. Steel grade is ASTM A992 (F_y=50 ksi, F_u=65 ksi). Select the lightest section for the beam including the self-weight in design. Consider also deflection limit ($L/360$) due to service loads and check your design.

$$\left(\text{Note:}\quad \Delta_{max} = \frac{5}{384}\frac{w_L L^4}{EI_x} \right)$$

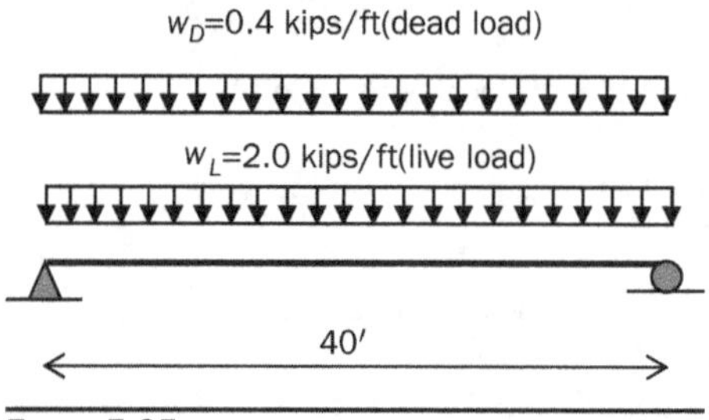

w_D=0.4 kips/ft(dead load)

w_L=2.0 kips/ft(live load)

40'

Figure 5.65

5.3 The simply supported beam with a concentrated load, P, in middle span in Fig. 5.66 is made up from a built-up section. The cross-section dimensions of this built-up section is also given in Fig. 5.66. Steel grade is A36 (F_y=36 ksi, F_u=58 ksi). Dead load (D) is 0.40P and live load (L) is 0.60P. Determine the maximum load, P, the beam can carry. Include self-weight.

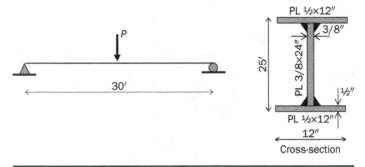

FIGURE 5.66

5.4 The simply supported beam in Fig. 5.67 is designed to transfer a concentrated load of P=250 kips. Dead load (D) is 0.30P and live load (L) is 0.70P. Steel grade is A36 (F_y=36 ksi, F_u=58 ksi). Check the design for bending, shear, local web yielding, and local web crippling. Redesign if necessary.

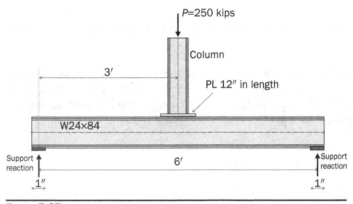

FIGURE 5.67

5.5 The simply supported W18×71 beam in Fig. 5.68 is subjected to a moving live load, P. Dead load (D) is 0.25P and live load (L) is 0.75P. Steel grade is A36 (F_y=36 ksi, F_u=58 ksi). Compute the maximum load, P, the beam can take.

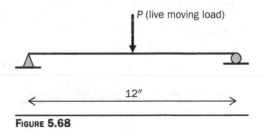

FIGURE 5.68

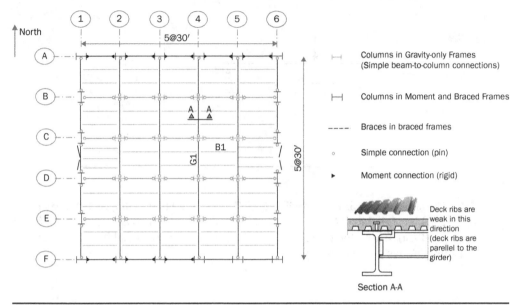

FIGURE **5.69**

5.6   Plan of a 4-story office building is given in Fig. 5.69. Design members B1 for bending, shear, and deflection. Steel grade is ASTM A992 (F_y=50 ksi, F_u=65 ksi). Assume dead load (D) of 100 psf including beam weight and live load (L) of 65 psf. Also assume that the floor deck provides full lateral stability to the top flange of B1. Ignore live load reduction.

Related to Lateral-Torsional Buckling:

5.7   Please answer the following questions:

   a. How many failure modes are there in a beam subject to bending?

   b. What does C_b coefficient take into account for?

   c. How do you increase the strength of a beam? Please list all possible options.

5.8   The simply supported W30×99 beam in Fig. 5.70 is subjected to a concentrated load, P, at middle span. Dead load (D) is 0.35P and live load (L) is 0.65P. Lateral supports are only provided at the ends (L_b=6'). Steel grade is A36 (F_y=36 ksi, F_u=58 ksi). Determine the maximum load, P, the beam can take.

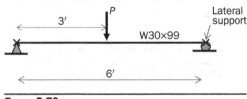

FIGURE **5.70**

5.9   The simply supported beam in Fig. 5.71 is subjected to uniformly distributed and concentrated dead and live loads. Dead load (D) is $0.80P$ and live load (L) is $0.20P$ due to concentrated loads at one-third of the span length. Select the lightest W-section under the following conditions:

   a. A36 steel; continuous lateral support

   b. A36 steel; lateral supports at ends only

   c. A36 steel; lateral supports at ends and at point A

   d. A992 steel; lateral supports at ends and at point A

   e. A992 steel; lateral supports at ends and at points A and B

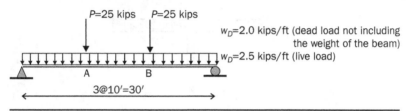

Figure 5.71

5.10   As shown in Fig. 5.72, the beam is a W16×40 of ASTM A992 ($F_y = 50$ ksi, $F_u = 65$ ksi). The uniform service dead load of 0.2 kip/ft (including the beam weight) is applied at overhang portions of the beam. The uniform service dead load of 0.4 kip/ft (including the beam weight) is applied between the two supports of the beam. The live service load, P, $2P$, and P, is applied at the left end, in the middle span, and the right end, respectively. The beam is laterally supported continuously ($L_b < L_p$). Calculate the maximum service load P. Only consider one limit state: flexural (bending) strength requirement.

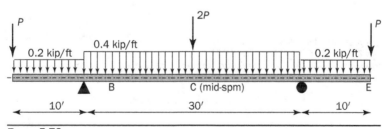

Figure 5.72

5.11   Select a W-shape section for the beam in Fig. 5.73 to carry uniform service dead load of $q_D = 0.6$ kip/ft (neglect beam self-weight), and movable service live load of $P_L = 6$ kips, over a beam with overhang, shown in Fig. 5.66. Note that P_L might be located at any point over the beam, with the value ranging from 0 to its maximum. Compression portions of the flanges and the tip of the overhang are properly laterally braced so that $L_b < L_p$. Use $F_y = 50$ ksi steel. Only consider strength requirements of bending and shear.

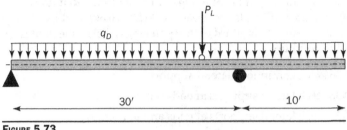

FIGURE 5.73

5.12   Determine the plastic moment capacity (M_p) of a W27×178 section in Fig. 5.74 $(M_p = F_y Z_x)$ (Steel Grade A992 Grade 50).

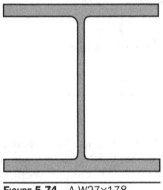

FIGURE 5.74   A W27×178 section.

5.13   Design the steel beam shown in Fig. 5.75. Consider bending and shear requirements. A992 steel $(F_y = 50$ ksi$)$ is used. Use a structural analysis software to analyze the statically indeterminate beam. Neglect the self-weight of the beams.

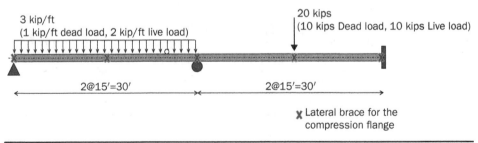

FIGURE 5.75

5.14   Select the lightest W-section for the steel beam shown in Fig. 5.76. Consider bending and shear requirements. A992 steel $(F_y = 50$ ksi$)$ is used. Use a structural analysis software to analyze the statically indeterminate beam. Neglect the self-weight of the beams.

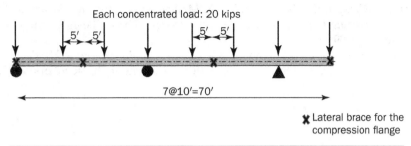

Each concentrated load: 20 kips

7@10'=70'

x Lateral brace for the
compression flange

FIGURE 5.76

5.15 For the floor framing given in Fig. 5.69, select the lightest W-shape for girder G1. Check bending, shear, and deflection with $L/240$ total load. Assume that G1 is laterally supported at the beam connections.

5.16 Write a python code that computes:

a. The nominal flexural strength of a W-shape

b. Shear strength of a compact, noncompact, and slender section

Related to Building Project in Chapter 1:

5.17 Please design the following structural members for the building given in the appendix in Chapter 1:

a. Secondary beams in the second story

b. Beams on the third floor

c. Girders on the third floor

Bibliography

Aghayere, A. O., and J. Vigil, *Structural Steel Design: A Practice Oriented Approach*, Pearson, New Jersey, NY, 2015.

AISC, *Seismic Design Manual*, American Institute of Steel Construction, Chicago, IL, 2018.

AISC 341, *Seismic Provisions for Structural Steel Buildings*, ANSI/AISC Standard 341-16, American Institute of Steel Construction, Chicago, IL, 2016.

AISC 360, *Specification for Structural Steel Buildings*, ANSI/AISC Standard 360-16, American Institute of Steel Construction, Chicago, IL, 2016.

AISC Manual, *Steel Construction Manual*, 15th ed., American Institute of Steel Construction, Chicago, IL, 2016.

ASCE 7, *Minimum Design Loads for Buildings and Other Structures*, ASCE/SEI 7-16, American Society of Civil Engineers, Reston, VA, 2016.

Bruneau, M., C. M. Uang, and R. Sabelli, *Ductile Design of Steel Structures*, 2nd ed., McGraw-Hill Education, New York, 2011.

Chajes, A., *Principles of Structural Stability Theory*, Waveland Pr Inc., 1993.

Hibbeler, R. C., *Structural Analysis*, 9th ed. in SI Units, Pearson, New Jersey, 2016.

Salmon, C. G., J. E. Johnson, and F. A. Malhas, *Steel Structures: Design and Behavior—Emphasizing Load and Resistance Factor Design*, 5th ed., Pearson International, New Jersey, 2009.

Shen, J., B. Akbas, O. Seker, and C. Carter, *Structural Engineering Handbook—Chapter 8: Design of Structural Steel Members*, 5th ed., McGraw-Hill, New York, 2020.

CHAPTER 6

Torsion

6.1   Introduction

Steel sections are very vulnerable to torsional moment, mainly because most steel members are made of thin-plated open sections (Fig. 6.1). Even though the most common structural shapes are quite efficient in bending, their torsional efficiencies are quite low. Contrary to structural shapes, solid structural shapes are not commonly used in steel building types of structures, and their torsional efficiency is quite high. On the other hand, thin-wall cross-sections are known for their high torsional and bending strength and commonly used in bridge construction. Torsion in a structural element occurs in cases where the loads are not acting through the shear center. Even though for most structural sections [W, C, L, HSS (hollow structural section), Z], shear centers coincide with the centroid of the cross-section, this is not the case for C-shapes and C-angles.

6.2   Accidental Torsion vs. Expected Torsion

Expected torsion can be defined as the torsion due to the structural system configuration, such as a cantilever beam with dead and live loads, spandrel beams supporting brick veneer, or beams subject to unequal loading on either side, etc. The most efficient approach or the best way to deal with torsional problems is to avoid exposing thin-plated open sections to any significant torsion and use closed sections if possible when significant torsion is inevitable. However, there are always cases where a secondary or unexpected torsional moment exists in structural members that causes accidental torsion. For example, as shown in Fig. 6.2, the beam is intended to carry the bending moment induced by the concentrated load, P. It is more often than not that the load might not run through the shear center of the beam section (located in the web), thus resulting in certain torsional moment. In this section, we will discuss issues related to the torsional impact on the steel members and how to take such an impact into account during the design and construction process.

6.3   Fundamentals in Elastic Torsional Analysis of Steel Members

State-of-practice in dealing with torsion: The behavior of steel members is limited to elastic behavior when torsional moment acts alone or in combination with other loadings

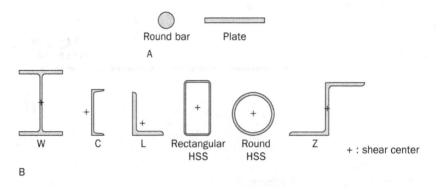

FIGURE 6.1   Steel section used in steel structures: (a) solid shapes, (b) structural shapes.

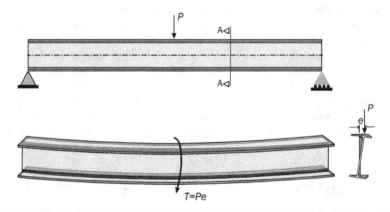

FIGURE 6.2   Possible accidental torsion due to eccentricity.

such as bending moment and axial force. The assumed elastic behavior is only possible when:

1. The impact of the torsional moment on the member is small in comparison to that of the bending moment and/or axial force.
2. The steel member remains elastic when subjected to the combination of all loadings (the bending moment, torsional moment, and axial force if any).

The elastic assumption enables us to use existing knowledge about torsion and bending to develop closed-form solutions for many practical loading and boundary cases. Note that all results are only meaningful when the member remains *ELASTIC!*

Torsion may be thought of as being composed of two parts:

1. Pure St. Venant's torsion, T_s

   In-plane rotational deformation

2. Warping torsion, T_w

   Out-of-plane rotational deformation, twisting prevented at the ends

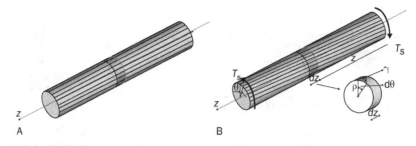

FIGURE 6.3 A circular shaft subject to pure torsion (no out-of-plane warping): (a) before deformation, (b) after deformation under pure torsion at free ends.

Thus, total torsion, T, can be expressed as the sum of pure torsion, T_s, and warping torsion, T_w, as follows:

$$T = T_s + T_w \tag{6.1}$$

6.3.1 Pure (St. Venant) Torsion, T_s

A cross-sectional plane before the application of torsion remains a plane, and only element rotation occurs during rotation (no end restraint, circular shaft in Fig. 6.3 is the only example of pure torsion). Contrary to a circular shaft, an I-shaped section subject to pure torsion with no end restraint will experience out-of-plane warping (Fig. 6.4).

Pure torsion refers to a torsional deformation pattern, in which the cross-section remains in the same plane before and after the twist. A simple, pure torsion case can be observed from a length of a circular member subjected to a torsion moment, T_s, in Fig. 6.3. The horizontal lines will twist, and the cross-section will remain circular under T_s acting at the two free ends. Since $\gamma dz = \rho d\theta$ from the geometry (Fig. 6.3a), γ can be stated as

$$\gamma = \rho \frac{d\theta}{dz} = \rho \phi \tag{6.2}$$

Using Hooke's law on shear strain (γ) and shear modulus (G), shear stress τ_s can be written as

$$\tau_s = \gamma G \tag{6.3}$$

Using equilibrium, we can write the torsion demand, T_s (or total resisting torsional moment), as

$$T_s = \int_{area} dT = \int_{area} \rho\tau dA = \int_{area} \rho(\gamma G)dA = \int_{area} \rho\left(\rho\frac{d\theta}{dz}G\right)dA = \int_{area} \rho^2\frac{d\theta}{dz}GdA \tag{6.4}$$

In Eq. (6.4), $\dfrac{d\theta}{dz}$ and G are constants. Thus, T_s becomes

$$T_s = \frac{d\theta}{dz}G\int_{area} \rho^2 dA = GJ\frac{d\theta}{dz} = GJ\theta' \tag{6.5}$$

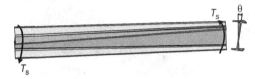

FIGURE 6.4 An I-shaped steel beam subject to pure torsion.

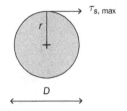

FIGURE 6.5 A circular cross-section with diameter D.

where

T_s = pure torsion, kip-in.
τ_s = shear stress due to pure torsion, ksi
θ = twist angle (increases as z increases)
$\phi = \dfrac{d\theta}{dz}$ torsional curvature (rate of twist)
γ = shear strain (strain angle)
G = shear modulus of elasticity of steel $\left(= \dfrac{E}{2(1+\mu)} \approx 11{,}200 \text{ ksi} \right)$
$\mu \approx 0.3$ Poisson ratio of steel

$J = \displaystyle\int_{\text{area}} \rho^2 dA$ torsional constant of the section, in.⁴

GJ = torsional rigidity

Pure torsional moment, T_s, in Eq. (6.5) states that it is equal to torsional rigidity times torsional curvature. Note the similarity in expressing moment due to flexure, which is equal to flexural rigidity (EI) times curvature. Using the derived relationships of $\tau = \gamma G = \left(\rho \dfrac{d\theta}{dz} \right) G$ from Eq. (6.3) and $\dfrac{d\theta}{dz} = \dfrac{T_s}{GJ}$ from Eq. (6.5), shear stress, τ_s, can be rewritten as

$$\tau_s = \rho \left(\frac{T_s}{GJ} \right) G = \frac{T_s \rho}{J} \qquad (6.6)$$

Example 6.1
Find the maximum shear stress for a circular cross-section with diameter D (Fig. 6.5).

Solution:
$\tau_{s,\max} = \dfrac{T_s r}{J}$ and $r = D/2$

For a circular section, torsional constant, J:

$$J = \frac{\pi D^4}{32}$$

Thus,

$$\tau_{s,\max} = \frac{T_s (D/2)}{\dfrac{\pi D^4}{32}} = \frac{16 T_s}{\pi D^3}$$

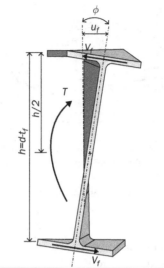

FIGURE 6.6 Twisting of an I-shaped beam subject to torsion.

6.3.2 Warping Torsion, T_w

Warping of a W-section is a condition in which the top and bottom flanges of the cross-section deflect in such a way that they are no longer parallel to each other. Warping plane will create normal stress, because the displacement is perpendicular to the section as well as shear stresses across the flange width. Warping torsion is the out-of-plane effect that arises when the flanges are laterally displaced during twisting, analogous to bending from laterally applied loads. For closed sections such as squares or circular tubes, each element of the section rotates without warping, i.e., the plane sections remain plane after rotation. However, for open structural sections, the warping torsion might introduce shear stress and normal stress in the top and bottom flanges. Such stresses are mainly due to the potential constraint of the warping deformation.

As illustrated in Fig. 6.6, it appears that the top and bottom flanges have a tendency to bend in their own plane (the plane perpendicular to the web). When such bending-induced deformation is somehow constrained, warping shear stress and normal stress exist in the same way as a beam subjected to a bending moment.

As shown in Fig. 6.6, flanges are subject to equivalent shear force, V_f, due to warping moment T_w. Let us look at the top flange and develop the relationship between warping torsional moment T_w and angle θ. First, we imagine that the effect of the torsional moment T_w on the flanges can be replaced by a couple of equivalent shear force, V_f. Then, a flange is treated as a "beam" bending in the plane of the flange. The warping torsion might introduce shear stress and normal stress in the top and bottom flanges. Web remains plane during rotation, i.e., flanges deflect an equal amount. u_f is considered to be a small deflection. Thus, T_w can be stated as

$$T_w = V_f h \qquad (6.7)$$

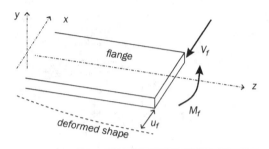

Figure 6.7   Lateral bending moment, M_f, and equivalent shear force, V_f, acting on one flange.

From geometry, lateral deflection of flange section, u_f, is proportional to twist angle as

$$u_f = \frac{h}{2}\theta \qquad (6.8)$$

If the top flange is treated as a beam subject to shear force V_f, one can use the elastic beam theory directly to relate u_f to V_f as (Fig. 6.7) follows:

$$\frac{d^2 u_f}{dz^2} = -\frac{M_f}{EI_f} \qquad (6.9)$$

and

$$V_f = \frac{dM_f}{dz} \rightarrow \frac{\partial^3 u_f}{dz^3} = -\frac{V_f}{EI_f} \qquad (6.10)$$

where

M_f = lateral bending moment on one flange, kip-in.
I_f = moment of inertia for one flange about the y-axis, in.[4]

Differentiating Eq. (6.8) three times, we have

$$\frac{d^3 u_f}{dz^3} = -\frac{h}{2}\frac{d^3 \theta}{dz^3} \qquad (6.11)$$

The combination of Eqs. (6.7), (6.10), (6.11) results in

$$V_f = -EI_f \frac{\partial^3 u_f}{dz^3} = -EI_f \left(\frac{h}{2}\right)\frac{\partial^3 \theta}{dz^3} \qquad (6.12)$$

$$T_w = V_f h = -EI_f \left(\frac{h}{2}\right)\frac{\partial^3 \theta}{dz^3} \qquad (6.13)$$

$$T_w = -EC_w \theta''' \qquad (6.14)$$

where

$C_w = I_f \left(\frac{h^2}{2}\right)$, warping torsional constant (in.[6]) (note that for T- and L-shaped sections, C_w is almost zero)

6.3.3   Total Torsion, T

Total torsion, T, in Eq. (6.1) is stated as the sum of pure torsion T_s, and warping torsion, T_w, as

$$T = T_s + T_w$$

Replacing T_s and T_w with Eqs. (6.5) and (6.14), respectively, we get

$$T = GJ\theta' - EC_w\theta'''$$ (6.15)

Rearranging Eq. (6.15) by dividing all terms with EC_w, the differential equation relating T to θ might be expressed as

$$\theta''' - \frac{1}{a^2}\theta' = -\frac{T}{EC_w}$$ (6.16)

where

$$a^2 = \frac{EC_w}{GJ}$$

The solution of the above third-order differential equation is the sum of homogenous and particular solutions. For given section properties, such as E, G, J, C_w, and torsion moment T, θ can be solved from Eq. (6.16) with proper boundary conditions. With θ known, T_s and T_w are available, and stresses associated with them will be evaluated. Some of the boundary conditions for torsion are shown in Fig. 6.8 (AISC Steel Design Guide Series 9, 2003).

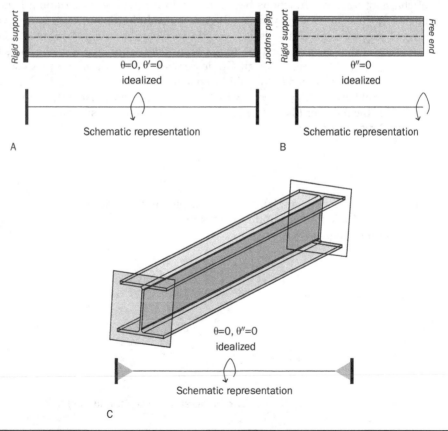

FIGURE 6.8 Boundary conditions for torsion: (a) torsionally fixed end, (b) free end, and (c) torsionally pinned end.

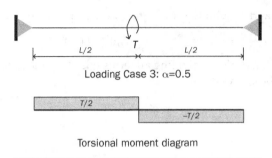

Loading Case 3: α=0.5

Torsional moment diagram

Figure 6.9 Concentrated torsional moment applied in the middle span (α = 0.5) of a torsionally pinned beam.

In practice, a structural engineer would first try to find the solution with the same boundary conditions and loading type as the charts indicate in Appendix B in AISC Steel Design Guide Series 9 (2003), which lists solutions to various cases in the form of charts of torsional functions based on Eq. (6.16), with corresponding loading and boundary conditions. In any case, when such a solution is not included in Appendix B (AISC Steel Design Guide Series 9, 2003), the solution to Eq. (6.16) would be necessary. As an example of solving Eq. (6.16), let us take a look at one of the cases listed in Appendix B in AISC Steel Design Guide Series 9 (2003): Case 3 with a concentrated torsional moment applied in the middle span (α = 0.5) (Fig. 6.9). The torsional moment is a constant between $z = 0$ and $z = L/2$, and $z = L/2$ and L. We only need to consider the half of the member ($0 < z < L/2$) since the deformation is symmetric.

For $0 < z < L/2$

$$\theta''' - \frac{1}{a^2}\theta' = -\frac{T/2}{EC_w} \tag{6.17}$$

At $z = 0$, $\theta = 0$ (no twist angle), $\theta''' = 0$ (warping is restrained).
At $z = L/2$, $\theta' = 0$ (the maximum twist occurs at $z = L/2$).
The solution to Eq. (6.17) is

$$\theta''' - \lambda^2\theta' = -\frac{T/2}{EC_w} \tag{6.18}$$

where

$$\lambda^2 = \frac{GJ}{EC_w}$$

$$\lambda = \frac{1}{a}$$

$$a = \sqrt{\frac{EC_w}{GJ}} \quad \text{(can be taken from section tables)}$$

The solution of Eq. (6.18), which is a third-order differential equation, is the sum of homogenous, θ_h, and particular solutions, θ_p, as

$$\theta = \theta_h + \theta_p \tag{6.19}$$

For the homogenous solution, the right-hand side of Eq. (6.18) is set to 0 and $\theta_h = Ae^{mz}$ is assumed. Thus, Eq. (6.18) becomes

$$\theta_h''' - \lambda^2 \theta_h' = 0 \tag{6.20}$$

Finding the first and third derivatives of θ_h as

$$\theta_h' = Ame^{mz} \tag{6.21a}$$

$$\theta_h'' = Am^2 e^{mz} \tag{6.21b}$$

$$\theta_h''' = Am^3 e^{mz} \tag{6.21c}$$

and replacing them in Eq. (6.20) results in

$$Ae^{mz}(m^3 - \lambda^2 m) = 0 \;\rightarrow\; m(m^2 - \lambda^2) = 0 \rightarrow \begin{cases} m_1 = 0 \\ m_{2,3} = \pm\lambda \end{cases} \tag{6.22}$$

Now, we have

$$\theta_h = A_1 e^{\lambda z} + A_2 e^{-\lambda z} + A_3 e^{0z} \tag{6.23}$$

Using hyperbolic function identities, we get

$$\theta_h = A\sinh(\lambda z) + B\cosh(\lambda z) + C \tag{6.24}$$

Assume $\theta_p = \dfrac{T/2}{GJ} z$. Now, we have the complete solution as

$$\theta = A\sinh(\lambda z) + B\cosh(\lambda z) + C + \theta_p \tag{6.25}$$

or

$$\theta = A\sinh\left(\frac{z}{a}\right) + B\cosh\left(\frac{z}{a}\right) + C + \frac{T/2}{GJ}z \tag{6.26}$$

where

A, B, C: constants determined by the three boundary conditions listed above

$$B = 0,\, C = 0,\, A = -\frac{Ta}{2GJ}\left(\frac{1}{\cosh(L/2a)}\right)$$

The solution is written as

$$\theta = \frac{Ta}{2GJ}\left[\frac{z}{a} - \frac{\sinh(z/a)}{\cosh(L/2a)}\right] \tag{6.27}$$

For easy use in design, charts for θ and its derivatives $\theta' = \dfrac{d\theta}{dz}$, $\theta'' = \dfrac{d^2\theta}{dz^2}$, and $\theta''' = \dfrac{\partial^3\theta}{dz^3}$ are given in Table 6.1 and plotted in Fig. 6.10, and also included in AISC Steel Design Guide Series 9 (2003).

Description of the case	Twist angle equation
Concentrated torque on member with pinned ends	

$0 \leq z \leq \alpha l$

$$\theta = \frac{Tl}{GJ}\left[(1.0 - \alpha)\frac{z}{l} + \frac{a}{l}\left(\frac{\sinh\frac{\alpha l}{a}}{\tanh\frac{l}{a}} - \cosh\frac{\alpha l}{a}\right)\sinh\frac{z}{a}\right]$$

$\alpha l \leq z \leq 1$

$$\theta = \frac{Tl}{GJ}\left[(1.0 - z)\frac{\alpha}{l} + \frac{a}{l}\left(\frac{\sinh\frac{\alpha l}{a}}{\tanh\frac{l}{a}} \times \sinh\frac{z}{a} - \sinh\frac{\alpha l}{a} \times \cosh\frac{z}{a}\right)\right]$$

| Concentrated torque on member with fixed ends | |

$0 \leq z \leq \alpha l$

$$\theta = \frac{Ta}{(H + 1)GJ}\left\{\left[H\left(\frac{1}{\sinh\frac{l}{a}} + \sinh\frac{\alpha l}{a} - \frac{\cosh\frac{\alpha l}{a}}{\tanh\frac{l}{a}}\right)\right.\right.$$

$$\left.\left. + \left(\sinh\frac{\alpha l}{a} - \frac{\cosh\frac{\alpha l}{a}}{\tanh\frac{l}{a}} + \frac{1}{\tanh\frac{l}{a}}\right)\right]\left[\cosh\frac{z}{a} - 1.0\right] - \sinh\frac{z}{a} + \frac{z}{a}\right\}$$

$\alpha l \leq z \leq l$

$$\theta = \frac{Ta}{(1 + \frac{1}{H})GJ}\left\{\left[\frac{\left(\cosh\frac{\alpha l}{a} - 1.0\right)}{H \times \sinh\frac{l}{a}} + \frac{\left(\cosh\frac{\alpha l}{a} - \cosh\frac{l}{a} + \frac{l}{a}\sinh\frac{l}{a}\right)}{\sinh\frac{l}{a}}\right] \right.$$

$$+ \cosh\frac{z}{a}\left[\frac{\left(1.0 - \cosh\frac{\alpha l}{a}\right)}{H \times \tanh\frac{l}{a}} + \frac{\left(1.0 - \cosh\frac{\alpha l}{a} \times \cosh\frac{l}{a}\right)}{\sinh\frac{l}{a}}\right]$$

$$\left. + \sinh\frac{z}{a}\left[\frac{\left(\cosh\frac{\alpha l}{a} - 1.0\right)}{H} + \cosh\frac{\alpha l}{a}\right] - \frac{z}{a}\right\}$$

TABLE 6.1 Torsion Function Equations (AISC Steel Design Guide)

where

$$H = \cfrac{\left[\cfrac{\left(1.0 - \cosh\dfrac{\alpha l}{a}\right)}{\tanh\dfrac{l}{a}} + \cfrac{\left(\cosh\dfrac{\alpha l}{a} - 1.0\right)}{\sinh\dfrac{l}{a}} + \sinh\dfrac{\alpha l}{a} - \dfrac{\alpha l}{a}\right]}{\left[\cfrac{\left[\cosh\dfrac{l}{a} + \cosh\dfrac{\alpha l}{a} \times \cosh\dfrac{l}{a} - \cosh\dfrac{\alpha l}{a} - 1.0\right]}{\sinh\dfrac{l}{a}} + \dfrac{l}{a}(\alpha - 1.0) \right] - \sinh\dfrac{\alpha l}{a}}$$

| Concentrated torque on member with fixed and free ends | |

$0 \le z \le \alpha l$

$$\theta = \frac{Ta}{GJ}\left[\left(\sinh\frac{\alpha l}{a} - \tanh\frac{l}{a} \times \cosh\frac{\alpha l}{a} + \tanh\frac{l}{a}\right)\left(\cosh\frac{z}{a} - 1.0\right)\right.$$

$$\left. - \sinh\frac{z}{a} + \frac{z}{a}\right]$$

$\alpha l \le z \le l$

$$\theta = \frac{Ta}{GJ}\left[\left(\tanh\frac{l}{a} \times \cosh\frac{\alpha l}{a} - \tanh\frac{l}{a} - \sinh\frac{\alpha l}{a}\right)\right.$$

$$\left. - \left(\cosh\frac{\alpha l}{a} - 1.0\right)\left(\tanh\frac{l}{a} \times \cosh\frac{z}{a}\right) + \left(\cosh\frac{\alpha l}{a} - 1.0\right) \times \sinh\frac{z}{a} + \frac{\alpha l}{a}\right]$$

TABLE 6.1 Torsion Function Equations (AISC Steel Design Guide) (*Continued*)

6.4   Stress Analysis and Design Considerations

There are three kinds of stresses in any I-shaped or channel section due to torsional loading:

1. Shear stresses τ_s in web and flanges due to the rotation of the elements of the cross-section (pure torsion case, T_s)

2. Shear stresses τ_{ws} in the flanges due to lateral bending (warping torsion case, T_w)

3. Normal stresses (tension and compression) σ_{ws} due to lateral bending of the flanges (lateral bending moment on flange, M_f) (warping torsion case, T_w)

It should be noted that normal stress, σ_b, and shear stresses, τ_b, due to flexure can be superimposed to the normal and shear stresses due to torsional loading.

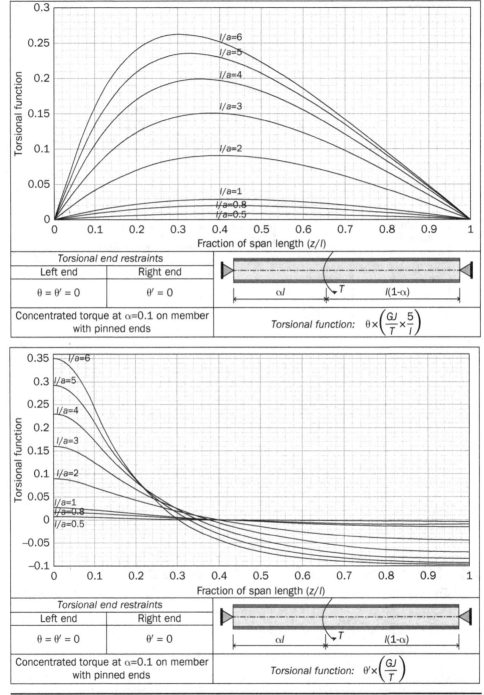

Figure 6.10 Torsional function curves.

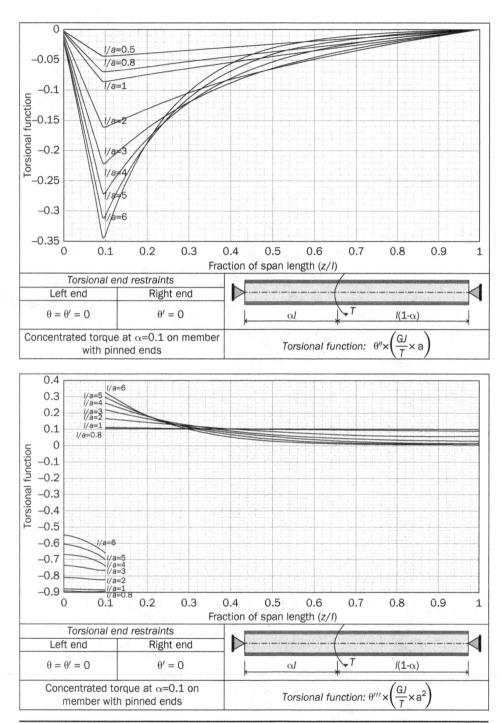

FIGURE 6.10   (Continued)

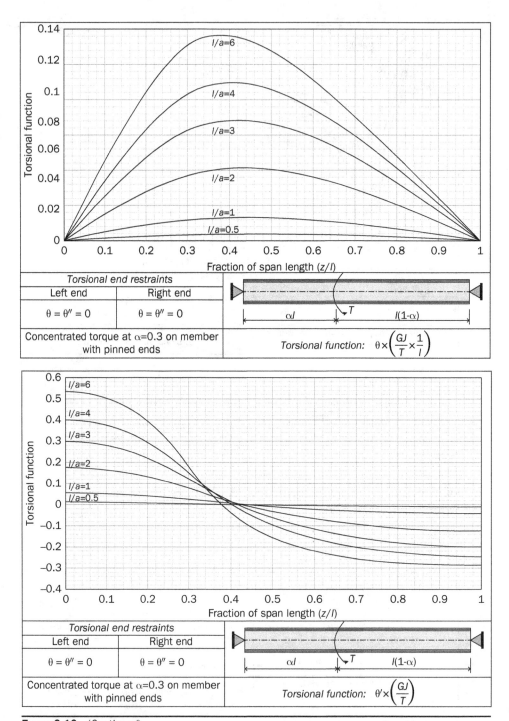

FIGURE 6.10 (Continued)

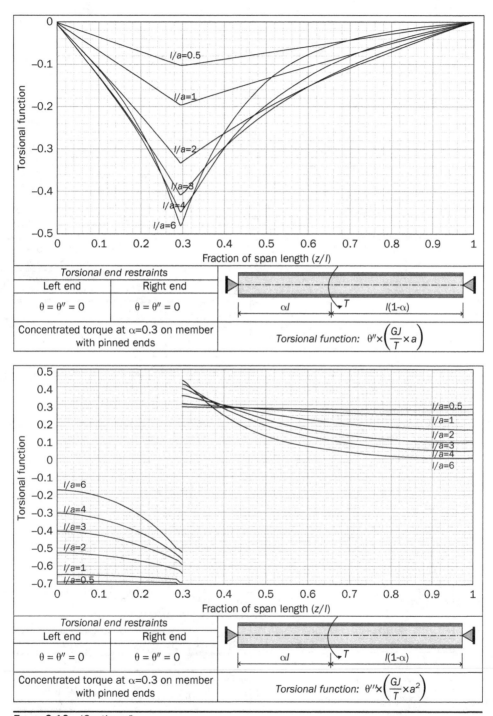

Figure 6.10 (Continued)

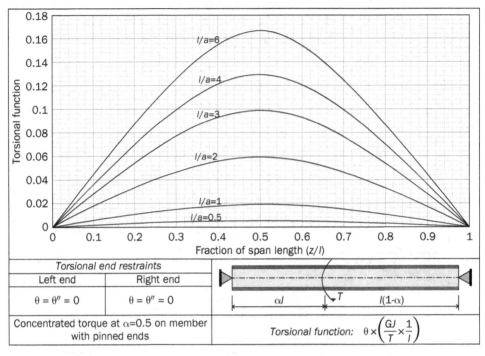

Torsional end restraints		
Left end	Right end	
$\theta = \theta'' = 0$	$\theta = \theta'' = 0$	
Concentrated torque at $\alpha=0.5$ on member with pinned ends		Torsional function: $\theta \times \left(\dfrac{GJ}{T} \times \dfrac{1}{l} \right)$

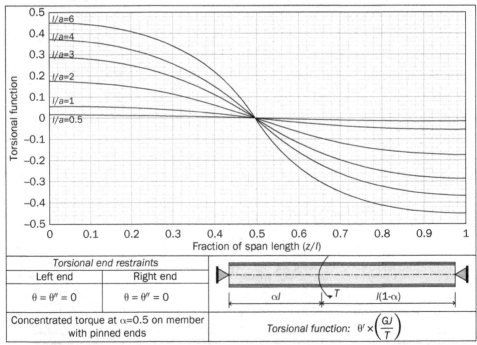

Torsional end restraints		
Left end	Right end	
$\theta = \theta'' = 0$	$\theta = \theta'' = 0$	
Concentrated torque at $\alpha=0.5$ on member with pinned ends		Torsional function: $\theta' \times \left(\dfrac{GJ}{T} \right)$

FIGURE 6.10 *(Continued)*

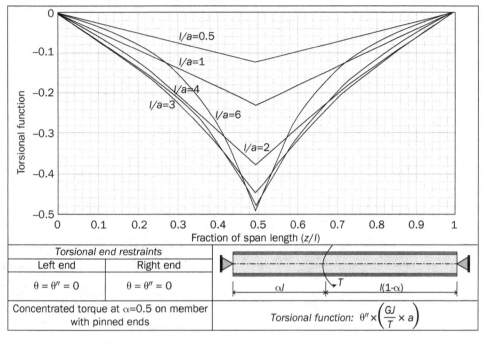

Torsional end restraints		
Left end	Right end	
$\theta = \theta'' = 0$	$\theta = \theta'' = 0$	
Concentrated torque at $\alpha=0.5$ on member with pinned ends	Torsional function: $\theta'' \times \left(\dfrac{GJ}{T} \times a\right)$	

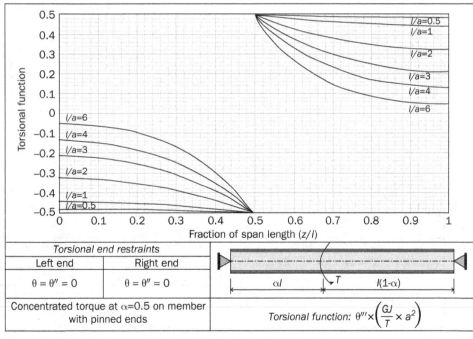

Torsional end restraints		
Left end	Right end	
$\theta = \theta'' = 0$	$\theta = \theta'' = 0$	
Concentrated torque at $\alpha=0.5$ on member with pinned ends	Torsional function: $\theta''' \times \left(\dfrac{GJ}{T} \times a^2\right)$	

FIGURE 6.10   *(Continued)*

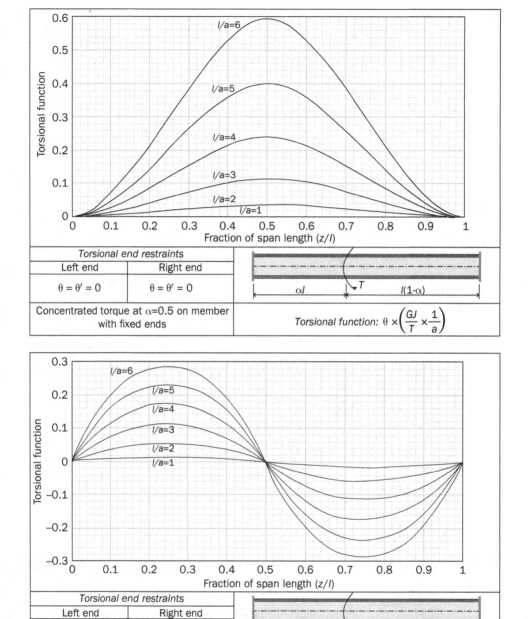

FIGURE 6.10   (Continued)

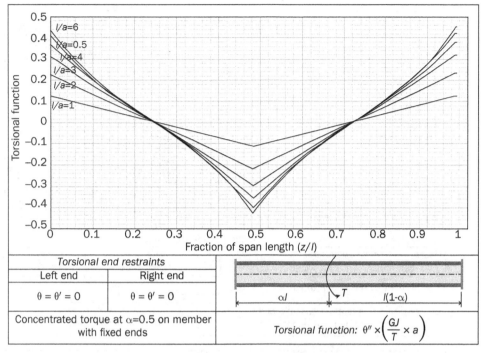

Torsional end restraints	
Left end	Right end
$\theta = \theta' = 0$	$\theta = \theta' = 0$

Concentrated torque at $\alpha=0.5$ on member with fixed ends	Torsional function: $\theta'' \times \left(\dfrac{GJ}{T} \times a \right)$

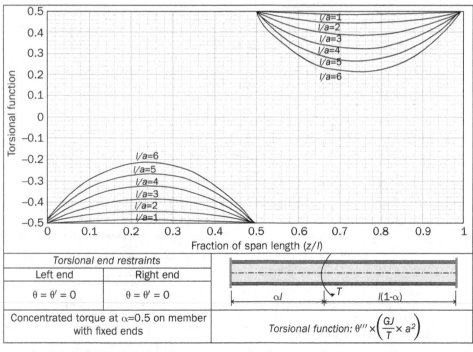

Torsional end restraints	
Left end	Right end
$\theta = \theta' = 0$	$\theta = \theta' = 0$

Concentrated torque at $\alpha=0.5$ on member with fixed ends	Torsional function: $\theta''' \times \left(\dfrac{GJ}{T} \times a^2 \right)$

FIGURE 6.10   *(Continued)*

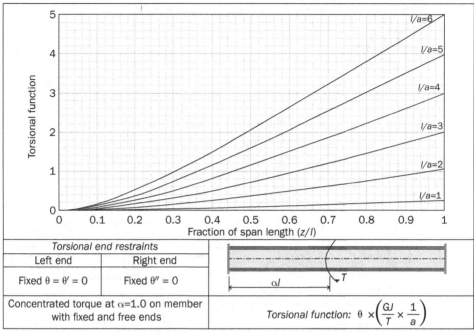

Torsional end restraints	
Left end	**Right end**
Fixed $\theta = \theta' = 0$	Fixed $\theta'' = 0$

Concentrated torque at α=1.0 on member with fixed and free ends	Torsional function: $\theta \times \left(\dfrac{GJ}{T} \times \dfrac{1}{a} \right)$

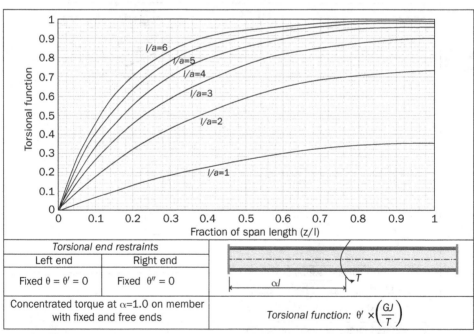

Torsional end restraints	
Left end	**Right end**
Fixed $\theta = \theta' = 0$	Fixed $\theta'' = 0$

Concentrated torque at α=1.0 on member with fixed and free ends	Torsional function: $\theta' \times \left(\dfrac{GJ}{T} \right)$

FIGURE 6.10 (Continued)

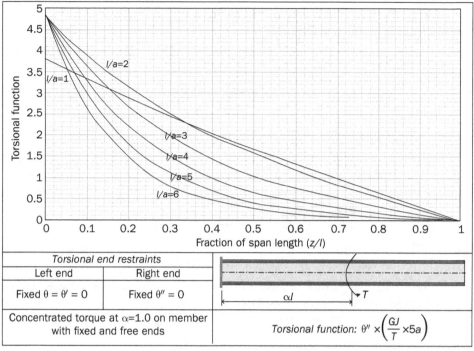

Torsional end restraints		
Left end	Right end	
Fixed $\theta = \theta' = 0$	Fixed $\theta'' = 0$	
Concentrated torque at $\alpha=1.0$ on member with fixed and free ends	Torsional function: $\theta'' \times \left(\dfrac{GJ}{T} \times 5a\right)$	

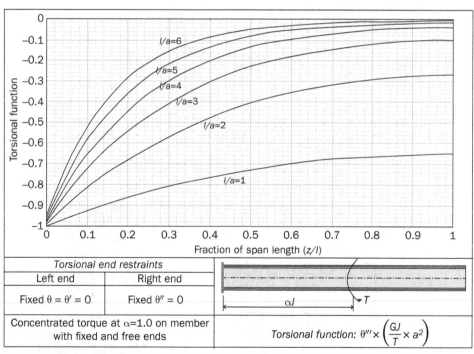

Torsional end restraints		
Left end	Right end	
Fixed $\theta = \theta' = 0$	Fixed $\theta'' = 0$	
Concentrated torque at $\alpha=1.0$ on member with fixed and free ends	Torsional function: $\theta''' \times \left(\dfrac{GJ}{T} \times a^2\right)$	

FIGURE 6.10   (Continued)

6.4.1   Pure Torsional Stress: Shear Stress, τ_s

a. Rectangular sections

Shear stress distribution for a rectangular section is given in Fig. 6.11. Shear strain, γ, from Fig. 6.11 is

$$\gamma \approx 2\frac{d\theta}{dz}\frac{t}{2} \;\rightarrow\; \gamma = \frac{d\theta}{dz}t \tag{6.28}$$

and

$$\tau_s = \gamma G \tag{6.29}$$

It should also be noted that $\tau_{s,max}$ in rectangular sections always occur in the middle of the long dimension. Using elasticity theory, the pure torsional maximum shear stress in a rectangular section subjected to torsion, as shown in Fig. 6.11, can be expressed as

$$\tau_{s,max} = \frac{k_1 T_S}{bt^2} \tag{6.30}$$

where

$$J = k_2 bt^3$$

Table 6.2 lists the maximum shear and normal stress points in an I-shaped section due to both torsional and flexural loading.

k_1 and k_2 values are given in Table 6.3. If $b/t > 5$, $k_1 = 3$, $k_2 = 1/3$, then

$$\tau_s = \frac{T_s}{\frac{1}{3}bt^2} \tag{6.31}$$

and

$$J = \frac{1}{3}bt^3$$

b. Steel sections

In steel sections, most elements such as flanges and webs, b/t ratio is far larger than 5.0, and k_1 and k_2 can be approximately assumed to be 3, 1/3, respectively (see Table 6.3). The maximum pure shear stress at the edge can be simply written as

$$\tau_s = \frac{T_s t}{J} = Gt\theta' \tag{6.32}$$

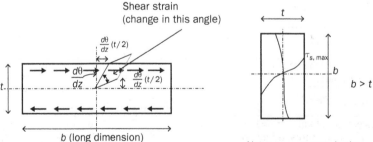

Note: $\tau_{s,\,max}$ occurs in the middle of the long dimension

Figure 6.11 Pure shear stress flow and distribution in a rectangular section.

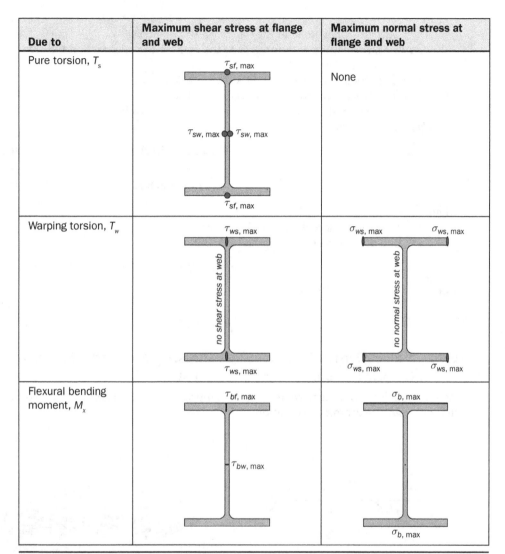

TABLE 6.2   Maximum Normal and Shear Stresses in an I-shaped Section

b/t	1.0	1.5	3.0	5.0	∞
k_1	4.81	4.33	3.75	3.44	3
k_2	0.141	0.196	0.263	0.291	1/3

TABLE 6.3 k_1 and k_2 Values for Rectangular Sections

where
τ_s = pure torsional stress at the flange or web edge, ksi
(Note that the stress is the largest at the edge of the section, and reduces linear to zero in the center)
τ_{sf} = pure torsional stress at the flange edge, ksi
τ_{ss} = pure torsional stress at the web edge, ksi
t = thickness of the element (t_f for flange and t_w for web), in.

For I-shaped, channel, and T-sections:

$$T_s = T_1 + T_2 + T_3 \text{ (see Fig. 6.12)} \tag{6.33}$$

$$J = \sum J_i = \sum \frac{1}{3} b_i t_i^3 \text{ (see Fig. 6.12)} \tag{6.34}$$

where
b = long dimension of rectangular element
t = thin dimension of rectangular element

$$\tau_{sw} = \frac{T_s t_w}{J} = G t_w \theta' \tag{6.35}$$

and

$$\tau_{sf} = \frac{T_s t_f}{J} = G t_f \theta' \tag{6.36}$$

6.4.2 Warping Torsion Stresses—Shear Stress, τ_{ws} and Normal Stress, σ_{ws}

a. Normal Stress due to warping, σ_{ws} (only at flange):

As shown in Fig. 6.13, the normal stress in compression and tension would result from the constrained warping deformation:

$$\sigma_{ws} = \frac{M_f}{I_f} x \tag{6.37}$$

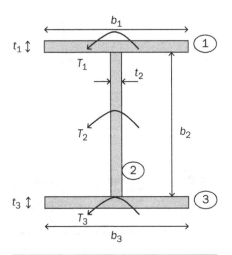

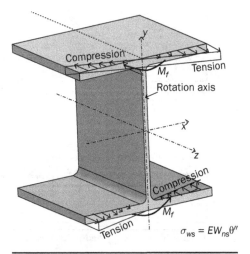

FIGURE 6.12 An approximate approach to compute torsional rigidity for steel sections.

FIGURE 6.13 Normal stresses in an I-shaped section due to warping (distributed linearly across the flange width).

Since $\dfrac{d^2 u_f}{dz^2} = -\dfrac{M_f}{EI_f}$

$$M_f = \frac{d^2 u_f}{dz^2} = EI_f\left(\frac{h}{2}\right)\frac{d^2\theta}{dz^2} = \frac{EC_w}{h}\frac{d^2\theta}{dz^2} \tag{6.38}$$

Note that (−) sign is dropped in Eq. (6.38) because tension occurs on one side and compression on the other side. Maximum normal stress occurs at $x=b_f/2$:

$$\sigma_{ws} = EI_f\left(\frac{h}{2}\right)\frac{d^2\theta}{dz^2}\left(\frac{b_f}{2I_f}\right) = \frac{Eb_f h}{4}\frac{d^2\theta}{dz^2} \tag{6.39}$$

or

$$\sigma_{ws} = EW_{ns}\theta'' \tag{6.40}$$

where

$C_w = I_f\left(\dfrac{h^2}{2}\right)$, warping torsional constant, in.[6]

$W_{ns} = \dfrac{b_f h}{4}$ normalized warping function at point s (Appendix A in Steel Design Guide Series 9, 2003), in.[2]

$\theta'' = \dfrac{d^2\theta}{dz^2}$ the second derivative of twist angle with respect to z

b. Shear stresses due to warping, τ_{ws} (only at flange)

τ_{ws} varies parabolically across the width of the rectangular flange (Fig. 6.14):

$$\tau_{ws} = \frac{V_f Q_f}{I_f t_f} \tag{6.41}$$

Remember that $V_f = -EI_f\dfrac{\partial^3 u_f}{dz^3} = -EI_f\left(\dfrac{h}{2}\right)\dfrac{\partial^3\theta}{dz^3}$

$$\tau_{ws} = -\frac{EI_f\left(\dfrac{h}{2}\right)\dfrac{\partial^3\theta}{dz^3}\dfrac{b_f^2 t_f}{8}}{I_f t_f} = -E\frac{b_f^2 h}{16}\frac{\partial^3\theta}{dz^3} \tag{6.42}$$

$$\tau_{ws} = -E\frac{S_w}{t_f}\theta''' \tag{6.43}$$

where

$Q_f = A\bar{x} = \dfrac{b_f t_f}{2}\left(\dfrac{b_f}{4}\right) = \dfrac{b_f^2 t_f}{8}$ statical moment of area of one-half flange about the y-axis, in.[3]

$S_w = \dfrac{b_f^2 h t_f}{16}$ warping statical moment, in.[4]

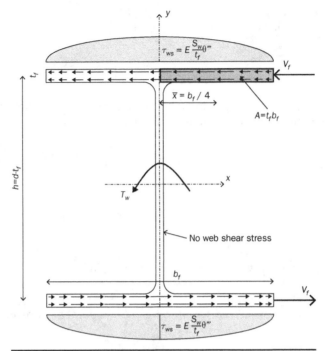

FIGURE 6.14   Shear stresses in an I-shaped section due to warping.

6.5   Behavior and Design

Behavior of members subject to torsion is always expected to be elastic, and analytical solutions are used due to lack of experimental work. Thus, a structural engineer should always avoid torsion in design. Structural members need to be checked under primary actions such as bending, compression, and tension. In cases where there is torsion, in addition to the torsional stresses, bending and shear stresses due to plane bending are normally present in the structural member. These stresses are determined by the following equations:

$$\sigma_b = \frac{M_u}{S} \tag{6.44}$$

and

$$\sigma_a = \frac{P}{A} \tag{6.45}$$

Design provisions:
Normal Stress (Fig. 6.15):

$$f_{un} \le \phi F_y \tag{6.46}$$

$$f_{un} = \sigma_a \pm \left(\sigma_{bx} + \sigma_{by} + \sigma_{ws} \right) \tag{6.47}$$

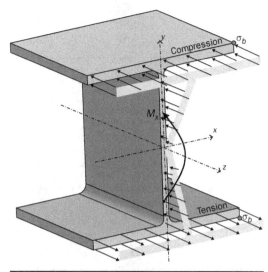

Figure 6.15 Normal stress distribution due to flexure in W-sections.

Shear Stress (Fig. 6.16):

$$f_{uv} \leq \phi 0.6 F_y \tag{6.48}$$

$$f_{uv} = \begin{cases} \tau_{sw} + \tau_{bw} & \text{for web} \\ \text{or} \\ \tau_{sf} + \tau_{ws} + \tau_{bf} & \text{for flange} \end{cases} \tag{6.49}$$

where

f_{un} = maximum factored normal stress due to axial load, bending, and torsion, ksi
f_{uv} = maximum factored shear stress due to pure torsion, warping, and bending, ksi
$\phi = 0.90$
σ_{ws} = maximum factored normal stress due to torsion (warping), ksi
σ_b = maximum factored normal stress due to bending about either x- or y-axis, ksi
σ_a = maximum factored normal stress due to axial load, ksi
τ_{bw} = shear stress due to flexure in web, ksi
τ_{bf} = shear stress due to flexure in flange, ksi
τ_{sw} = shear stress due to pure torsion in web, ksi
τ_{sf} = shear stress due to pure torsion in flange, ksi
τ_{ws} = shear stress due to torsion (warping) in flange, ksi
M_u = maximum factored bending moment, kip-in.
A = cross-section area, in.²
S = elastic section modulus about either x- or y-axis in.³

In Eqs. (6.46) and (6.48), f_{un} and f_{uv} are the demand side of the design equation, whereas ϕF_y and $0.6\phi F_y$ represent the capacity. Note that f_{uv} is usually far less than permitted.

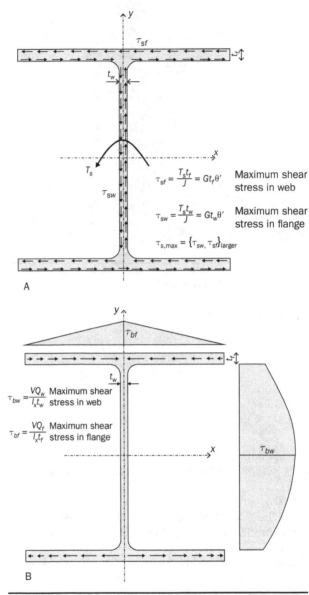

FIGURE 6.16   Shear stress due to (a) pure torsion and
(b) flexure in W-sections.

The normal strength of a section subject to torsion or torsion combined with flexure is
not clearly determined. Shear stresses are assumed to be small and can be neglected.

Example 6.2

Determine the torsional properties (J, C_w, a, W_{ns}, S_w) and statical moments (Q_f, Q_w) of
W21 × 57. ($E = 29{,}000$ ksi, $\mu=0.3$ Poisson ratio for steel).

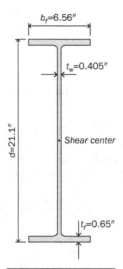

FIGURE 6.17 W21×57 cross-section.

Cross-section properties (from *AISC Manual*):
$t_f = 0.65$ in., $t_w = 0.405$ in. (Fig. 6.17)
$S_x = 111$ in.^{3}, $I_x = 1{,}170$ in.^{4}

Solution:
a. Torsional properties:

$$J = k_2 bt^3$$

b/t ratios for both flange and web:
$b_f/t_f = 6.56''/0.65'' = 10.1$ (for flange) >5 and $(d-2t_f)/t_w) = (21.1'' - 2 \times 0.65'')/0.405'' = 48.9$
$> 5 \rightarrow k_2 = 1/3$

$$J \approx \sum \frac{1}{3}bt^3 = \frac{1}{3}\left[2\left(6.56''\right)\left(0.65''\right)^3\right) + \left(21.1'' - 2 \times 0.65''\right)\left(0.405''\right)^3\right] = 1.64 \text{ in.}^4$$

$$C_w = I_f\left(\frac{h^2}{2}\right) = \frac{t_f b_f^3}{12}\left(\frac{(d-t_f)^2}{2}\right) = \frac{(0.65'')(6.56'')^3}{12}\left(\frac{(20.45'')^2}{2}\right) = 3{,}197 \text{ in.}^6$$

$$h = 21.1'' - 0.65'' = 20.45 \text{ in.}$$

$$G = \frac{E}{2(1+\mu)} = \frac{29{,}000^{ksi}}{2(1+0.3)} = 11{,}154 \text{ ksi } (\mu = 0.3 \text{ Poisson ratio for steel})$$

$$a = \sqrt{\frac{EC_w}{GJ}} = \sqrt{\frac{29{,}000^{ksi} \times 3{,}197^{in.^6}}{11{,}154^{ksi} \times 1.64^{in.^4}}} = 71.2 \text{ in.}$$

$$W_{ns} = \frac{b_f h}{4} = \frac{6.56'' \times 20.45''}{4} = 33.5 \text{ in.}^2$$

$$S_w = \frac{b_f^2 h t_f}{16} = \frac{(6.56'')^2 (20.45'')(0.65'')}{16} = 35.8 \text{ in.}^4$$

b. Statical moments:

$$Q_f = \left(\frac{6.56''}{2}\right)(0.65'')\frac{(21.1'' - 0.65'')}{2} = 21.8 \text{ in.}^3 \text{ (at the midwidth of the flange)}$$

$$Q_f = \left(\frac{6.56'' - 0.405''}{2}\right)(0.65'')\frac{(21.1'' - 0.65'')}{2} = 20.5 \text{ in.}^3 \text{ [at the face of the web (more}$$

correct to calculate the maximum flange shear stress); AISC Steel Design Guide Series 9 (2003) provides values for the point at the intersection of the web and flange centerlines]

$$Q_w = (6.56'')(0.65'')\left(\frac{21.1'' - 0.65''}{2}\right) + \frac{(21.1'' - 2 \times 0.65'')}{2}(0.405'')$$

$$\frac{(21.1'' - 2 \times 0.65'')}{4} = 62.1 \text{ in.}^3$$

c. Comparison with the given values in AISC Steel Design Guide Series 9 (2003)

Torsional properties and statical moments of W21×57 from AISC Steel Design Guide Series 9 (2003):
Torsional properties: $J=1.77 \text{ in.}^4$, $C_w=3.190 \text{ in.}^6$, $a=68.3 \text{ in.}$, $W_{ns}=33.4 \text{ in.}^2$, $S_w=35.6 \text{ in.}^4$
Statical moments: $Q_f = 20.9 \text{ in.}^3$, $Q_w = 64.3 \text{ in.}^3$

Please note the small differences of the torsional properties and statical moments between the hand calculation and the AISC Steel Design Guide. The differences are due to the effect of fillets at the junction of the flange to the web, which is ignored in the hand calculation. However, hand calculated torsional properties and statical moments, assuming rectangular webs and flange is within the acceptable limits and can be used for design purposes.

Example 6.3
A W12×50 beam spans 12 ft (144 in.) and supports an 18-kip factored load (12-kip service load) at midspan that acts at an 8-in. eccentricity with respect to the shear center. Check if the design is adequate (Fig. 6.18).

Solution:
Cross-section and torsional properties of W12×50:
Torsional properties: $J = 1.71 \text{ in.}^4$, $C_w = 1.88 \text{ in.}^6$, $a = 52.3 \text{ in.}$, $W_{no} = 23.3 \text{ in.}^2$, $S_{w1} = 30.2 \text{ in.}^4$

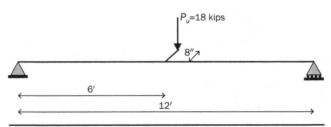

FIGURE 6.18 Simply supported beam with eccentric loading at midspan.

Statical moments: $Q_f = 14.7$ in.^{3}, $Q_w = 36.2$ in.^{3}
Bending properties: $S_x = 64.2$ in.^{3}, $I_x = 391$ in.^{4}
Other properties: $t_f = 0.64$ in., $t_w = 0.37$ in.

The end conditions are assumed to be flexurally and torsionally pinned. Moment diagrams for bending and torsion are given in Fig. 6.19.

Case 1: Bending (Fig. 6.20)

1. Normal stress

$$\sigma_b = \frac{M_b}{S_x} = \frac{54^{\text{kip-ft}} \times 12^{\text{in./ft}}}{64.2^{\text{in.}^3}} = 10.1 \text{ ksi}$$

2. Shear stress (web)

$$\tau_{bw} = \frac{V_{z=7.5'} \times Q_w}{I_x \times t_w} = \frac{9.0^{\text{kips}} \times 36.2^{\text{in.}^3}}{391^{\text{in.}^4} \times 0.37^{\text{in.}}} = 2.25 \text{ ksi}$$

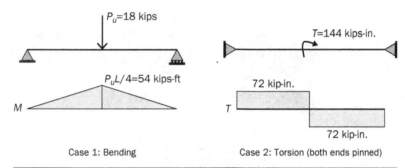

FIGURE 6.19   Moment diagrams for bending and torsion.

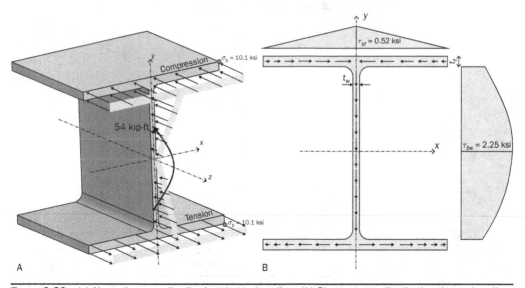

FIGURE 6.20   (*a*) Normal stress distribution due to bending. (*b*) Shear stress distribution due to bending.

3. Shear stress (flange)

$$\tau_{bf} = \frac{V_{z=7.5'} \times Q_f}{I_x \times t_f} = \frac{9.0^{kips} \times 14.7^{in.^3}}{391^{in.^4} \times 0.64^{in.}} = 0.52 \text{ ksi}$$

Case 2: Torsion (Fig. 6.21)

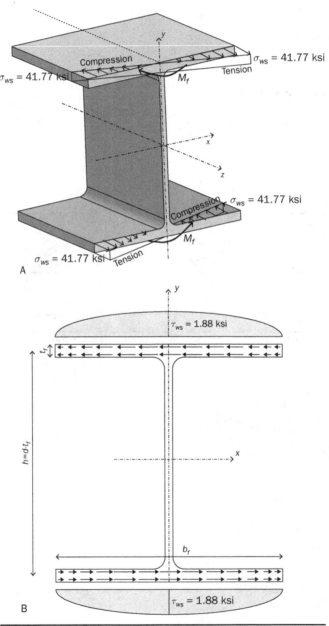

FIGURE 6.21 (a) Normal stress distribution due to warping. (b) Shear stress distribution due to warping.

Since both ends pinned

$$\alpha = 0.5$$

$L/a = 144''/52.3'' = 2.75; \; z/L = 0.5$

$$\theta \times \left(\frac{GJ}{T} \times \frac{1}{l}\right) = +0.09 \rightarrow \theta = \frac{0.09 \times 144^{in.} \times 144^{kip\text{-}in.}}{11,200^{ksi} \times 1.71^{in.^4}} = 0.0974$$

$$\theta' \times \left(\frac{GJ}{T}\right) = 0 \rightarrow \theta' = 0$$

$$\theta'' \times \left(\frac{GJ}{T} \times a\right) = -0.43 \rightarrow \theta'' = \frac{(-0.43) \times 144^{kip\text{-}in.}}{11,200^{ksi} \times 1.71^{in.^4} \times 52.3^{in.}} = -6.1818 \times 10^{-5}$$

$$\theta''' \times \left(\frac{GJ}{T} \times a^2\right) = -0.5 \rightarrow \theta''' = \frac{(-0.5) \times 144^{kip\text{-}in.}}{11,200^{ksi} \times 1.71^{in.^4} \times \left(52.3^{in.}\right)^2}$$

$$= -1.3744 \times 10^{-6}$$

1. Pure torsion (shear stress)

$$\tau_s = \frac{T_s t}{J} = Gt\theta'$$

Web: $\tau_{sw} = \dfrac{T_s t_w}{J} = Gt_w \theta' = 11,200^{ksi} \times 0.37^{in.} \times 0 = 0$

Flange: $\tau_{sf} = \dfrac{T_s t_f}{J} = Gt_f \theta' = 11,200^{ksi} \times 0.64^{in.} \times 0 = 0$

2. Warping stress (normal and shear stresses)

$$\sigma_{ws} = EW_{ns}\theta'' = 29,000^{ksi} \times 23.3^{in.^2} \times \left(-6.1818 \times 10^{-5}\right) = -41.77 \text{ ksi}$$

$$\tau_{ws} = -\frac{ES_{ws}\theta'''}{t_f} = -\frac{29,000^{ksi} \times 30.2^{in.^4} \times \left(-1.3744 \times 10^{-6}\right)}{0.64^{in.}} = 1.88 \text{ ksi}$$

Superposition of stresses:

1. Shear stress

Shear stresses due to pure torsion are equal to zero (web and flange) for the midspan $(z = 6.0')$.

2. Normal stress

Normal stress in the web due to bending is neglected.

Superposition of Stresses: Summary

	σ_b	σ_{ws}	f_{un}	τ_s	τ_{ws}	τ_b	f_{uv}
Flange	10.1 ksi	41.77 ksi	51.9 ksi	–	1.88 ksi	0.52 ksi	2.4 ksi
Web	–	–	–	–	–	2.25 ksi	2.25 ksi

$f_{un} = 51.9 \text{ ksi} < \phi F_y = 0.9 \times 50^{ksi} = 45 \text{ ksi (Not safe...)}$

$f_{uv} = (2.4 \text{ ksi}; 2.25 \text{ ksi})_{max} = 2.4 \text{ ksi} < 0.6\phi F_y = 0.6 \times 0.9 \times 50^{ksi} = 27 \text{ ksi}$

The design is not adequate.

Example 6.4

For a 5.0-kip-ft pure torsional moment, compute maximum torsional shear stress, τ_s, for L8×6×3/4 (Fig. 6.22).

Solution:

If we treat the angle as a plate (Fig. 6.23), maximum shear stress can be determined as $b/t = 14''/\frac{3}{4}'' = 19 > 5$, then

$$\tau_s = \frac{T_s}{\frac{1}{3}bt^2} = \frac{5.0^{\text{kip-ft}} \times 12^{\text{in./ft}}}{\frac{1}{3}14''\left(\frac{3}{4}''\right)^2} = 22.86 \text{ ksi}$$

Example 6.5

A cantilever beam is loaded by a P (100% live load). During the service, P is moved to the edge of the top flange, resulting in an eccentricity of half of the flange. Check if the beam is adequate. Consider normal stresses only. The beam is laterally braced (no lateral-torsional buckling). $F_y = 50$ ksi steel (Fig. 6.24).

Solution:

1. Structural analysis models for bending and torsion moments

For this example, governing load combination is: $1.2D + 1.6L$.
Factored load P:

$$P_u = 1.6 \times 20^{\text{kips}} = 32 \text{ kips}$$

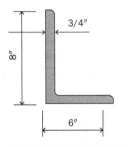

FIGURE 6.22 L8×6×3/4 cross-section.

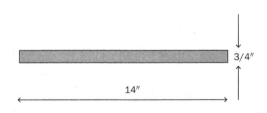

FIGURE 6.23 A plate equivalent to L8×6×3/4 in approximate area.

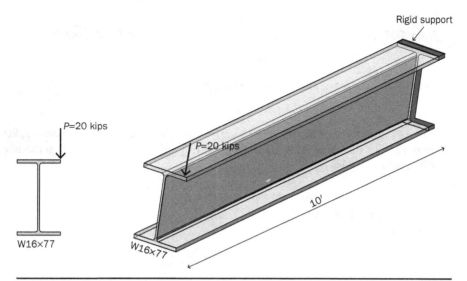

FIGURE 6.24 Cantilever beam with eccentric loading at the end.

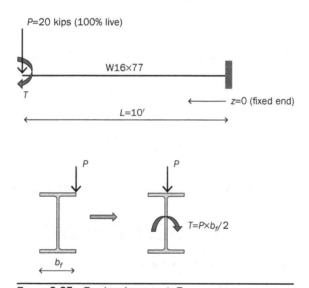

FIGURE 6.25 Torsional moment, T.

Factored maximum bending moment (at the fixed end):

$$M_u = P_u \times L = 320 \text{ kip-ft} = 3{,}840 \text{ kip-in.}$$

Factored torsional moment (uniformly distributed over the beam length; Fig. 6.25):

$$T_u = P_u \times b_f/2$$
$$= 32^{kips} \times (10.3''/2) = 164.8 \text{ kip-in.}$$

(The maximum normal stress induced by this torsional moment is in the flanges of the W16×77 at the fixed end.)

2. Torsion-induced normal stress, σ_{ws}

From AISC Steel Design Guide Series 9 (2003) (on Page 36 for W16×77):

$$W_{ns} = 40.6 \text{ in.}^2, J = 3.57 \text{ in.}^4, a = 78.9 \text{ in.}$$

Torsion-induced normal stress in the flanges of W-section:

$$\sigma_{ws} = EW_{ns}\theta''$$

where θ'' can be found from Fig. 6.10 (or in AISC Steel Design Guide Series 9, 2003). As shown in Fig. 6.26, this is the same as concentrated torque at $\alpha = 1.0$ on member with fixed and free ends:

$$L/a = 10' \times 12''/78.9'' = 1.52$$

At $z/L = 0$ (the fixed end) (Fig. 6.10)
when $L/a = 1.0$,

$$\theta'' = \left(\frac{GJ}{T} \times 5a\right) = 3.8$$

when $L/a = 2.0$,

$$\theta'' = \left(\frac{GJ}{T} \times 5a\right) = 4.8$$

For $L/a = 1.52$, we can linearly interpolate the values between $L/a = 1$ and 2 as

$$\theta'' = \left(\frac{GJ}{T} \times 5a\right) = 4.3$$

or

$$\theta'' = \left(\frac{4.3T}{5GJa}\right)$$

Using T_u in place of T, and $G = 11,200$ ksi as shear modulus of steel, we have

$$\theta'' = \left(\frac{4.3T_u}{5GJa}\right) = \frac{4.3 \times 164.8^{\text{kip-in.}}}{5 \times 11,200^{\text{ksi}} \times 3.57^{\text{in.}^4} \times 78.9^{\text{in.}}} = 4.49\times10^{-5}$$

Then, we can get the normal stress due to torsion at the fixed end:

$$\sigma_{ws} = EW_{ns}\theta'' = 29,000^{\text{ksi}} \times 40.6^{\text{in.}^2} \times (4.49 \times 10^{-5}) = 52.86 \text{ ksi}$$

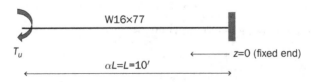

FIGURE 6.26   Concentrated torque at the end of the cantilever beam.

Note: the torsion-induced normal stress alone has already exceeded the design stress of $\phi F_y = 0.9 \times 50^{ksi} = 45$ ksi!

3. Bending-induced normal stress, σ_{bx}

The maximum bending stress occurs in the flanges of the section at the fixed end:

$$\sigma_{bx} = \frac{M_u}{S_x} = \frac{3,840^{kip\text{-}in.}}{134^{in.^3}} = 28.66 \text{ ksi}$$

where S_x = elastic section modulus of W16×77 from *AISC Manual*, Table 1-1.
Note: The beam is adequate for bending only, $\sigma_{bx} = 28.66$ ksi $< \phi F_y = 0.9 \times 50$ ksi $= 45$ ksi. In other words, the original design without considering the eccentricity-induced torsion would be safe (even a little more conservative).

4. Final check of the design when P moves to the edge of the flange

The maximum normal stress occurs in the flanges of the section at the fixed end because both torsional and bending normal stresses are at the same place:

$$f_{un} = \sigma_{ws} + \sigma_{bx} = 52.86^{ksi} + 28.66^{ksi} = 82^{ksi} > \phi F_y = 0.9 \times 50^{ksi} = 45 \text{ ksi}$$

The beam is unsafe under the combined bending and torsional moment.
Discussion: The beam was originally designed for a bending moment of 320 kip-ft, due to a seemingly small eccentricity-induced torsional moment of 164.8 kip-in. or 14 kip-ft, which is only 4% of the bending moment. However, such a tiny torsional moment would produce almost twice as much normal stress as the mighty bending moment does! AND, this example is not very rare in practice. There is only ONE reason: open sections (W-section, channel, angle, etc.) are very vulnerable to torsional moment. We need to pay special attention to any possible torsion in structural engineering practice!!!

Example 6.6
The 30-ft-long W21×93 beam AC, with fixed support at its ends (for bending and torsion), A and C, is supporting a cantilever beam BD. BD is perpendicular to beam AC and is rigidly connected to AC at B. The load P (50% dead and 50% live loads) is applied with 7 ft from the centerline of beam AC. Check if the beam design is adequate. Consider normal stresses only. The beam is laterally braced (no lateral-torsional buckling). $F_y = 50$ ksi steel (Fig. 6.27).

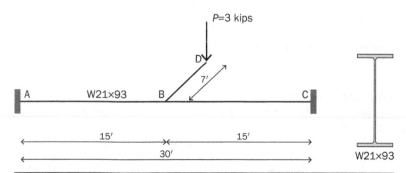

FIGURE 6.27 W21×93 beam with fixed support at its ends and subjected to torsion at midspan.

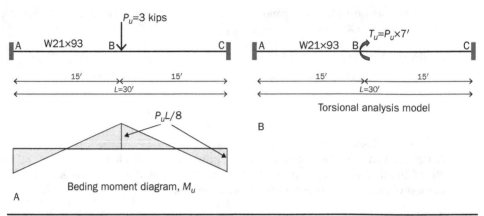

FIGURE 6.28 Loading for (a) bending, (b) torsion.

Solution:

1. Structural analysis models for bending and torsion moments

For this example, the governing load combination is: $1.2D+1.6L$
Factored load Pu:

$$P_u = (1.2 \times 0.5 + 1.6 \times 0.5)(3^{kips}) = 4.2 \text{ kips}$$

The beam can be treated as two loading cases: the bending moment and the torsional moment (Fig. 6.28):

The factored maximum bending moment, M_u, is the same at A, B, and C:

$$M_u = P_u L/8 = 4.2^{kips} \times 30'/8 = 15.75 \text{ kip-ft}$$

Factored torsional moment (uniformly distributed over the beam length):

$$T_u = P_u \times 7^{ft} = 4.2^{kips} \times 7' = 29.4 \text{ kip-ft}$$

2. Torsion-induced normal stress, σ_{ws}

From AISC Steel Design Guide Series 9 (2003) (on Page 36 for W21×93):

$$W_{ns} = 43.6 \text{ in.}^2, J = 6.83 \text{ in.}^4, a = 65.3 \text{ in.}$$

Torsion-induced normal stress in the flanges of W-section:

$$\sigma_{ws} = EW_{ns}\theta''$$

where θ'' can be found from Fig. 6.10.
As shown in the figure above, this is the same as concentrated torque at $\alpha = 0.5$ on member with fixed ends (Fig. 6.10):
$L/a = 30' \times 12/65.3'' = 5.5$, from the top chart of Fig. 6.10,

$\theta'' = \left(\dfrac{GJ}{T} \times a\right)$ has the same peak value $(+/-)$ at $z/L = 0$ (section A), 0.5 (section B), and 1.0 (section C), when $L/a = 4.0$,

$$\theta'' = \left(\frac{GJ}{T} \times a\right) = \pm0.38$$

when $L/a = 6.0$,

$$\theta'' = \left(\frac{GJ}{T} \times a\right) = \pm0.45$$

For $L/a = 5.5$, we can linearly interpolate the values between $L/a = 4.0$ and 6.0 as

$$\theta'' = \left(\frac{GJ}{T} \times a\right) = \pm0.4325$$

or,

$$\theta'' = \frac{\pm0.4325T}{GJa}$$

In the above equation, if we use T_u in place of T, $G = 11{,}200$ ksi as shear modulus of steel, and drop $+/-$ sign, we have

$$\theta'' = \frac{\pm0.4325T}{GJa} = \frac{0.4325 \times 29.4^{\text{kip-ft}} \times 12^{\text{in./ft}}}{11{,}200^{\text{ksi}} \times 6.83^{\text{in.}^4} \times 65.3^{\text{in.}}} = 3.05{\times}10^{-5}$$

Then, we can get the normal stress due to torsion at the fixed ends (A and C) and middle section (B):

$$\sigma_{ws} = EW_{ns}\theta'' = 29{,}000^{\text{ksi}} \times 43.6^{\text{in.}^2} \times \left(3.05 \times 10^{-5}\right) = 38.62 \text{ ksi}$$

Note: the torsion-induced normal stress is near but less than $\phi F_y = 0.9{\times}50\text{ksi} = 45$ ksi!

3. Bending-induced normal stress, σ_{bx}

The maximum bending stress occurs in the flanges of the section at the fixed end:

$$\sigma_{bx} = \frac{M_u}{S_x} = \frac{15.75^{\text{kip-ft}} \times 12^{\text{in./ft}}}{192^{\text{in.}^3}} = 0.98 \text{ ksi}$$

where $S_x =$ sectional modulus of W21×93 from *AISC Manual*, Table 1-1.
Note: The beam is adequate for bending only, $\sigma_{bx} = 28.66$ ksi $< \phi F_y = 0.9 \times 50$ ksi $= 45$ ksi. In other words, the original design without considering the eccentricity-induced torsion would be safe (even a little more conservative).

4. Final check of the design when P moves to the edge of the flange

The maximum normal stress occurs in the flanges of the section at the fixed end because both torsional and bending normal stresses are at the same place:

$$f_{un} = \sigma_{ws} + \sigma_{bx} = 38.62^{\text{ksi}} + 0.98^{\text{ksi}} = 39.6^{\text{ksi}} > \phi F_y = 0.9 \times 50^{\text{ksi}} = 45 \text{ ksi}$$

The beam is safe under the combined bending and torsional moment.
Discussion: The beam was originally designed to support the cantilever with anticipated torsion moment as the major loading; thus, a proper section was selected apparently based on the torsional analysis. In other words, we can design for anticipated torsion.

6.6   Problems

6.1   For a 3.0-kip-ft pure torsional moment, compute maximum torsional stress, τ_s, and draw stress distribution for each of the sections in Fig. 6.29.

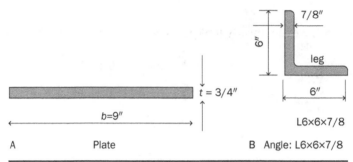

A   Plate

B   Angle: L6×6×7/8

FIGURE 6.29   (Hint: treat the angle as a plate with $b = 5'' + 5'' = 10''$ and $t = \frac{3}{4}''$).

6.2   A cantilever beam is loaded by a P (100% dead load). During the construction, P was misplaced at the edge of the top flange, resulting in an eccentricity of half of the flange. Check if the beam is adequate. Only consider normal stresses. The beam is laterally braced (no lateral-torsional buckling) (Fig. 6.30).

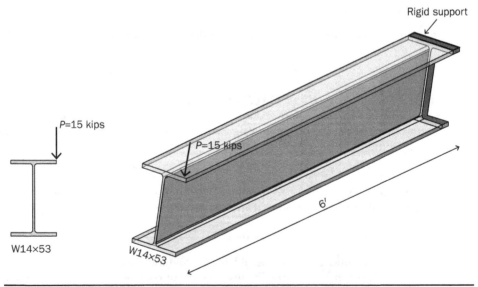

FIGURE 6.30   W14×53 cantilever beam subject to eccentric loading at the end.

6.3 The sketch in Fig. 6.31 shows the detail of a W18×97 beam rigidly connected to supporting columns for torsion and bending. The load P (30% dead and 70% live loads) is applied with 71 in. from the edge of the girder flange. Check if the beam design is adequate (only consider normal stresses). Assume that the beam is supported with fixed ends in torsion and bending, and neglect any deformation in the columns. The beam is laterally braced (no lateral-torsional buckling).

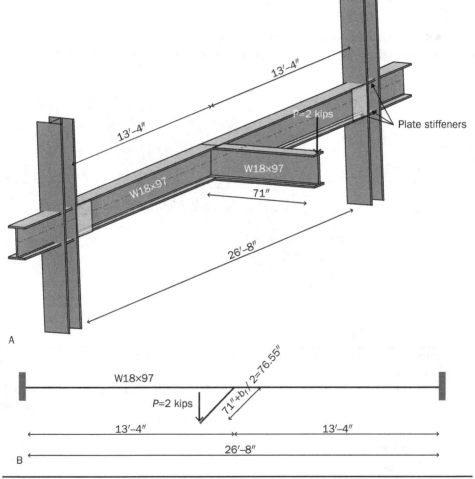

A

B

FIGURE 6.31 W18×97 beam rigidly connected beam: (*a*) loading for bending and torsion, (*b*) torsional analysis model.

6.4 Select the most efficient (the lightest) cross-section from the cross-sections in Fig. 6.32 to carry 2.5-kip-ft pure torsion (maximum shear stress should not be greater 15 ksi).

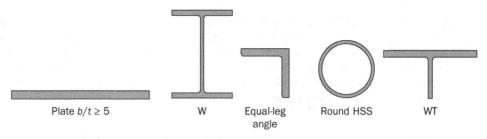

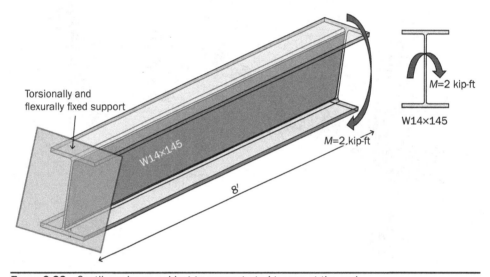

FIGURE 6.32 Various shapes under pure torsion.

6.5 Calculate shear stress and normal stress distributions over the beam cross-sections at $z = 0$ and $z = 1.5$ m for the cantilever beam in Fig. 6.33 (steel grade: A36).

FIGURE 6.33 Cantilever beam subject to concentrated torque at the end.

6.6 The drawings in Fig. 6.34 show details of a floor framing and loading of beams and girders: design beam B1 and girder G1 using A36 steel. After the design, calculate the rotation of the cross-section at midspan of the girder.

Dead load = 65 psf

Live load = 50 psf

Wall weight = 13 kip/in.

6.7 Write a python program that computes

    a. The torsional properties and statical moments of a W-shape

    b. Torsional stresses due to pure torsion and warping torsion

    c. Shear and normal stresses due to torsion and flexure

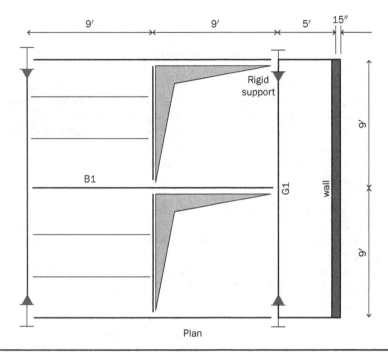

Figure 6.34 Floor plan.

6.8 A tee-shaped girder is subjected to service load P (20% dead load and 80% live load) (steel grade A36; Fig. 6.35).

    a. Calculate the maximum value of P.

    b. For the value of P calculated in Part (a) calculate the displacement of point A in the vertical direction.

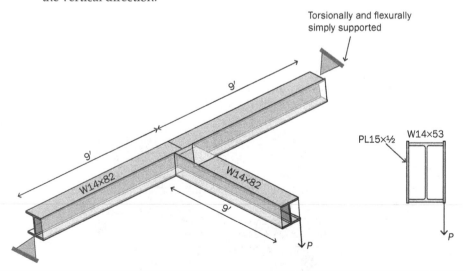

Figure 6.35 Simply supported beam subject to torsion at midspan.

Bibliography

Aghayere, A. O., and J. Vigil, *Structural Steel Design: A Practice Oriented Approach*, Pearson Education, Inc., 2015.

AISC, *Seismic Design Manual*, American Institute of Steel Construction, Chicago, IL, 2018.

AISC Steel Design Guide Series 9, *Torsional Analysis of Structural Steel Members*, American Institute of Steel Construction (AISC), Chicago, IL, 2003.

AISC 360, *Specification for Structural Steel Buildings*, ANSI/AISC Standard 360-16, American Institute of Steel Construction, Chicago, IL, 2016.

AISC Manual, *Steel Construction Manual*, 15th ed., American Institute of Steel Construction, Chicago, IL, 2016.

ASCE 7, *Minimum Design Loads for Buildings and Other Structures*, ASCE/SEI 7-16, American Society of Civil Engineers, Reston, VA, 2016.

Salmon, C. G., J. E. Johnson, and F. A. Malhas, *Steel Structures: Design and Behavior—Emphasizing Load and Resistance Factor Design*, 5th ed., Pearson Education, Inc., New Jersey, 2009.

Shen, J., B. Akbas, O. Seker, and C. Carter, *Structural Engineering Handbook—Chapter 8: Design of Structural Steel Members*, 5th ed., McGraw-Hill, New York, 2020.

CHAPTER 7

Beam-Columns

7.1 Introduction

In real structures, there is no single member subjected to only axial load or bending moment, but both. In most practical situations, the effect of either one of them is neglected, and the member is designed as a beam or an axially loaded column or tension member. If the effect of both axial load and bending moment cannot be neglected, the element is called a *beam-column* member that is a structural member subjected to combined bending moment and axial force. As such, the rules for laterally supported beams and lateral-torsional buckling (LTB) of beams set forth in previous chapters apply for beam-columns.

In general, bending moments about the x-axis (strong axis) and the y-axis (weak axis), and axial force are considered in the beam-column design, as shown in Fig. 7.1. A beam-column member has a general member form of beam or column when some components (P, M_x, M_y) become negligible in comparison to a dominant one. For example, it becomes a column, when $M_x = 0$, $M_y = 0$, and it becomes a beam bent about its strong axis when $P = 0$, $M_y = 0$. Note that M_y will be equal to zero in engineering practice because a moment connection to a W-section about the weak axis is rarely used. Interaction between P and M is the main character of the beam-column member, which affects both sides of the design equation, capacity, and demand.

7.2 Demand and Capacity Evaluation on Beam-Columns

7.2.1 Capacity Side

As shown in Fig. 7.2, the bending moment capacity reduces significantly as P increases.

7.2.2 Demand Side (P-δ Effect)

Demand comes from the structural analysis of a numerical model (Figs. 7.3 and 7.4). P-δ effect considers element deformation on the internal forces.

$$M_r = \underbrace{\frac{ql^2}{8}}_{M_0} + \underbrace{P\delta_{max}}_{M^*} = M_0 + M^* = \left(1 + \frac{M^*}{M_0}\right)M_0 = B_1 M_0 \tag{7.1}$$

where

> $M_r = M_u$ = required flexural strength when q and P are considered, kip-in.
> M_0 = bending moment due to lateral loads (q only), and calculated by first-order analysis, kip-in.
> M^* = additional bending moment due to P-δ interaction, kip-in.

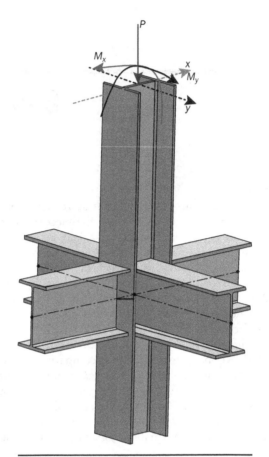

FIGURE 7.1 A column subjected to axial force P, and bi-axial bending moments M_x and M_y.

$B_1 = 1 + \dfrac{M^*}{M_0}$, a factor to consider the magnifying effect of P-δ interaction on M_0.

Note that δ_{max} is caused by both lateral load q and axial load P in an interactive fashion. Let us imagine a possible loading-deformation process:

a. The member is first loaded with lateral load q only, and has its primary bending moment and deformation as shown in Fig. 7.3a.

b. The axial force P is applied at the end of primary deformation, and creates additional bending moment M_0, which in turn increases the deformation from δ_0 to δ_{max} (Fig. 7.4b).

Note that δ_{max} depends on both q and P, and the exact calculation of M_r with P-δ effect/interaction includes lengthy solution even for an elastic beam-column, which is of little application in the actual design. However, such amplification of design moment due to P-δ effect might cause problems and should be accounted for, at least, approximately.

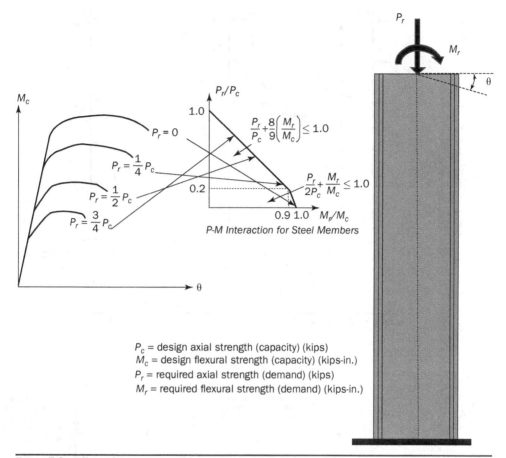

P_c = design axial strength (capacity) (kips)
M_c = design flexural strength (capacity) (kips-in.)
P_r = required axial strength (demand) (kips)
M_r = required flexural strength (demand) (kips-in.)

FIGURE 7.2   Effect of *P-M* interaction on bending strength.

7.2.3 Demand Side (*P-Δ* Effect)

With axial force present, the bending moment becomes larger and larger as the lateral deformation increases (Fig. 7.5).

$$M_r = FL \qquad\qquad \text{first-order moment} \qquad\qquad (7.2a)$$

$$M_r = FL + P\Delta = B_2(FL) \quad \text{second-order moment} \qquad (7.2b)$$

A short reminder of the first-order and second-order structural analysis would be helpful at this point. A first-order linear elastic structural analysis is based on a simple relationship given in Eq. (7.3a). The external forces $\{F\}$ acting at the joints of a structure should be equal to the multiplication of the structure's stiffness matrix, $[K]$, and the joint deformations $\{u\}$. However, in a second-order analysis, the geometric nonlinearity matrix, $[K_g]$, should be added to the structure's stiffness matrix $[K]$ [Eq. (7.3b)]. It should be noted that the terms of the $[K_g]$ involve axial forces developed within the

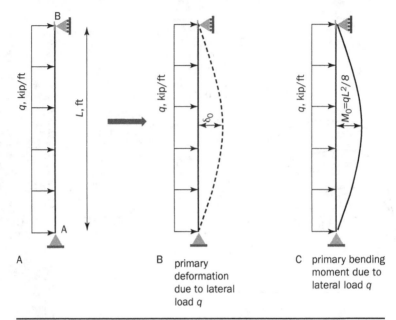

A

B primary
deformation
due to lateral
load q

C primary bending
moment due to
lateral load q

FIGURE 7.3 A Simply supported column subject to lateral force q.

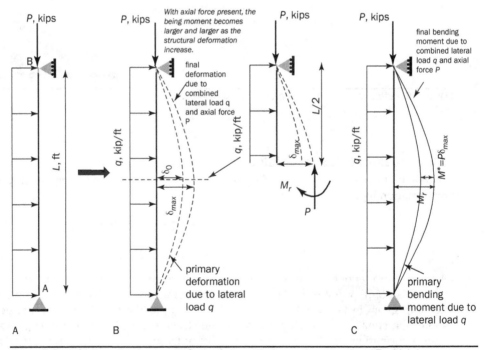

A

B

C

FIGURE 7.4 $P\text{-}\delta$ interaction increases flexural strength demand on member AB which has no lateral translation (nt).

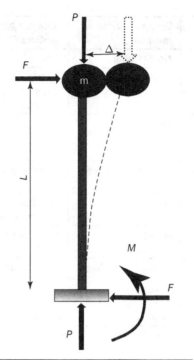

FIGURE 7.5 A cantilever column subject to vertical and horizontal loads.

element. Thus, to carry out a second-order analysis, axial forces in the elements should be known.

$$\{F\} = [K]\{u\} \tag{7.3a}$$

$$\{F\} = \{[K] + [K_g]\}\{u\} \tag{7.3b}$$

7.3 Design for Stability

Beam-column members are primarily used in a moment frame (MF). In general, moment frames are more flexible than other commonly used structural systems such as braced frames and shear walls. A building with MF as its lateral load–resisting system tends to have lateral displacement significantly enough to lead to its lateral instability. To provide stability for buildings with moment frames, any moment frame as a whole and all of its elements shall be designed considering the destabilizing impact of lateral displacement in the moment frame, as follows:

a. Flexural, shear, and axial member deformations, and all other component and connections deformations that contribute to the displacements of the structure

b. Second-order effects (including P-Δ and P-δ effects)

	Direct analysis method	Moment amplification method	First-order analysis
Limitations	No limitation	Yes	Yes
First-order or second-order analysis	Second-order	Approximate second-order	First-order
Computer-aided analysis for second-order analysis	Required	–	–
Nominal or reduced member stiffness	Reduced EI and EA	Nominal EI and EA	Nominal EI and EA
Notional loads	Yes	Only for gravity-only load combinations	As an additional lateral load
Effective length factor, K	$K = 1$	Alignment charts are used	$K = 1$
Geometric nonlinearities, imperfections, and inelasticity	Directly considered	Not considered directly	Not considered directly

TABLE 7.1 Comparison of the Stability Design Methods

    c. Geometric imperfections

    d. Stiffness reductions due to inelasticity, including the effect of partial yielding of the cross-section which may be accentuated by the presence of residual stresses

    e. Uncertainty in system, member, and connection strength and stiffness

Any rational method of design for stability that considers all of the listed effects is permitted. AISC 360, Chapter C offers three methods for the design for stability (Table 7.1):

    1. Direct analysis method

This method is intended to be used for a computer-based design process. It has to be used together with a practical design method for conceptual and preliminary design.

    2. Moment amplification method [or approximate second-order analysis (or effective length method)]

This method has been used since the 1960s. This method can be used alone to design moment frames in typical buildings and can be used together with the direct analysis method for large-scale structures where computer-aided design becomes necessary.

    3. First-order analysis method

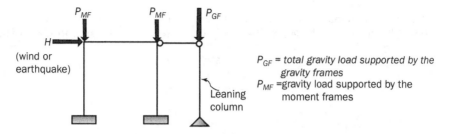

FIGURE 7.6 Numerical model incorporating leaning column.

7.3.1 Direct Analysis Method

Direct analysis method (DAM) of design requires structural engineering software and is becoming a standard method for stability analysis among structural engineers. The main characteristics of the direct analysis method are as follows:

1. No limitations on use of this method.

2. Computer-aided analysis is required to perform a second-order (P-Δ) analysis to obtain the second-order moments and axial loads.

3. Geometric nonlinearities, imperfections, and inelasticity are all taken into account.

4. All axial loads on any gravity-only columns should be included in the structural model.

5. Notional lateral load (a percentage of the gravity loads on the frame), N_i, is applied to the structural model to consider the geometric imperfections and should be added to the only gravity-only load combinations except the case where the ratio of the maximum second-order drift to maximum first-order drift ($\Delta_{2nd}/\Delta_{1st}$) exceeds 1.7 in which notional loads should also be added to the combinations that include other lateral loads (earthquake and wind).

6. Reduced EI and EA are used to account for inelastic behavior. Stiffness reduction factor, τ_b, is used to adjust the flexural stiffness of all members.

To apply the DAM properly, the following steps are followed:

Step 1: A realistic model of the lateral force-resisting system with gravity-only columns (or leaning columns) should be constructed to introduce the impact of gravity frames on second-order effects. A leaning column consisting of rigid truss elements is incorporated in the numerical model. In any stability analysis, it is necessary to capture the destabilizing effect of columns that rely on the lateral frame for stability but are not a part of the lateral frame. These columns with pinned ends are commonly referred to as "leaning columns" (Fig. 7.6).

Step 2: The stiffness of the lateral force-resisting system should be reduced.
Flexural and axial stiffness should be reduced by 20%, assuming $\tau_b = 1.0$ [Eq. (7.4b)]. This assumption needs to be checked after determining the actual τ_b based on the level

of axial force demand. An additional factor, τ_b, is to be introduced for flexural stiffness $(0.8\tau_b EI)$ adjustments based on the level of axial force-to-capacity ratio [Eq. (7.5)].

$$\text{Axial stiffness: } 0.8EA \tag{7.4a}$$

$$\text{Flexural stiffness: } 0.8EI \tag{7.4b}$$

$$\tau_b = 1 \text{ when } \frac{P_r}{P_y} \leq 0.5 \tag{7.5a}$$

$$\tau_b = 4\left(\frac{P_r}{P_y}\right)\left[1 - \left(\frac{P_r}{P_y}\right)\right] \text{ when } \frac{P_r}{P_y} > 0.5 \tag{7.5b}$$

where

P_r = required compressive strength, kips
P_y = yield strength, kips

Step 3: Notional loads as a percentage of gravity loads or notional displacements to account for initial imperfections should be applied.

The notional load, N_i, is taken as 0.002 times the total gravity load at each story level, which is based on an initial story out-of-straightness of 1/500 for all stories, i.e., 0.002 coefficient is the permitted erection tolerance (Fig. 7.7).

$$N_i = 0.002Y_i \tag{7.6}$$

where

Y_i = the total gravity load applied at story level *i* including the loads supported by gravity frames, kips

AISC suggests that notional loads need to be added to load combinations in which the notional load is larger than the lateral load in the frames. Thus, notional loads usually can be ignored in all but the gravity-only load combinations. However, if a designer wishes to simplify the design process, it is always conservative to include notional loads in all load combinations. The load combinations, including notional load effects, are given in Table 7.2.

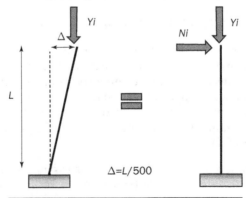

FIGURE 7.7 Notional load.

Step 4: A second-order analysis that takes P-Δ and P-δ effects into account should be conducted. Capacities of the structural members should be determined assuming $K = 1.0$. The ratio of the second-order drift to first-order drifts $(\Delta_{2nd}/\Delta_{1st})$ should be checked.

AISC 360 requires that $\Delta_{2nd}/\Delta_{1st}$ ratio is to be examined when determining the load combinations in which the notional loads are to be included. If $\Delta_{2nd}/\Delta_{1st}$ ratio is less than 1.7 for all combinations in all stories, it is not required to include the notional loads in the combinations with other lateral loads (Fig. 7.8).

Step 5: Capacities of the structural members should be determined assuming $K = 1.0$ and adequacy of the members should be checked (see Sec. 7.6). It should be noted that reduced stiffness is only used in strength analyses. For serviceability checks listed below, nominal stiffness of the elements should be used:

- Drift limits for wind and seismic loads
- Determination of building periods
- Vibration control

Combination	Description
Comb#1	$1.4D \pm (1.4ND)$
Comb#2	$1.2D + 1.6L + 0.5L_r \pm (1.2ND + 1.6NL + 0.5NL_r)$
Comb#3	$1.2D + 1.0L + 1.6L_r \pm (1.2ND + 1.0NL + 1.6NL_r)$
Comb#4	$1.2D + 1.6L_r \pm 0.5W$
Comb#5	$1.2D \pm 1.0W + L + 0.5L_r$
Comb#6	$0.9D \pm 1.0W$

ND = notional dead load, NL = notional live load, NL_r = notional roof live load.

TABLE 7.2 Load Combinations Including Notional Load Effects

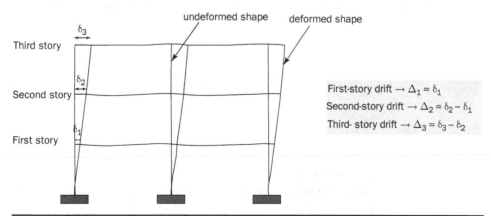

FIGURE 7.8 Lateral displacements of a three-story moment frame due to lateral loads (wind or earthquake).

7.3.2 Moment Amplification Method

This is the most commonly used method to consider second-order effects on members.

1. An alternative method to second-order analysis.
2. Nominal EI and EA are used in the analysis.

This method is covered in detail in the following section.

7.3.3 First-Order Analysis Method

This method is not very common due to its limitations which are as follows:

1. Second-order analysis is not required.
2. Nominal EI and EA are used in the analysis.
3. $K = 1.0$ for all members.
4. The factored axial compression loads should not be greater than 50% of the yield strength for all members whose flexural stiffness contributes to the lateral stiffness of the frames.

7.4 Approximate Second-Order (Amplified First-Order) Analysis Method (Moment Amplification Method)

7.4.1 Moment Amplifiers, B_1 and B_2, and Required Design Moment

For a conventional design, an approximate and simple approach would be sufficient to consider the impact of P-δ and P-Δ effect on required design bending strength, M_r.

7.4.2 Beam-Column Members with No Joint Translation (nt)—Braced Frames

Beam-column members in braced frames with no lateral translation (*nt*) at their two-member ends are in this category. One such example is shown in Fig. 7.9. In this case, the two ends do not have any lateral displacement (no movement in y-direction in Fig. 7.9), and the lateral displacement of the beam-column member is restricted between its two ends.

Required second-order flexural strength, M_r, and axial strength, P_r, due to P-δ effect is given as follows:

$$M_r = B_1 M_{nt} \tag{7.7a}$$

$$P_r = P_{nt} \tag{7.7b}$$

where

$$B_1 = \frac{C_m}{1 - \dfrac{P_r}{P_{e1}}} \geq 1$$

B_1 = multiplier to account for P-δ effects

$B_1 \cong 1$ if there is no transverse load acting between the beam-column element ends

M_{nt} = first-order moment obtained from the structural analysis with no lateral translation, kip-in.

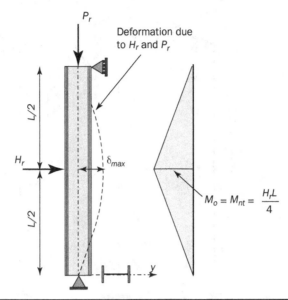

FIGURE 7.9 No lateral translation of a member at its two ends.

P_{nt} = first-order axial force obtained from the structural analysis with no lateral translation, kips

C_m = coefficient assuming no lateral translation of the frame
  = $0.6 - 0.4(M_1/M_2)$ if there is no transverse loading between supports (Fig. 7.10)
  = 1.0 if the transverse load is acting between supports (Fig. 7.10)

M_1, M_2 = smaller and larger moments at the ends of the member, respectively, and obtained from first-order analysis [M_1/M_2 is positive for reverse (double) curvature, negative for single curvature]

$P_{e1} = \dfrac{\pi^2 EI^*}{(K_1 L)^2}$ elastic critical buckling strength of the member, kips

EI^* = flexural rigidity
  = $0.8\tau_b EI$ for the direct analysis method
  = EI for effective length and first-order analysis methods

K_1 = effective length factor (= 1.0 unless a smaller value is justified by analysis)

Example 7.1

Find C_m values for columns A and B for the single-story frame in Fig. 7.11.

Solution (Fig. 7.12):

$$\dfrac{|M_1|}{|M_2|} = +0.5 \rightarrow C_m = 0.6 - 0.4(0.5) = 0.4 \text{ for column A}$$

$$\dfrac{|M_1|}{|M_2|} = -0.5 \rightarrow C_m = 0.6 - 0.4(-0.5) = 0.8 \text{ for column B}$$

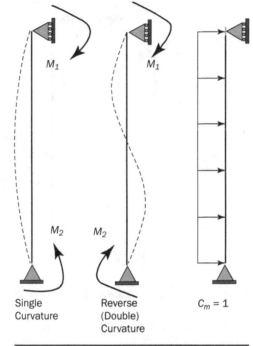

Single Curvature

Reverse (Double) Curvature

$C_m = 1$

FIGURE 7.10 C_m values.

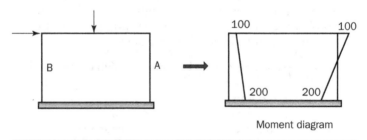

Moment diagram

FIGURE 7.11 Single-story frame subject to gravity and lateral loads, and the corresponding moment diagram.

7.4.3 Beam-Column Members with Lateral Translation (*lt*): Unbraced Frames

Required second-order flexural strength, M_r, and axial strength, P_r, due to P-δ and P-Δ effects are given as follows:

$$M_r = B_1 M_{nt} + B_2 M_{lt} \qquad (7.8a)$$

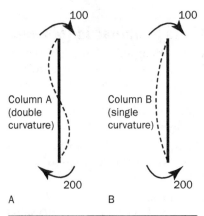

FIGURE 7.12   (a) Column A. (b) Column B.

$$P_r = P_{nt} + B_2 P_{lt} \qquad (7.8b)$$

where

M_{lt} = first-order moment obtained from the structural analysis with lateral translation, kip-in.

P_{lt} = first-order axial force obtained from the structural analysis with lateral translation, kips

P_{story} = total vertical load supported the story including the load on the gravity-only columns, kips

$P_{e\,story}$ = elastic critical buckling strength for the story (not easy to determine)

$$= R_m \frac{HL}{\Delta_H}$$

R_m = 0.85 for moment frames, or 0 for braced frames

L = height of the story, in.

Δ_H = first-order drift due to lateral forces (drift limit imposed by the specification can be used), in.

H = story shear, kips

B_2 = multiplier to account for P-Δ effects determined for each story

$$B_2 = \frac{1}{1 - \dfrac{P_{story}}{P_{e\,story}}} \geq 1 \text{ or}$$

$$B_2 = \frac{1}{1 - \dfrac{P_{story}}{R_m \dfrac{HL}{\Delta_H}}} \geq 1 \text{ this equation is more convenient to use}$$

From the definition of B_2, we need to know the column and beam sizes to determine B_2.

7.5   Design of Members in Unbraced (Moment) Frames

Application of the moment amplification method on unbraced frames is summarized in Figs. 7.13 and 7.14. Steps to determine the required flexural strength, M_r, and axial strength, P_r, due to P-δ and P-Δ effects in an unbraced frame are [Eq. (7.8)]:

1. Only apply gravity loads to the frame, and obtain bending moment (M_{nt}) and axial force (P_{nt}) by first-order structural analysis.

2. Only apply lateral loads to the frame and obtain bending moment (M_{lt}), axial force (P_{lt}), and total lateral translation (story drift) by first-order structural analysis.

3. Calculate moment amplifiers B_1 and B_2 to include P-δ and P-Δ effects on bending moment and axial force. B_1 is the same as in the braced frame.

Example 7.2

Determine B_2, moment magnification factor, for the columns on the first story of the four-story moment frame with earthquake and gravity loads given in Fig. 7.15 (gravity loads include the loads from the interior gravity frames).

Solution:

$$B_2 = \frac{1}{1 - \dfrac{P_{story}}{R_m \dfrac{HL}{\Delta_H}}} \geq 1$$

$$R_m = 0.85$$

$$\frac{\Delta_H}{L} = \frac{1}{200} \quad \text{drift limit for wind}$$

$$H = 80^{kips} + 60^{kips} + 40^{kips} + 20^{kips} = 200 \text{ kips}$$

Load combinations that include earthquake loads (ignore vertical seismic load effect) are as follows:

$$1.2D + 1.0E + L + 0.2S$$

$$0.9D + 1.0E \text{ not critical for calculating } B_2$$

$$1.2D = 1.2[3.0^{kip/ft} + 6.0^{kip/ft} + 6.0^{kip/ft} + 6.0^{kip/ft}] \times 90^{ft} = 2{,}268.0 \text{ kips}$$
$$L = [2.5^{kip/ft} + 2.5^{kip/ft} + 2.5^{kip/ft}] \times 90^{ft} = 675.0 \text{ kips}$$
$$0.2S = 0.2[2.0^{kip/ft}] \times 90^{ft} = 36.0 \text{ kips}$$

$$P_{story} = 2{,}268^{kips} + 675^{kips} + 36^{kips} = 2{,}979.0 \text{ kips}$$

$$B_2 = \frac{1}{1 - \dfrac{1.0(2{,}979^{kips})}{0.85\dfrac{200(200)}{1}}} = 1.1 < 1.5 \quad \text{O.K.}$$

Note that direct analysis method should be used for $B_2 > 1.5$.

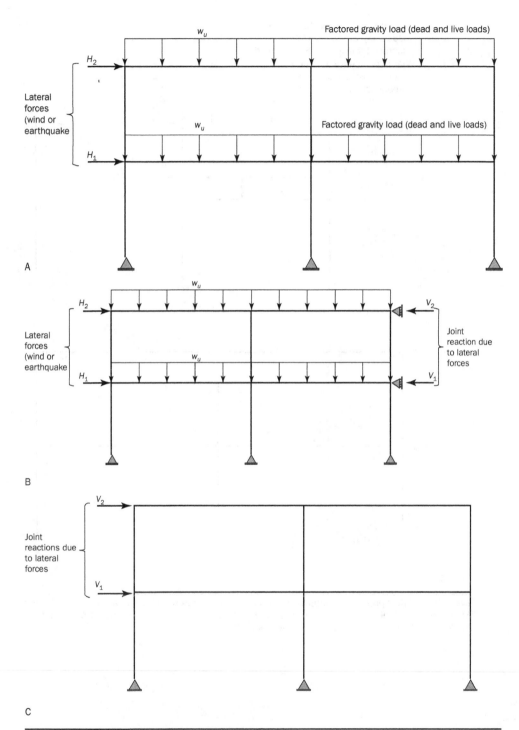

FIGURE 7.13 Application of moment amplification method on an unbraced frame: (*a*) numerical model, (*b*) modified numerical model to obtain P_{nt} and M_{nt} (M_{nt} is amplified by B_1), (*c*) modified numerical model to obtain P_{lt} and M_{lt} (both P_{lt} and M_{nt} are amplified by B_2).

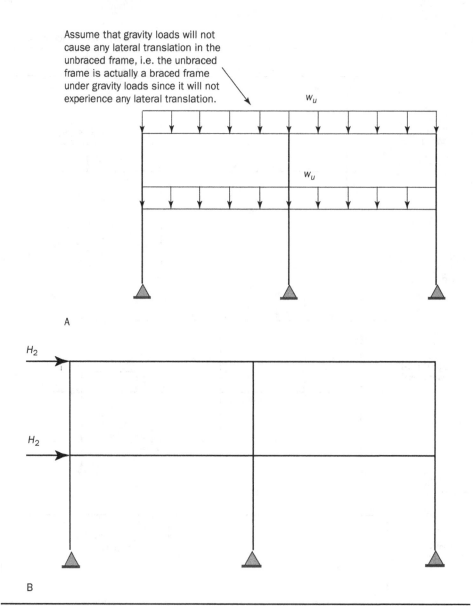

Assume that gravity loads will not cause any lateral translation in the unbraced frame, i.e. the unbraced frame is actually a braced frame under gravity loads since it will not experience any lateral translation.

w_u

w_u

A

H_2

H_2

B

Figure 7.14 An approximate alternative approach to design members in an unbraced frame: (a) Modified numerical model to obtain P_{nt} and M_{nt} (M_{nt} is amplified by B_1), (b) modified numerical model to obtain P_{lt} and M_{lt} (both P_{lt} and M_{nt} are amplified by B_2).

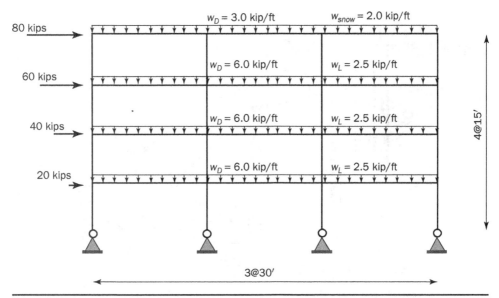

FIGURE 7.15

7.6   Design for Combined Axial Force and Flexure (AISC 360, Section H)

Strength interaction curve for beam-column design:

To consider *P-M* interaction effect on capacity of W-shapes, AISC specification pre-scribes interaction equations for compression and bending moment about both strong axis (*x-x*) and weak axis (*y-y*) (Fig. 7.16). In a moment frame, bending moment of a W-shape about the weak axis is often zero. Thus, we will only have to bend about the strong axis of the column. And the interaction equations are as follows:

a. For $\dfrac{P_r}{P_c} \geq 0.2$

$$\frac{P_r}{P_c} + \frac{8}{9}\left(\frac{M_r}{M_c}\right) \leq 1.0 \tag{7.9a}$$

b. For $\dfrac{P_r}{P_c} < 0.2$

$$\frac{P_r}{2P_c} + \frac{M_r}{M_c} \leq 1.0 \tag{7.9b}$$

where

$P_c = \phi_c P_n$
   = design axial strength, kips
$M_c = \phi_b M_n$
   = design flexural strength, kip-in.
$\phi_c = 0.90$
$\phi_b = 0.90$

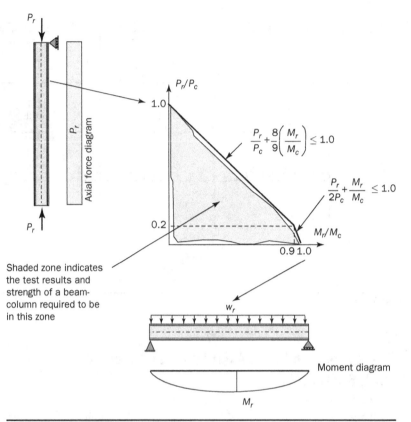

FIGURE 7.16 *P-M* interaction diagram for steel members.

7.7 Basic Steps in Beam-Column Design

A beam-column design problem is generally based on (a) the evaluation of the design adequacy or determination of maximum load, and (b) selection of a beam-column section.

Type 1: Evaluation of the Design Adequacy or Determination of Maximum Load

For Type 1 problems, W-shape, P_c, M_c, B_1, and B_2 are known and the adequacy of the design for given P_r, M_{lt}, M_{nt} or finding the maximum allowable load (P_r, M_r) is required. Solution to the problem is to use either Eq. (7.9a) or (7.9b) to check the design or find P_r and/or M_r.

Type 2: Selection of the Section

For Type 2 problems, P_r, M_{lt}, M_{nt}, etc., are given or obtained through structural analysis, and selection of the beam-column section is required. A solution to the problem consists of two steps:

> Step 1: "Guess" a trial section.
> Step 2: With the trial section, Type 1 procedure is followed to check the adequacy of the trial section. If necessary, the trial section is modified, and Steps 1 and 2 are repeated.

A reasonable initial guess of the trial section might reduce or even eliminate repeated effort. Such an initial guess appears to be less complicated if the depth of the W-shape is fixed (say to W14). Many designers might try a section based on previous design experience.

It should also be noted as a general rule of thumb whether or not LTB is prevented when:

1. $L_b < L_p$ or
2. $L_p < L_b < L_r$ and $C_b \geq 1.67$

Example 7.3

As shown in Fig. 7.17, the member is loaded with a lateral distributed load W (kip/ft) and axial force P (kips), and braced in both in-plane and out-of-plane of the bending at its ends (A and C). In addition, lateral support is provided in the weak direction at the mid-height (point B). Neglect the self-weight. Determine the maximum load W (kip/ft).

$P_L = 200$ kips, $P_D = 100$ kips

W (50%Live, 50%Dead) (kip/ft)

W12×120, $F_y = 50$ ksi, $A_g = 35.3$ in.², $Z_x = 186$ in.³, $S_x = 163$ in.³

Solution:

The member has no translation at its ends.

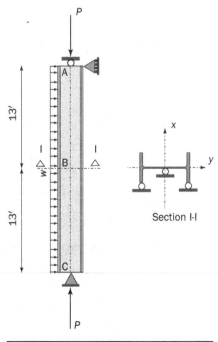

FIGURE 7.17

a. First-order analysis

Since this is a braced-frame case (with no translation), results from a first-order (linear) structural analysis include M_{nt} and P_{nt}:

$$P_{nt} = 1.2P_D + 1.6P_L = 1.2 \times 200^{kips} + 1.6 \times 100^{kips} = 400 \text{ kips}$$

$$W_r = 1.2W_D + 1.6W_L = 1.2(50\%W) + 1.6(50\%W) = 1.4W \text{ kip/ft}$$

$$M_{nt} = \frac{W_r L^2}{8} = \frac{(1.4W)(26')^2}{8} = 118.3W \text{ kip-ft}$$

$$M_r = B_1 M_{nt} \quad (B_2 = 0)$$

$$P_r = P_{nt} \quad (B_2 = 0)$$

b. Column strength

$$K_x = K_y = 1.0, L_x = 26', L_y = 13'$$

$$\text{W12}\times120, F_y = 50 \text{ ksi}, r_y = 3.13 \text{ in.}, r_x = 5.51 \text{ in.}, r_x/r_y = 1.76$$

$$\left(\frac{KL}{r}\right)_y = \frac{1.0 \times 13' \times 12^{in./ft}}{3.13''} = 49.8$$

$$\left(\frac{KL}{r}\right)_x = \frac{1.0 \times 26' \times 12^{in./ft}}{5.51''} = 56.6$$

$$\left(\frac{KL}{r}\right) = \text{the larger of } \left(\frac{KL}{r}\right)_x \text{ and } \left(\frac{KL}{r}\right)_y = \left(\frac{KL}{r}\right)_x = 56.6$$

The column buckles about the x-axis.

$$\left(\frac{KL}{r}\right)_x = 56.6 < 4.71\sqrt{\frac{E}{F_y}} = 113$$

$$F_e = \frac{\pi^2 E}{\left(\frac{KL}{r}\right)^2} = 89.3 \text{ ksi}$$

$$F_{cr} = \left[0.658^{F_y/F_e}\right]F_y = 39.6 \text{ ksi}$$

$$P_n = A_g F_{cr} = 35.3^{in.^2} \times 39.6^{ksi} = 1{,}398 \text{ kips}$$

$$P_c = \phi_c P_n = 0.9 \times 1{,}398^{kips} = 1{,}258 \text{ kips}$$

Alternative solution for column strength: Use column tables in *AISC Manual*, Table 4-1 as follows:

$$(KL)_y = 13'$$

$$\frac{(KL)_x}{(r_x/r_y)} = \frac{26}{1.76} = 14.8' \text{ Controls}$$

$$(KL) = \text{the larger of } (KL)_y = 13' \text{ and } \frac{(KL)_x}{(r_x/r_y)} = 14.8' = 14.8'$$

Enter the *AISC Manual*, Part 4, Table 4-1 for W12×120, $F_y = 50$ ksi

$$P_c = \phi P_n = 1{,}248 \text{ kips}$$

c. Beam strength

Points A, B, and C are considered lateral bracing points. $L_b = 13'$ (AB portion) (Fig. 7.18)

$$L_p = 1.76 r_y \sqrt{\frac{E}{F_y}} = 1.76 \times 3.13'' \sqrt{\frac{29{,}000^{ksi}}{50^{ksi}}} = 11.1 \text{ ft (or from } AISC \text{ Manual, Part 3, Table}$$

3-6 for $F_y = 50$ ksi)

$L_r = 56.5$ ft (from *AISC Manual*, Part 3, Table 3-6 for $F_y = 50$ ksi)

$$L_p < L_b < L_r$$

$$C_b = \frac{12.5 \times \left(\frac{1}{8} q L^2\right)}{2.5 \times \left(\frac{1}{8} q L^2\right) + 3 \times \left(\frac{7}{128} q L^2\right) + 4 \times \left(\frac{3}{32} q L^2\right) + 4 \times \left(\frac{15}{128} q L^2\right)} = 1.30 \text{ (or from}$$

Table 5.5)

$$M_n = C_b \left[M_p - \left(M_p - 0.7 F_y S_x\right)\left(\frac{L_b - L_p}{L_r - L_p}\right) \right] \le M_p = F_y Z_x$$

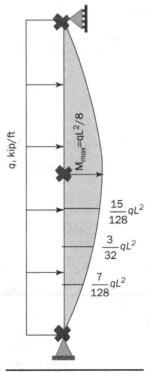

Figure 7.18 Finding C_b.

$$M_n = 1.30\left[775^{kip-ft} - \left(775^{kip-ft} - 0.7 \times \frac{50^{ksi} \times 163^{in.^3}}{12^{in./ft}}\right)\left(\frac{13' - 11.1'}{56.5' - 11.1}\right)\right]$$

$$\leq M_p = \frac{50^{ksi} \times 186^{in.^3}}{12^{in./ft}}$$

$$= 990 \text{ kip-ft} > 775 \text{ kip-ft} \rightarrow M_n = M_p = 775 \text{ kip-ft}$$

$$M_c = \phi_b M_n = 0.9(775^{kip-ft}) = 698 \text{ kip-ft}$$

d. Moment magnification factor B_1 (B_2 is not needed for the braced beam-column member)

$$P_{e1} = \frac{\pi^2 EI^*}{(K_1 L)^2} = \frac{\pi^2 EA_g}{(KL/r)_x^2} = \frac{\pi^2 \times 29,000^{ksi} \times 31.3^{in.^2}}{(1.0 \times 26' \times 12^{in./ft}/5.51'')^2} = 3,148 \text{ kips (or use } AISC \text{ } Manual,$$

Table 4-1 for W12×120, $F_y = 50$ ksi)

$C_m = 1.0$ (transverse load between two ends of the member)

$$B_1 = \frac{C_m}{1 - \dfrac{P_r}{P_{e1}}} = \frac{1.0}{1 - \dfrac{400^{kips}}{3,148^{kips}}} = 1.15 \geq 1$$

$$M_r = B_1 M_{nt} = (1.15)(118.3W) = 136W \text{ kip-ft}$$

e. Strength interaction curve

$$\frac{P_r}{P_c} = \frac{400^{kips}}{1,258^{kips}} = 0.32 > 0.2 \text{ , use Eq. (7.9a)}$$

$$\frac{P_r}{P_c} + \frac{8}{9}\left(\frac{M_r}{M_c}\right) \leq 1.0$$

$$0.32 + \frac{8}{9}\left(\frac{136W}{698^{kip-ft}}\right) \leq 1.0 \text{ (Fig. 7.19)}$$

$$W \leq 3.94 \text{ kip/ft} \rightarrow W_{max} = 3.94 \text{ kip/ft}$$

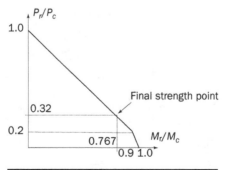

FIGURE 7.19

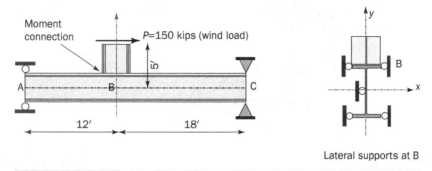

FIGURE 7.20

Example 7.4

As shown in Fig. 7.20, member ABC is simply supported at its ends. A horizontal load P due to wind is applied at the end of a W-shape member welded to the beam at B. Lateral braces are provided at A, B, and C. Select the lightest W14-section using $F_y = 50$-ksi steel. Neglect structural self-weight. Select the lightest W14-section.

Solution:

The member has no translation at points A, B, and C.

    a. First-order analysis (demand analysis)

Since this is a braced frame case (with no translation), results from a first-order (linear) structural analysis include M_{nt} and P_{nt}. The free-body diagram of member ABC is shown in Fig. 7.21. Using the basic equilibrium equations, we get the reaction forces as follows:

$$\sum F_X = 0 \rightarrow \text{horizontal reaction at C} = P\ (\leftarrow)$$
$$\sum M_A = 0 \rightarrow \text{vertical reaction at C} = P/6\ (\uparrow)$$
$$\sum M_C = 0 \rightarrow \text{vertical reaction at A} = P/6\ (\downarrow)$$

    Bending moment and axial force diagrams are shown in Fig. 7.21, which indicates clearly that the BC portion is subject to larger axial compressive force and bending moment than the AB portion (with no axial force and smaller bending moment). BC is a typical "beam-column" member with "no translation" at its two ends, B and C. The required strength for the member is

$$P_{nt} = 240 \text{ kips}$$
$$M_{nt} = 3P = 3 \times 240^{\text{kips}} = 720 \text{ kip-ft}$$

$$M_r = B_1 M_{nt}\ (B_2 = 0)$$
$$P_r = P_{nt}\ (B_2 = 0)$$

    b. Select a trial W14-section

BC is loaded with axial compression force and bending moment. A typical method to design a beam-column member is to get a trial section first by any means, and then

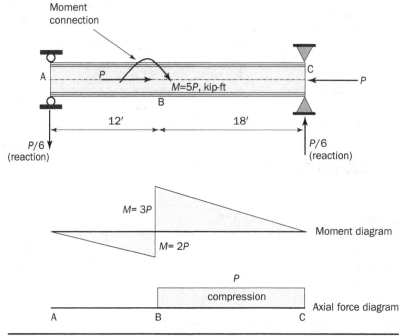

Moment connection

A

P

$M=5P$, kip-ft

B

C

P

12'   18'

$P/6$
(reaction)

$P/6$
(reaction)

$M= 3P$

Moment diagram

$M= 2P$

P

compression

Axial force diagram

A   B   C

Figure 7.21

check the trial section to see if it meets the interaction equation [Eq. (7.9)]. There are a few ways to estimate a trial section, including:

1. Assuming it is a "beam" subject to bending moment only, but with consideration of the axial force when actually selecting the trial section

2. Assuming it is a "column" subject to axial compression force only, but with consideration of the bending moment when actually selecting the trial section

3. Trying a section used in a similar structure

In any case, it will be simple if the depth of the W-shape is predetermined. For this example, let us first assume it is a "beam" only subject to $M_{nt} = 720$ kip-ft. For a quick preliminary selection, assume $M_n = M_p$ with $F_y = 50$ ksi.

$$\text{For W14}\times120 \rightarrow \phi_b M_n = 795 \text{ kip-ft} > M_r = 720 \text{ kip-ft}$$
$$\text{For W14}\times132 \rightarrow \phi_b M_n = 878 \text{ kip-ft} > M_r = 720 \text{ kip-ft}$$

It seems that W14×120 would be the lightest W14 for the bending moment requirement. With consideration of the axial force, let us try a heavier section, W14×132.

c. Check W14×132: column strength of BC

$K_x = K_y = 1.0$, $L_x = 18'$, $L_y = 18'$
W14×132, $F_y = 50$ ksi, $r_y = 3.76$ in., $r_x = 6.28$ in., $r_x/r_y = 1.67$, $S_x = 209$ in.^{3}, $Z_x = 234$ in.^{3}

Since $\left(\dfrac{KL}{r}\right)_y > \left(\dfrac{KL}{r}\right)_x$, the column buckles about y-axis.

$$\left(\frac{KL}{r}\right)_y = \frac{1.0 \times 18' \times 12^{\text{in./ft}}}{3.76''} = 57 < 4.71\sqrt{\frac{E}{F_y}} = 113$$

$$F_e = \frac{\pi^2 E}{\left(\dfrac{KL}{r}\right)^2} = \frac{\pi^2 \times 29{,}000^{\text{ksi}}}{(57)^2} = 88 \text{ ksi}$$

$$F_{cr} = \left(0.658^{F_y/F_e}\right)F_y = 39.4 \text{ ksi}$$

$$P_n = A_g F_{cr} = 38.8^{\text{in.}^2} \times 39.4^{\text{ksi}} = 1{,}530 \text{ kips}$$

$$P_c = \phi_c P_n = 0.9 \times 1{,}530^{\text{kips}} = 1{,}380 \text{ kips}$$

Alternative solution for column strength: Use column tables in *AISC Manual*, Table 4-1 as follows:

$$(KL)_y = 18'$$

$$\frac{(KL)_x}{(r_x/r_y)} = \frac{18'}{1.67} = 11'$$

$(KL) =$ the larger of $(KL)_y = 18'$ and $\dfrac{(KL)_x}{(r_x/r_y)} = 11' = 11'$

Enter the *AISC Manual*, Part 4, Table 4-1 for W14×132, $F_y = 50$ ksi

$$P_c = \phi P_n = 1{,}370 \text{ kips}$$

d. Beam strength: W14×132

$$L_b = 18'$$

From *AISC Manual*, Part 3, Table 3-6 for $F_y = 50$ ksi, $L_p = 13.3$ ft, $L_r = 55.8$ ft:

$$L_p < L_b < L_r$$

$$C_b = \frac{12.5M}{2.5M + 3(0.75M) + 4(0.5M) + 4(0.25M)} = 1.67$$

$$M_n = C_b\left[M_p - (M_p - 0.7F_y S_x)\left(\frac{L_b - L_p}{L_r - L_p}\right)\right] \le M_p = F_y Z_x$$

$$M_n = 1.67\left[975^{\text{kip-ft}} - \left(975^{\text{kip-ft}} - 0.7 \times \frac{50^{\text{ksi}} \times 209^{\text{in.}^3}}{12^{\text{in./ft}}}\right)\left(\frac{18' - 13.3'}{55.8' - 13.3'}\right)\right]$$

$$\le M_p = \frac{50^{\text{ksi}} \times 234^{\text{in.}^3}}{12^{\text{in./ft}}}$$

$$= 1{,}561 \text{ kip-ft} > 975 \text{ kip-ft} \rightarrow M_n = M_p = 975 \text{ kip-ft}$$

$$M_c = \phi_b M_n = 0.9(975^{\text{kip-ft}}) = 878 \text{ kip-ft}$$

e. Moment magnification factor B_1 (B_2 is not needed for the beam-column member without lateral translation)

$$P_{e1} = \frac{\pi^2 EI^*}{(K_1 L)^2} = \frac{\pi^2 EA_g}{(KL/r)_x^2} = \frac{\pi^2 \times 29{,}000^{ksi} \times 38.8^{in.^2}}{\left(1.0 \times 18' \times 12^{in./ft} / 6.28''\right)^2} = 9{,}380 \text{ kips (or use } AISC$$

Manual, Table 4-1 for W14×132, $F_y = 50$ ksi)

The member without lateral (transverse) loads between its two ends, B and C:

$$C_m = 0.6 - 0.4\left(\frac{M_1}{M_2}\right) = 0.6$$

where

$$M_1 = 0 < M_2 = 3P$$

$$B_1 = \frac{C_m}{1 - \dfrac{P_r}{P_{e1}}} = \frac{0.6}{1 - \dfrac{240^{kips}}{9{,}980^{kips}}} = 0.61 < 1 \rightarrow B_1 = 1.0$$

$$M_r = B_1 M_{nt} = (1.0)(720^{kip\text{-}ft}) = 720 \text{ kip-ft}$$

f. Strength interaction curve

$$\frac{P_r}{P_c} = \frac{240^{kips}}{1{,}370^{kips}} = 0.175 < 0.2 \text{ , use Eq. (7.9b)}$$

$$\frac{P_r}{2P_c} + \left(\frac{M_r}{M_c}\right) \le 1.0$$

$$\frac{0.318}{2} + \left(\frac{720^{kip\text{-}ft}}{878^{kip\text{-}ft}}\right) = 0.088 + 0.820 = 0.91 \le 1.0 \text{ (Fig. 7.22)}$$

Use W14×132, $F_y = 50$ ksi

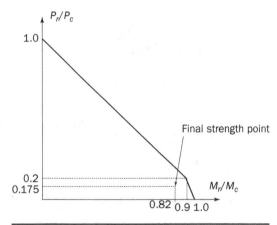

FIGURE 7.22

Example 7.5

A steel frame is shown in Fig. 7.23. A992 ($F_y = 50$ ksi) steel is used. Dead load P is applied at the top of two columns. The wind load $H = 0.5P$ is applied at both sides of the roof level. The frame is laterally braced at all joints, the middle beam span, and supports. The beam-to-column connections are rigid connections. Determine the maximum service load P. In this problem, assume $B_1 = 1.0$ and $B_2 = 1.3$.

Solution:

a. Analysis for M_r and P_r

$$M_r = B_1 M_{nt} + B_2 M_{lt}$$
$$P_r = P_{nt} + B_2 P_{lt}$$

Case 1: No translation (dead load) (Fig. 7.24)
Note: M_{nt} is zero.

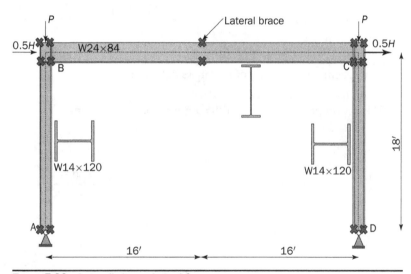

FIGURE 7.23 Laterally braced steel frame.

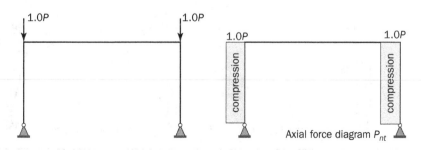

FIGURE 7.24 Dead loads and corresponding axial force diagram

Case 2: Lateral translation (wind load) (Fig. 7.25)

Member AB (beam-column): Load combination of $1.2D + 1.0W + L$ governs.

$$M_{nt} = 0$$
$$M_{lt} = 1.2M_D + 1.0M_W + L = 1.0 \times 9P$$
$$= 9P \text{ kip-ft}$$
$$P_{nt} = 1.2P_D + 1.0P_W + P_L = 1.2 \times (1.0P)$$
$$= 1.2P \text{ kips}$$
$$P_{lt} = 1.2P_D + 1.0P_W + P_L = 1.0 \times (9P / 16)$$
$$= 9P / 16 \text{ kips}$$

$$M_r = B_1M_{nt} + B_2M_{lt} = B_2M_{lt} = 1.3 \times 9P$$
$$= 11.7P \text{ kip-ft}$$
$$P_r = P_{nt} + B_2P_{lt} = 1.2P + 1.3 \times 9P / 16$$
$$= 1.93125P \text{ kips}$$

Member BC (beam): Load combination of $1.2D + 1.0W + L$ governs.

$$M_{nt} = 0$$
$$P_{nt} = 0$$
$$P_{lt} = 0$$
$$M_{lt} = 1.2M_D + 1.0M_W + L = 1.0 \times 9P$$
$$= 9P \text{ kip-ft}$$

$$M_r = B_1M_{nt} + B_2M_{lt} = B_2M_{lt} = 1.3 \times 9P$$
$$= 11.7P \text{ kip-ft}$$

   b. Capacities (M_c and P_c) of members AB and BC

Member AB (beam-column)—W14×120:

$F_y = 50$ ksi (Compact section for both compression and bending)

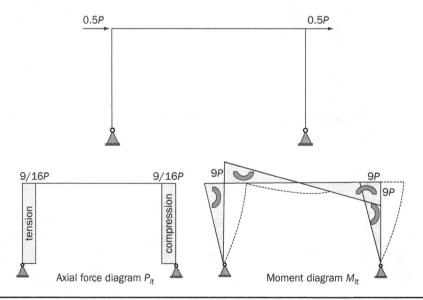

FIGURE 7.25 Lateral loads and corresponding axial force and moment diagrams.

- Column capacity

$$G_A = 10 \text{ (Pin)}$$

$$\left. G_B = \frac{\sum (E_c I_c / L_c)}{\sum (E_b I_b / L_b)} = \frac{1380 / 18}{2370 / 32} = 1.04 \right\} K_x = 1.90 \text{ (Fig. C-A-7.2, Page 16.1-513)}$$

Method 1: Use *AISC Manual*, Table 4-1 (for $F_y = 50$ ksi)

$$K_x \cong 1.90 \quad K_y = 1.0 \quad r_x = 6.24 \text{ in.} \quad r_y = 3.74 \text{ in.}$$

$$K_x / K_y = 1.9 > r_x / r_y = 1.67$$

$\rightarrow$ Column will buckle about x-axis

$$KL = \text{the larger of } \left\{ (KL)_y, \ (KL)_x / \left(r_x / r_y \right) \right\}$$

$$KL = (KL)_x = \frac{K_x L_x}{r_x / r_y} = \frac{1.9 \times 18'}{1.67} = 20.48 \text{ ft}$$

Enter Table 4-1 for W14×120 with $KL = 20.48$ ft $\rightarrow \phi_c P_n = 1{,}160$ kips

Method 2: Use formulas given by AISC 360 (a general method for any steel)

$$(KL/r)_x = \frac{1.9 \times (18' \times 12^{\text{in./ft}})}{6.24^{\text{in.}}} = 65.77 > (KL/r)_y = \frac{1.0 \times (18' \times 12^{\text{in./ft}})}{3.74^{\text{in.}}} = 57.75$$

$$KL = \text{the larger of } \left\{ (KL)_y, \ (KL)_x / \left(r_x / r_y \right) \right\}$$

$$KL = (KL/r)_x = 65.77 < 4.71\sqrt{E/F_y} = 113 \text{ (Inelastic Buckling, Use Eq. E3-2)}$$

$$F_e = \frac{\pi^2 E}{(KL / r)_x^2} = \frac{\pi^2 \times 29{,}000^{\text{ksi}}}{65.77^2} = 66.17 \ ksi$$

$$F_{cr} = (0.658^{F_y / F_e}) F_y = (0.658^{50^{\text{ksi}} / 66.17^{\text{ksi}}}) \times 50^{\text{ksi}} = 36.44 \ ksi$$

$$P_n = A_g F_{cr} = 35.3^{\text{in.}^2} \times 36.44^{\text{ksi}} = 1{,}286.4 \text{ kips}$$

$$P_c = \phi P_n = 0.9 \times 1286.4^{\text{kips}} = 1{,}158 \text{ kips (Round-off error)}$$

- Beam capacity

$L_p = 13.2$ ft $< L_b = 18$ ft $< L_r = 51.9$ ft Inelastic LTB might be possible.

Since $L_p < L_b < L_r$ and $C_b = 1.67 \rightarrow M_n = M_p = Z_x F_y = 212^{\text{in.}^3} \times 50^{\text{ksi}} / 12^{\text{in./ft}} = 883$ kip-ft

$M_c = \phi M_n = 0.9 \times 883^{\text{kip-ft}} = 795$ kip-ft

- Check *P-M* interaction

$$M_r = B_1 M_{nt} + B_2 M_{lt} = B_2 M_{lt} = 1.3 \times 9P = 11.7P \text{ kip-ft} \quad \left| \ M_c = 795 \text{ kip-ft} \right.$$

$$P_r = P_{nt} + B_2 P_{lt} = 1.2P + 1.3 \times 9P / 16 = 1.93125P \text{ kips} \quad \left| \ P_c = 1{,}160 \text{ kips} \right.$$

Assume $\dfrac{P_r}{P_c} < 0.2; \ \dfrac{P_r}{2P_c} + \dfrac{M_r}{M_c} \leq 1.0 \Rightarrow \dfrac{1.93125P}{2 \times 1160} + \dfrac{11.7P}{795} \leq 1.0 \Rightarrow P_{\text{max, AB}} = 64.31 \text{ kips}$

Check the assumption;

$$\frac{P_r}{P_c} = \frac{1.93125 \times 64.31}{1160} = 0.107 < 0.2 \rightarrow \text{The assumption is correct!}$$

Member BC (beam)—W24×84:

$F_y = 50$ ksi (Compact section for bending)

$$M_r = B_1 M_{nt} + B_2 M_{lt} = B_2 M_{lt} = 1.3 \times 9P = 11.7P \text{ kip-ft}$$

$L_p = 6.89$ ft. $< L_b = 16$ ft. $< L_r = 20.3$ ft. Inelastic LTB might be possible.
Since $C_b = 1.67 \rightarrow M_n = M_p = 933$ kip-ft

$$M_c \geq M_r$$
$$\Rightarrow M_c = \phi M_n = 0.9 \times 933 = 840 \text{ kip-ft} \geq M_r = 11.7P \text{ kip-ft}$$
$$P_{max,BC} = 71.8 \text{ kips}$$

- $P_{max} = (P_{max,AB}; P_{max,BC})_{min} = 64.31$ kips

Example 7.6

Figure 7.26 shows a 30-ft-long vertical member AC supporting a horizontal cantilever member BD. P has an 80% dead load and 20% live load. $F_y = 50$-ksi steel is used.

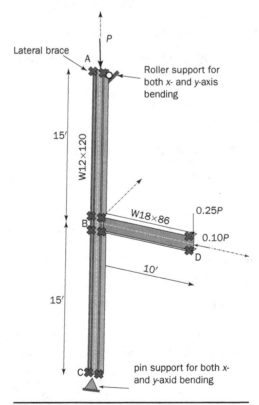

FIGURE 7.26

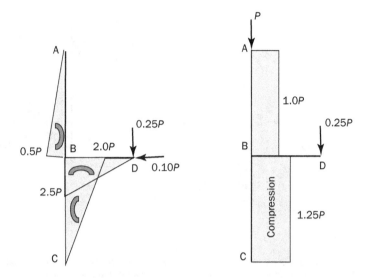

Bending moment diagram, M_D, of AC for M_{nt} (kip-ft)
(drawn on compression side)

Axial force diagram, P_D, of AC for P_{nt} (kips)

Determine the maximum load P. Assume $B_1 = 1.0$ and $B_2 = 1.0$ for member AC, and $B_1 = 1.0$ and $B_2 = 1.2$ for member BD.

Solution:
   a. Analysis for M_r and P_r

$$M_r = B_1M_{nt} + B_2M_{lt}$$
$$P_r = P_{nt} + B_2P_{lt}$$

Member AC has no translation at A and C. $M_{lt} = 0$ and $P_{lt} = 0.0$

$$M_r = B_1M_{nt} + B_2M_{lt} = B_1M_{nt}$$
$$P_r = P_{nt} + B_2P_{lt} = P_{nt}$$

Member AC: Load combination of $1.2D + 1.6L$ governs for 80% dead load, 20% live load (Fig. 7.27).
   (BC has a larger moment and axial force)

$$M_{nt} = 1.2M_D + 1.6M_L = 1.2 \times (0.8 \times 2P) + 1.6 \times (0.2 \times 2P) = 2.56P \text{ kip-ft}$$

$$P_{nt} = 1.2P_D + 1.6P_L = 1.2 \times (0.8 \times 1.25P) + 1.6 \times (0.2 \times 1.25P) = 1.6P \text{ kips}$$

$$M_r = B_1M_{nt} + B_2M_{lt} = B_1M_{nt} = 1.0 \times 2.56P = 2.56P \text{ kip-ft}$$

$$P_r = P_{nt} + B_2P_{lt} = P_{nt} = 1.6P \text{ kips}$$

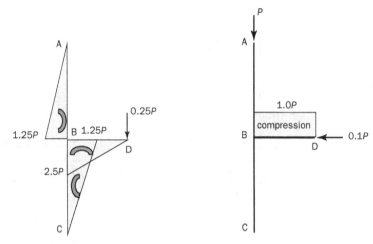

Bending moment diagram, M_D, of BD with lateral translation for M_{lt} (kip-ft) ($P_{lt} = 0$ in BD)

Axial force diagram, P_D, of BD under 0.1P without lateral translation for P_{nt} (kips) ($M_{nt} = 0$)

FIGURE 7.28

Member BD has a lateral translation at D. We can look at two loads separately. Case 1: Only 0.25P creating lateral translation $M_{nt} = 0$ and $P_{lt} = 0.0$

$$M_r = B_1 M_{nt} + B_2 M_{lt} = B_1 M_{nt}$$
$$P_r = P_{nt} + B_2 P_{lt} = P_{nt}$$

Member BD: Load combination of $1.2D + 1.6L$ governs for 80% dead load, 20% live load (Fig. 7.28).

$M_{nt} = 0.0$

$M_{lt} = 1.2 M_D + 1.6 M_L = 1.2 \times (0.8 \times 2.5P) + 1.6 \times (0.2 \times 2.5P) = 3.2P$ kip-ft

$P_{nt} = 1.2 P_D + 1.6 P_L = 1.2 \times (0.8 \times 0.1P) + 1.6 \times (0.2 \times 0.1P) = 0.128P$ kips

$P_{lt} = 0.0$

$M_r = B_1 M_{nt} + B_2 M_{lt} = B_2 M_{lt} = 1.2 \times 3.2P = 3.84P$ kip-ft

$P_r = P_{nt} + B_2 P_{lt} = P_{nt} = 0.128P$ kips

b. Capacities (M_c and P_c) of members AC and BD

Member AC (W12×120, $F_y = 50$ ksi):

W12×120, $F_y = 50$ ksi is a compact section; thus, there is no local buckling.

- *Column capacity, P_c*

$$(KL)_x = 30 \text{ ft} \quad (KL)_y = 15 \text{ ft} \quad r_x = 5.51 \text{ in.} \quad r_y = 3.13 \text{ in.}$$
$$(KL)_x / (KL)_y = 2.0 > r_x / r_y = 1.76$$
$$\rightarrow \text{Column will buckle about } x\text{-axis (Fig. 7.29)}$$

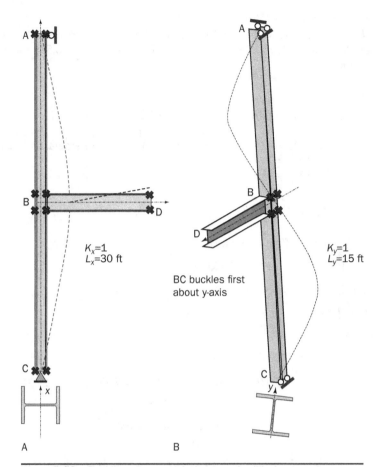

FIGURE 7.29 Buckling shapes: (*a*) about *x*-axis, (*b*) about *y*-axis.

Method 1: Use *AISC Manual*, Table 4-1 (only when $F_y = 50$ ksi)

Enter Table 4-1 for W12×106 with $KL = $ the larger of $\{(KL)_y, (KL)_x/(r_x / r_y)\}$

$KL = \{15 \text{ ft}, 17.05 \text{ ft}\} = 17.05 \text{ ft} \rightarrow \phi_c P_n = 1{,}158$ kips (Using Linear Interpolation)

Method 2: Use formulas given by AISC 360 (a general method for any F_y steel)

$$(KL/r)_{max} = (KL/r)_x = \frac{30' \times 12^{in./ft}}{5.51^{in.}} = 65.3 < 4.71\sqrt{E/F_y} = 113 \text{ Inelastic Buckling}$$

$$\rightarrow \text{ Use AISC 360-10, Eq. E3-2}$$

$$F_e = \frac{\pi^2 E}{(KL/r)_y^2} = \frac{\pi^2 \times 29{,}000^{ksi}}{65.3^2} = 67.12 \text{ ksi}$$

$$F_{cr} = (0.658^{F_y/F_e})F_y = (0.658^{50^{ksi}/67.12^{ksi}}) \times 50^{ksi} = 36.6 \text{ ksi}$$

$$P_n = A_g F_{cr} = 35.2^{in.^2} \times 36.6^{ksi} = 1{,}288 \text{ kips}$$

$$P_c = \phi P_n = 0.9 \times 1288 = 1159 \text{ kips}$$

- *Beam capacity, M_c*

Since $F_y = 50$ ksi, *AISC Manual*, Table 3-2 can be used to determine L_p and L_r values. *AISC Manual*, Table 3-2: W12×120 ➔ $L_p = 11.1$ ft. $< L_b = 15$ ft. $< L_r = 56.5$ ft. Inelastic LTB might be possible.

When

$$L_p < L_b \leq L_r \Rightarrow \begin{cases} M_n = \text{The smaller of } \{M_p^*, M_p\} \\ M_p^* = C_b\left[M_p - (M_p - 0.7F_yS_x)\left(\dfrac{L_b - L_p}{L_r - L_p}\right)\right] \\ M_p = Z_xF_y \end{cases} \text{(Inelastic LTB)} \rightarrow \text{AISC 360}$$

Since $L_p < L_b < L_r$ and $C_b = 1.67 \rightarrow M_n = M_p = Z_xF_y = 186 \times 50 / 12 = 775$ kip-ft

$M_c = \phi M_n = 0.9 \times 775 = 697$ kip-ft

Member BD (W18×86, $F_y = 50$ ksi):

$F_y = 50$ ksi: Compact section.

- *Column capacity, P_c*

$$G_B = 0$$
$$G_D = \infty \text{ (free end)}$$

$$K_x = 2.0 \text{ (Fig. 7.30)} \quad K_y = 1.0 \quad r_x = 7.77 \text{ in.} \quad r_y = 2.63 \text{ in.}$$
$$K_x / K_y = 2.0 < r_x / r_y = 2.95$$
$$\rightarrow \text{Column will buckle about } y\text{-axis}$$

$$(KL/r)_y = \frac{10' \times 12^{\text{in./ft}}}{2.63''} = 45.62 < 4.71\sqrt{E/F_y} = 113 \text{ (Inelastic Buckling, Use Eq. E3-2)}$$

$$F_e = \frac{\pi^2 E}{(KL/r)_y^2} = \frac{\pi^2 \times 29{,}000^{\text{ksi}}}{45.62^2} = 137.53 \text{ ksi}$$

$$F_{cr} = (0.658^{F_y/F_e})F_y = (0.658^{50^{\text{ksi}}/137.53^{\text{ksi}}}) \times 50^{\text{ksi}} = 42.94 \text{ ksi}$$

$$P_n = A_g F_{cr} = 25.3^{\text{in.}^2} \times 42.94^{\text{ksi}} = 1086 \text{ kips}$$
$$P_c = \phi P_n = 0.9 \times 1{,}086^{\text{kips}} = 977 \text{ kips}$$

- *Beam capacity, M_c*

From *AISC Manual*, Table 3-2: W18×86, $F_y = 50$ ksi ➔ $L_p = 9.29$ ft, $L_r = 28.6$ ft

BD: $C_b = 1.67$ (from the bending moment diagram), and $L_b = 10$ ft. So, we have $L_p < L_b < L_r$, and inelastic LTB would occur if $C_b = 1.0$, or C_b is slightly larger than 1.0. Since $C_b = 1.67$ is substantially larger than 1.0, LTB is prevented, and $\rightarrow M_n = M_p = Z_xF_y = 186^{\text{in.}^3} \times 50^{\text{ksi}} / 12^{\text{in./ft}} = 775$ kip-ft

$M_c = \phi M_n = 0.9 \times 775^{\text{kip-ft}} = 697$ kip-ft

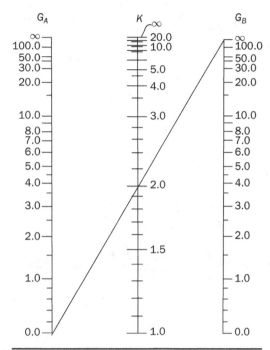

G_A K G_B

FIGURE 7.30 Alignment chart for unbraced frames.

c. Interaction curves for the maximum value of P

Based on member AC

From (a) and (b):

$M_r = 2.56P$ kip-ft; $M_c = 697$ kip-ft

$P_r = 1.60P$ kips, $P_c = 1,158$ kips

Assume $\dfrac{P_r}{P_c} < 0.2$;

$$\frac{P_r}{2P_c} + \frac{M_r}{M_c} \leq 1.0 \Rightarrow \frac{1.60P}{2 \times 1158} + \frac{2.56P}{697} \leq 1.0 \Rightarrow P_{max1}$$
$$= 229.1 \text{ kips (based on member } AC)$$

Check the assumption:

$$\frac{P_r}{P_c} = \frac{1.60 \times 229.1}{1158} = 0.31 > 0.2 \rightarrow \text{The assumption is wrong!}$$

Let's change the assumption: $\dfrac{P_r}{P_c} > 0.2$;

$$\frac{P_r}{P_c} + \frac{8}{9}\frac{M_r}{M_c} \leq 1.0 \Rightarrow \frac{1.60P}{1158} + \frac{8}{9}\frac{2.56P}{697} \leq 1.0 \Rightarrow P_{max1}$$
$$= 215.2 \text{ kips (based on member } AC)$$

Check the assumption:

$$\frac{P_r}{P_c} = \frac{1.60 \times 215.2}{1158} = 0.29 > 0.2 \;\rightarrow\; \text{The assumption is correct!}$$

Based on Member BD

From (a) and (b):

$$M_r = 3.84P \text{ kip-ft}; \quad M_c = 697 \text{ kip-ft}$$
$$P_r = 0.128P \text{ kips}, \quad P_c = 977 \text{ kips}$$

Assume: $\dfrac{P_r}{P_c} < 0.2$, $\quad \dfrac{P_r}{2P_c} + \dfrac{M_r}{M_c} \le 1.0 \Rightarrow \dfrac{0.128P}{2 \times 977} + \dfrac{3.84P}{697} \le 1.0 \Rightarrow P_{\text{max},2} = 179.4 \text{ kips}$

Check the assumption: $\dfrac{P_r}{P_c} = \dfrac{0.128 \times 179.4}{977} = 0.02 < 0.2 \rightarrow$ The assumption is correct!

Final conclusion: $P_{\text{max}} = (P_{\text{max},1};\, P_{\text{max},2})_{\text{min}} = 179.4 \text{ kips (member } BD \text{ controls).}$

7.8 Application of Design Tables in AISC Manual for Beam-Columns

An initial trial section has to be "guessed" at the beginning of the design before a relatively lengthy process of checking the interaction equation takes place. It is fairly straightforward to guess a section that is close to the final design when the depth of W-shape is fixed. On the other hand, it will not be so easy when we intend to select the lightest W-shape among all W-shapes (almost 300 of them available) with any depth, which might range from W8 to W44. For each trial section, we need to calculate column and beam nominal strengths, respectively, before checking the interaction equation. To help with the effort, the *AISC Manual* provides tabulated values of beam and column strengths (Table 6-1 in Part 6, *AISC Manual*), which allows to directly obtain the beam and column strengths as soon as we have a trial W-shape with depths between W8 and W44 with $F_y = 50$ ksi.

It should be noted that the architect and/or building owner always tend to like a shallow W-section (W8, W12, or W14) to save the floor area. However, in some cases (for example, the seismic load is significant) when a beam-column member is subjected to a large bending moment, the engineer always finds a deeper section to be a better solution: lighter and easier for connection.

7.8.1 Background of Design Tables

Let's rewrite Eq. (7.9a) considering the bending about weak axis (*y*-axis) and strong axis (*x*-axis):

For $\dfrac{P_r}{P_c} \ge 0.2$

$$\frac{P_r}{P_c} + \frac{8}{9}\left(\frac{M_{rx}}{M_{cx}} + \frac{M_{ry}}{M_{cy}}\right) \le 1.0 \tag{7.10}$$

Let $\quad p = \dfrac{1}{P_c}$ (kips)$^{-1}$, where $P_c = \phi_c P_n$

$\qquad b_x = \dfrac{8}{9M_{cx}}$, (kip-ft)$^{-1}$, where $M_{cx} = \phi_c M_{nx}$

$\qquad b_y = \dfrac{8}{9M_{cy}}$, (kip-ft)$^{-1}$, where $M_{cy} = \phi_c M_{ny}$

Eq. (7.10) becomes

For $\dfrac{P_r}{P_c} = pP_r \geq 0.2$

$$pP_r + b_x M_{rx} + b_y M_{ry} \leq 1.0 \qquad\qquad (7.11)$$

Similarly, for $\dfrac{P_r}{P_c} = pP_r < 0.2$

$$0.5pP_r + \frac{9}{8}\left(b_x M_{rx} + b_y M_{ry}\right) \leq 1.0 \qquad\qquad (7.12)$$

Table 6-1 in *AISC Manual* lists p, b_x, and b_y values for W-shapes with $F_y = 50$-ksi steel. The local and lateral-torsional buckling has been considered in the table. The tabulated values of b_x have included cases of $L_p < L_b \leq L_r$, but assume that $C_b = 1.0$. One can modify b_x values when actual $C_b > 1.0$ as we do in the beam design using AISC 360 Specification, Section F1. There is no need to check local and lateral-torsional buckling since they are included in the tabulated values.

Example 7.7
The member with W14×193 as shown in Fig. 7.31 is subjected to combined bending moment and axial load from the dead load (30%) and live load (70%). Assume the member is laterally braced at the two ends only ($L_b = 28$ ft), and pin supports at the two ends, and additional lateral support at B for out-of-plane buckling ($L_x = 28$ ft and $L_y = 14$ ft). Check whether W14×193 with $F_y = 50$ ksi is adequate.

Solution:
a. Loads

$$P_{nt} = 1.2P_D + 1.6P_L = \left[1.2 \times 30\% + 1.6 \times 70\%\right](P) = 1.48 \times 440^{\text{kips}} = 650 \text{ kips}$$

$$M_{nt} = 1.48(M) = 1.48 \times 620^{\text{kip-ft}} = 918 \text{ kip-ft}$$

$$M_r = B_1 M_{nt} \ (B_2 = 0)$$

$$P_r = P_{nt} \ (B_2 = 0)$$

b. Moment magnification factor B_1 (B_2 can be considered equal to zero for the beam-column member without lateral translation at its ends)

From Table 4-1 in *AISC Manual* (for any F_y) for W14×193,

$$P_{e1} = P_{ex} = \underbrace{\frac{\pi^2 EA_g}{(KL/r)_x^2}}_{\left(\pi^2 EA_g r_x^2\right)} = 68{,}700 \times 10^4 \,/\, \left(28' \times 12^{\text{in./ft}}\right)^2 = 6{,}085 \text{ kips}$$

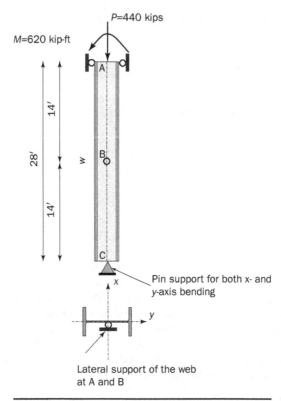

P=440 kips

M=620 kip-ft

Pin support for both x- and y-axis bending

Lateral support of the web at A and B

FIGURE 7.31

The member without lateral (transverse) loads between its two ends, B and C:

$$C_m = 0.6 - 0.4\left(\frac{M_1}{M_2}\right) = 0.6$$

where

$$M_1 = 0 < M_2 = 620 \text{ kip-ft}$$

$$B_1 = \frac{C_m}{1 - \dfrac{P_r}{P_{e1}}} = \frac{0.6}{1 - \dfrac{650^{\text{kips}}}{6,085^{\text{kips}}}} = 0.67 < 1 \rightarrow B_1 = 1.0$$

$$M_r = B_1 M_{nt} = (1.0)(918^{\text{kip-ft}}) = 918 \text{ kip-ft}$$

$$P_r = P_{nt} = 650 \text{ kips}$$

    c. Obtain column and beam strengths for W14×193 from Table 6-1 of *AISC Manual*
We have

$$(KL)_y = 14', (KL)_x/(r_x/r_y) = 28'/1.60 = 17.5 \text{ ft.}, L_b = 28', F_y = 50 \text{ ksi}$$

Enter Table 6-1 of *AISC Manual* for W14×193 with $KL = 17.5'$ [the larger of $(KL)_y$ and $(KL)_x/(r_x/r_y)$] and $L_b = 28'$,

$$p = 0.477 \times 10^{-3} \text{ (kips)}^{-1}$$

$$b_x = 0.727 \times 10^{-3} \text{ (kip-ft)}^{-1}$$

d. Check the section using the interaction equation [Eq. (7.9)]

$$pP_r + b_x M_{rx} + b_y M_{ry} \leq 1.0 \qquad \text{for } \frac{P_r}{P_c} = pP_r \geq 0.2$$

$$0.5 pP_r + \frac{9}{8}(b_x M_{rx} + b_y M_{ry}) \leq 1.0 \qquad \text{for } \frac{P_r}{P_c} = pP_r < 0.2$$

$\frac{P_r}{P_c} = pP_r = (0.477 \times 10^{-3})(650^{\text{kips}}) = 0.310 \geq 0.2 \rightarrow$ Eq. (7.9a) applies. Note that $M_{rx} = M_r$ and $M_{ry} = 0$. Thus

$$pP_r + b_x M_{rx} + b_y M_{ry} = pP_r + b_x M_{rx} = 0.310 + (0.727 \times 10^{-3})(918^{\text{kip-ft}}) = 0.977 \leq 1.0$$

W14×193, $F_y = 50$ ksi is adequate for the member.

Example 7.8

Figure 7.32 shows the elevation of identical frames on Line 1 and Line 5 (Fig. 7.32*a*) and a typical plan of an office building. There is a braced frame (whose elevation is not shown) on Lines A and D, respectively. All connections between columns and girders in the whole buildings are indicated clearly in the building plan (Fig. 7.32*b*). The building has dead (*D*) and live (*L*) loads on each floor and roof, and design wind (*W*) loads. Governing load combination: $1.2D + 1.0L + 1.0W$.

Given:

- $F_y = 50$-ksi steel is used.
- Preliminary design has the following member sizes in the moment frames on Lines 1 and 5:

Level	Column on Line A, B, C, or D	Girder
3	W14×82	W27×114
2	W14×82	W30×173
1	W14×109	W30×173

- Gravity loads on each floor and roof: Dead load = 95 psf; live load = 100 psf.
- The lateral loads F_1, F_2, and F_3 are design wind loads for ONE Frame either on Line 1 or Line 5.
- Gravity load effect on Moment Frame on Line 1: The analysis under gravity loads for the moment frame on Line 1: (1) axial forces (P_{nt}) for the columns in this frame is given in Fig. 7.33*a*; (2) M_{nt} is very small and will be neglected in the design.
- Lateral load effect on Moment Frame on Line 1: P_{lt} and M_{lt} are given in Fig. 7.33*b* and *c*.

Questions:

1. Draw deformation shape of affected members when Column *fj*, buckles about its *y*-axis, and determine value for K_y and L_y of Column *fj* (see Fig. 7.32*a* and *b*).

2. Draw deformation shape of affected members when Column *fj* buckles about its *x*-axis, and determine value for K_x and L_x of Column *fj* (see Fig. 7.32*a* and *b*).

3. Determine unbraced length, L_b, and C_b of Column *fj*.

4. Determine nominal axial and bending strength, P_n and M_n, respectively, for Column *fj*.

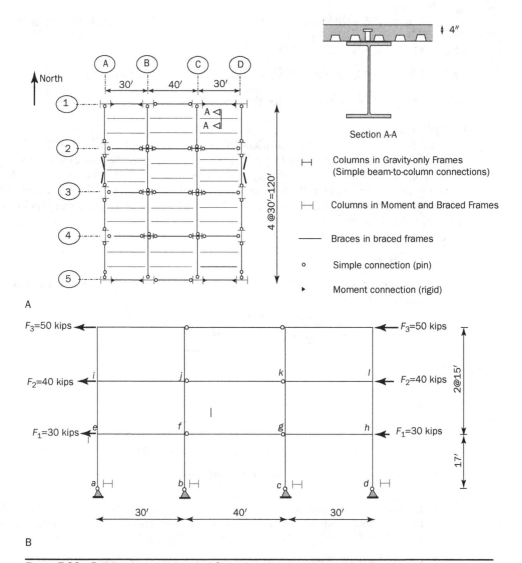

FIGURE 7.32 Building layout: (*a*) typical floor plan; (*b*) elevation on Line 5.

5. Determine required design axial and bending strength, P_r and M_r, respectively, for Column fj. Use $B_1 = 1.0$; but, calculate B_2 based on the floor and roof dead and live loads, and the lateral displacement at the first floor is 1.0 in., and second-floor level is 1.9 in. due to the wind loads.

6. Is the preliminary section for Column fj adequate? If not, select another W14 to make Column fj safe.

7. Check the safety of Column cg in Fig. 7.32b. Given: $K_x = 1.60$; $K_y = 1.0$; $B_1 = 1.0$; $B_2 = 1.5$.

Solution:

a. Draw deformation shape of affected members when Column fj buckles about its y-axis, and determine the value for K_y and L_y of Column fj (Fig. 7.34).

$$K_y = 1.0 \text{ and } L_y = 15 \text{ ft}$$

b. Draw deformation shape of affected members when Column fj buckles about its x-axis (Fig. 7.35), and determine the value for K_x and L_x of Column fj (Fig. 7.36).

$$\left.\begin{array}{l} G_j = \dfrac{\sum(E_c I_c / L_c)}{\sum(E_b I_b / L_b)} = \dfrac{2 \times 881 / 15'}{8230 / 30'} = 0.43 \\[3mm] G_f = \dfrac{\sum(E_c I_c / L_c)}{\sum(E_b I_b / L_b)} = \dfrac{881 / 15' + 1240 / 17'}{8230 / 30'} = 0.48 \end{array}\right\} K_x \cong 1.17$$

From AISC Section Tables:

W14×82: $I_x = 881$ in.⁴
W14×109: $I_x = 1240$ in.⁴
W30×173: $I_x = 8230$ in.⁴

Note: Since beams fg and jk have pin (shear) connections, they have no resistance to rotations at the joints. Therefore, beams fg and jk should not be taken into account for the calculation of K_x.

$$K_x = 1.17 \text{ and } L_x = 15 \text{ ft.}$$

c. Determine the unbraced length, L_b, and C_b of Column fj.

$$C_b = \frac{12.5 M_{max}}{2.5 M_{max} + 3 M_A + 4 M_B + 3 M_C}$$

$$\left.\begin{array}{l} M_{max} = 650 \text{ kip-ft} \\ M_A = 250 \text{ kip-ft} \\ M_B = 50 \text{ kip-ft} \\ M_C = 350 \text{ kip-ft} \end{array}\right\} C_b = \frac{12.5 \times 650^{kip\text{-}ft}}{2.5 \times 650^{kip\text{-}ft} + 3 \times 250^{kip\text{-}ft} + 4 \times 50^{kip\text{-}ft} + 3 \times 350^{kip\text{-}ft}} = 2.24$$

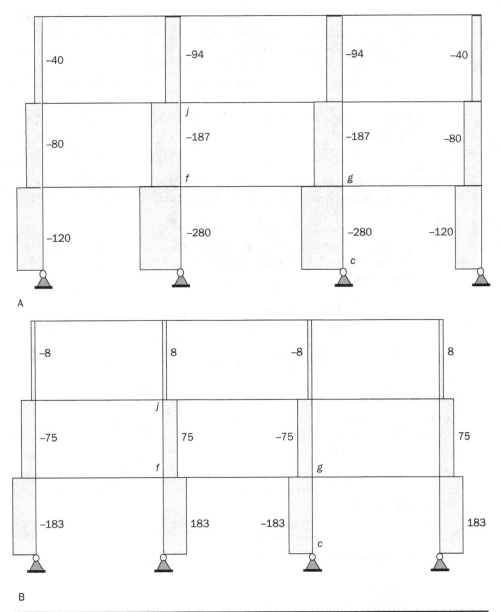

Figure 7.33 Axial force and bending moment in moment frame on Line 1 under different load cases. (Note: M_{nt} is too small to be included in this problem.) (a) Moment frame: Column axial forces under gravity load only, P_{nt} (kips). (Note: negative sign means "compression.") (b) Moment frame: Column axial forces under lateral load only, P_{lt} (kips). (Note: negative sign means "compression.") (c) Moment frame on Line 1: Column bending moment under lateral load only, M_{lt} (kip-ft) (drawn on tension side).

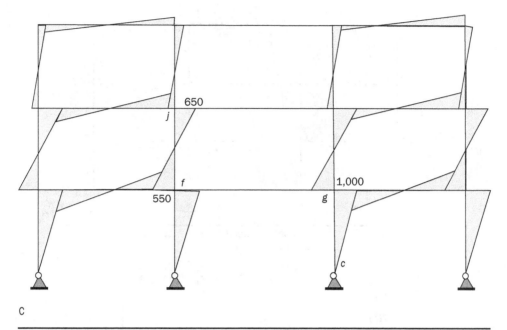

c

Figure 7.33   (*Continued*)

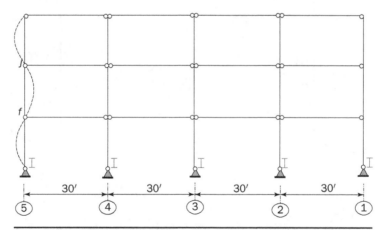

Figure 7.34   Deformed shape when Column *fj* buckles about the *y*-axis (frame along Line B).

Note that

    a. $C_b < 1.67$ for a single-curvature bending moment

    b. $C_b = 1.67$ for a right triangle bending moment

    c. $C_b > 1.67$ for a double-curvature bending moment (as shown in Fig. 7.37)

Since $M_n = M_p$ when $L_p < L_b < L_r$ and $C_b \geq 1.67$, there is no need to calculate the exact value, if $L_p < L_b < L_r$ with a right triangle. Note that bending moment shapes of columns

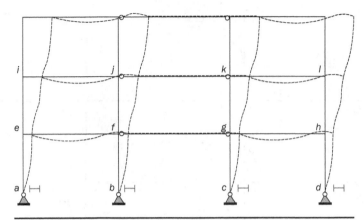

FIGURE 7.35 Deformed shape when Column *fj* buckles about the x-axis (frame along Line 5).

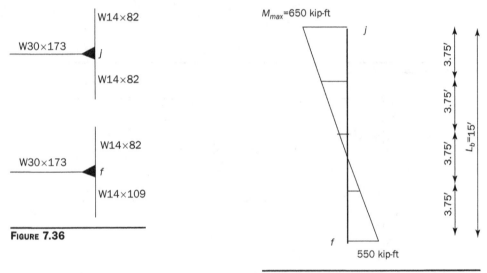

FIGURE 7.36

FIGURE 7.37

in moment frames are all right triangle, except the first floor when pin supporting footing is used in which case $C_b = 1.67$.

d. Determine nominal axial and bending strength, P_n and M_n, respectively, for Column *fj*.

- *Determine P_n*

W14×82, $F_y = 50$ ksi (Compact section for compression)

$$K_x/K_y = 1.17 < r_x/r_y = 2.44 \quad \rightarrow \text{Column will buckle about } y\text{-axis}$$

$$(KL/r)_y = \frac{1.0 \times (15' \times 12^{\text{in./ft}})}{2.48''} = 72.58 < 4.71\sqrt{29,000^{\text{ksi}}/50^{\text{ksi}}}$$

$$= 113 \rightarrow \text{Inelastic Buckling}$$

$$F_e = \frac{\pi^2 E}{(KL/r)_y^2} = \frac{\pi^2 \times 29{,}000^{ksi}}{72.58^2} = 54.33 \text{ ksi}$$

$$F_{cr} = (0.658^{F_y/F_e})F_y = (0.658^{50^{ksi}/54.33^{ksi}}) \times 50^{ksi} = 34.01 \text{ ksi}$$

$$P_n = A_g F_{cr} = 24^{in.^2} \times 34.01^{ksi} = 816 \text{ kips}$$

- Determine M_n

W14×82, $F_y = 50$ ksi (Compact section for bending)

AISC Manual, Table 3-2: $L_p = 8.76$ ft. $< L_b = 15$ ft. $< L_r = 33.2$ ft. (Inelastic LTB occurs when $C_b = 1.0$)

There is no LTB, since $C_b = 2.24 \rightarrow M_n = M_p = Z_x F_y = 579$ kip-ft

e. Determine the required design axial and bending strength, P_r and M_r, respectively, for Column fj. Use $B_1 = 1.0$; but calculate B_2 based on the floor and roof dead and live loads, and the lateral displacement at the first floor is 1.0 in., and the second-floor level is 1.9 in. due to the wind loads.

$$P_{story} = \left(\frac{1.2 \times 95^{psf} + 1.0 \times 100^{psf}}{1{,}000} \right)\left(\frac{100' \times 120'}{2} \right) \times (2) = 2{,}568 \text{ kips}$$

$$(1.2D + 1.0W + L \text{ governs})$$

$\Delta_H = 1.9 - 1.0 = 0.9$ in. (Story Drift)

$H = 2 \times (50^{kips} + 40^{kips}) = 180$ kips (Story shear force due to Wind Load)

$R_m = 0.85$ for Moment Frames

$$B_2 = \frac{1}{1 - \dfrac{P_{story}}{R_m \dfrac{HL}{\Delta_H}}} = \frac{1}{1 - \dfrac{(2{,}568^{kips})}{(0.85)\dfrac{(180^{kips})(15' \times 12^{in./ft})}{0.9^{in.}}}} = 1.092$$

$\left. \begin{array}{l} P_{nt} = 187 \text{ kips} \\ P_{lt} = 75 \\ M_{nt} = 0 \\ M_{lt} = 650 \text{ kip-ft} \\ B_1 = 1.0 \\ B_2 = 1.092 \end{array} \right\}$ $\left| \begin{array}{l} M_r = B_1 \times M_{nt} + B_2 \times M_{lt} = 0 + 1.092 \times 650^{kip\text{-}ft} = 736.8 \text{ kip-ft} \\ P_r = P_{nt} + B_2 \times P_{lt} = 187^{kips} + 1.092 \times 75^{kips} = 268.9 \text{ kips} \end{array} \right|$

f. Is the preliminary section for Column fj adequate? If not, select another W14 to make Column fj safe.

Method 1: Use AISC 360 formula

$$P_c = \phi P_n = 0.9 \times 816^{kips} = 735 \text{ kips}$$

$$M_c = \phi M_n = 0.9 \times 579^{kip\text{-}ft} = 521 \text{ kip-ft}$$

$$\frac{P_r}{P_c} = \frac{268.9^{kips}}{735^{kips}} = 0.36 > 0.2 \rightarrow \frac{P_r}{P_c} + \frac{8}{9}\frac{M_r}{M_c} \leq 1.0$$

$$\Rightarrow \frac{268.9^{kips}}{735^{kips}} + \frac{8}{9}\frac{736.8^{kip\text{-}ft}}{521^{kip\text{-}ft}} = 1.62 > 1.0 \text{ N.G. !}$$

Try W14×120, $F_y = 50$ ksi (Compact section for bending and compression)

- *Determine P_c*

Assume $K_x = 1.17$ → $K_x/K_y = 1.17 < r_x / r_y = 1.67$ → Column will buckle about y-axis

Enter Table 4-1 in *AISC Manual* with $KL = 15$ ft. → $P_c = \phi P_n = 1{,}340$ kips

- *Determine M_c*

AISC Manual, Table 3-2: $L_p = 13.2$ ft. $< L_b = 15$ ft. $< L_r = 51.9$ ft. (Inelastic LTB occurs if $C_b = 1.0$)

Since $C_b = 2.24$, LTB is prevented → $M_n = M_p$

$$M_c = \phi Z_x F_y = 0.9 \times 212^{\text{in.}^3} \times 50^{\text{ksi}}/12^{\text{in./ft}} = 795 \text{ kip-ft}$$

- *Check P-M Interaction*

$$\frac{P_r}{P_c} = \frac{268.9^{\text{kips}}}{1340^{\text{kips}}} = 0.201 \geq 0.2 \rightarrow \frac{P_r}{P_c} + \frac{8}{9}\left(\frac{M_r}{M_c}\right) \leq 1.0$$

$$\Rightarrow \frac{268.9^{\text{kips}}}{1340^{\text{kips}}} + \frac{8}{9}\left(\frac{736.8^{\text{kip-ft}}}{795^{\text{kip-ft}}}\right) = 1.02 \text{ (within the reasonable margin of 5\%)}$$

Method 2: Use Table 6-1 of AISC Manual

$p \times 10^{-3} = 1.36$ (kips)$^{-1}$ for $KL=15$ ft

$b_x \times 10^{-3} = 1.71$ (kip-ft)$^{-1}$ for $L_b=0$ ft (b values for $L_b=0$ can be used when $C_b > 1.67$)

$pP_r = 1.36 \times 10^{-3} \times 268.9^{\text{kips}} = 0.366 > 0.2$

Thus,

$pP_r + b_x M_r = 1.36 \times 10^{-3} \times 268.9^{\text{kips}} + 1.71 \times 10^{-3} \times 736.8^{\text{kip-ft}} = 1.62 > 1.0$ NG!

Try W14×120

$$p \times 10^{-3} = 0.746 \text{ (kips)}^{-1} \text{ for } KL=15 \text{ ft}$$
$$b_x \times 10^{-3} = 1.12 \text{ (kip-ft)}^{-1} \text{ for } L_b=0 \text{ ft}$$

$pP_r = 0.746 \times 10^{-3} \times 268.9^{\text{kips}} = 0.199 < 0.2$

Thus,

$$\frac{1}{2}pP_r + \frac{9}{8}b_x M_r = 0.746 \times 10^{-3} \times 268.9^{\text{kips}} + 1.12 \times 10^{-3} \times 736.8^{\text{kip-ft}}$$
$$= 1.02 \text{ (within the reasonable margin of 5\%)}$$

g. Check the safety of Column *cg*. Given: $K_x = 1.60$; $K_y = 1.0$; $B_1 = 1.0$; $B_2 = 1.5$.

Method 1: Use formulas given in AISC 360

W14×109, $F_y = 50$ ksi (Compact section for bending and compression)

- *Determine P_c*

$K_x / K_y = 1.60 < r_x / r_y = 1.67$ → Column will buckle about x-axis

Enter Table 4-1 with $\dfrac{(KL)_x}{r_x/r_y} = \dfrac{1.60 \times 17'}{1.67} = 16.29 \rightarrow P_c = \phi P_n = 1{,}181.3$ kips

- *Determine M_c*

AISC Manual, Table 3-2: $L_p = 13.2$ ft. $< L_b = 17$ ft. $< L_r = 48.5$ ft. (Inelastic LTB occurs if $C_b = 1.0$)

Since $C_b = 1.67 \rightarrow M_n = M_p$

$$M_c = \phi Z_x F_y = 0.9 \times 192^{in.^3} \times 50^{ksi} / 12^{in./ft} = 720 \text{ kip-ft}$$

- *Determine P_r and M_r*

$$
\left.
\begin{aligned}
P_{nt} &= 280 \text{ kips} \\
P_{lt} &= 183 \text{ kips} \\
M_{nt} &= 0 \\
M_{lt} &= 1{,}000 \text{ k-ft} \\
B_1 &= 1.0 \\
B_2 &= 1.5
\end{aligned}
\right\}
\quad
\left|
\begin{aligned}
M_r &= B_1 \times M_{nt} + B_2 \times M_{lt} = 0 + 1.5 \times 1{,}000^{kip\text{-}ft} = 1{,}500 \text{ kip-ft} \\
P_r &= P_{nt} + B_2 \times P_{lt} = 280^{kips} + 1.5 \times 183^{kips} = 554.5 \text{ kips}
\end{aligned}
\right|
$$

Check P-M Interaction

$$\frac{P_r}{P_c} = \frac{554.5^{kips}}{1{,}181.3^{kips}} = 0.47 > 0.2 \rightarrow \frac{P_r}{P_c} + \frac{8}{9}\frac{M_r}{M_c} \leq 1.0$$

$$\Rightarrow \frac{554.5^{kips}}{1{,}181.3^{kips}} + \frac{8}{9}\frac{1{,}500^{kip\text{-}ft}}{720^{kip\text{-}ft}} = 2.32 > 1.0 \text{ N.G. !}$$

Method 2: Use Table 6-1 of AISC Manual

$p \times 10^{-3} = 0.8491 \text{ (kips)}^{-1}$ for $KL=16.29$ ft

$b_x \times 10^{-3} = 1.23 \text{ (kip-ft)}^{-1}$ for $L_b=0$ ft (b values for $L_b=0$ can be used when $C_b > 1.5$)

$pP_r = 0.8491 \times 10^{-3} \times 554.5^{kips} = 0.471 > 0.2$

Thus,

$pP_r + b_x M_r = 0.8491 \times 10^{-3} \times 54.5^{kips} + 1.23 \times 10^{-3} \times 1{,}500^{kip\text{-}ft} = 2.32 > 1.0 \text{ Not Good!}$

Example 7.9

Figure 7.38 shows plan of a typical three-story steel building. The elevation of the identical moment frames on Lines 1 and 4 and the preliminary members made of A992 steel ($F_y = 50$ ksi) are also given in Fig. 7.38. The structure is subjected to dead (D), live (L), and wind (W) loads. Gravity loads on each floor: Dead load = 80 psf; live load = 40 psf. Gravity loads on roof: Dead load = 48 psf; live load = 16 psf. Using the tributary areas, gravity and wind load distributions on the frame are determined as given in Fig. 7.39. A preliminary design is given with engineering judgment. Determine the required strength of the second story interior column (Column B1 or B4) in the building shown in Fig. 7.38 using the direct analysis method (DAM). Check if the selected shape for column B1 is adequate.

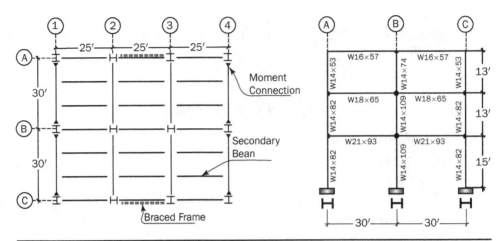

FIGURE 7.38   A typical plan for floors and roof (left) and elevation on Lines A and C (right).

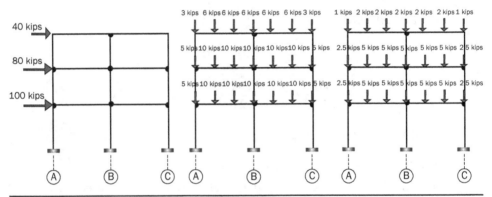

FIGURE 7.39   Wind (left), dead (middle), and live (right) load distribution on the frame.

Solution:

Step I: Create a realistic model with the leaning columns

A 2D numerical model of the moment frame shown in Fig. 7.40a is created using rigidly connected elastic beam elements. The columns in the moment frame are assumed to be fixed at the base. Each column in the frame is divided into five elements to take P-δ effect into account. To accurately account for the second-order effects, a leaning column (i.e., P-Δ column) consisting of rigid truss elements is incorporated in the numerical model. The column is loaded with the vertical gravity loads supported by the columns that are not part of the moment frame. Based on the tributary area (Fig. 7.40b) supported by the gravity-only columns, the concentrated dead loads acting on the leaning column at each story level are computed as follows:

$$P_{\text{leaning}}^{\text{floor}} = (25^{\text{ft}} \times 60^{\text{ft}}) \times 80^{\text{psf}}/1000^{\text{lb/kip}} = 120 \text{ kips}$$

$$P_{\text{leaning}}^{\text{roof}} = (25^{\text{ft}} \times 60^{\text{ft}}) \times 48^{\text{psf}}/1000^{\text{lb/kip}} = 72 \text{ kips}$$

Dead loads supported by the gravity columns at each story level

The loads acting on the leaning (i.e., P-Δ) columns for the other load cases can be computed in a similar manner using the same tributary area.

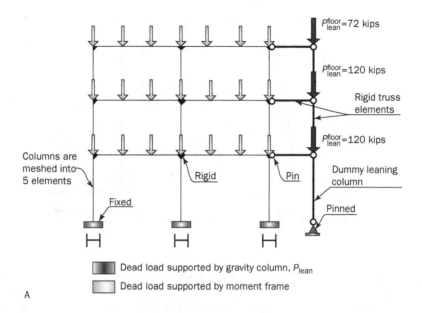

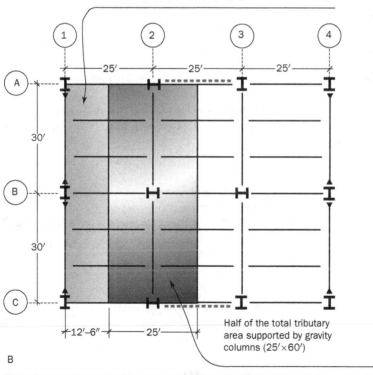

Figure 7.40 2D numerical model. (*a*) A representative numerical model with dead loads. (*b*) Tributary area for the dummy leaning columns.

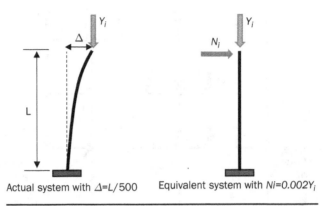

Actual system with $\Delta = L/500$ Equivalent system with $Ni = 0.002Y_i$

FIGURE 7.41 Equivalent lateral notional load to account for the permitted erection tolerance.

Step II: Reduce stiffness of all members that contribute to the stability

To account for the potential reduction in stiffness, flexural stiffness is reduced initially by 20%. Note that an additional reduction in the flexural stiffness can be made based on the axial force level. A stiffness reduction factor, τ_b, is introduced for the flexural stiffness when the axial force in a member is greater than 50% of its axial load–carrying capacity.

$$\tau_b = 1.0 \text{ when } \frac{\alpha P_r}{P_y} \le 0.5 \quad \text{and} \quad \tau_b = 4(\alpha P_r/P_y)[1 - (\alpha P_r/P_y)] \text{ when } \frac{\alpha P_r}{P_y} > 0.5$$

Step III: Apply notional loads in terms of the factored gravity loads or notional displacements to account for initial imperfections

As demonstrated in Fig. 7.41, the notional load, N_i, is taken as 0.002 times the total gravity load at each story level, which is based on an initial story out-of-straightness of $1/500$ for all stories.

$$N_i = 0.002\alpha Y_i$$

where $\alpha = 1.0$ (LRFD) and Y_i are the total gravity load distributed on floors and roof, including the loads supported by gravity frames.

Example notional load calculation for the dead loads is as follows (Fig. 7.42):

$$ND_1 = ND_2 = 0.002\alpha DL^{\text{floor}} = 0.002 \times 1.0 \times (120^{\text{kips}} + 5 \times 10^{\text{kips}} + 2 \times 5^{\text{kips}}) = 0.36 \text{ kips}$$

$$ND_3 = 0.002\alpha DL^{\text{roof}} = 0.002 \times 1.0 \times (72^{\text{kips}} + 5 \times 6^{\text{kips}} + 2 \times 3^{\text{kips}}) = 0.22 \text{ kips}$$

The load combinations adopted for the stability analysis are summarized in Table 7.2. Factored notional loads, which are indicated in parenthesis (Table 7.2) are initially incorporated in the load combinations without lateral loads (i.e., Combinations 1 through 3).

Step IV: Verify $\Delta_{2nd}/\Delta_{1st}$

To determine whether or not the notional loads need to be considered in the combinations with the lateral loads, such as $1.2D + 1.0W + L + 0.5L_r$, *the Specification* requires the ratio of second-order drifts to first-order drifts to be examined. If the ratio of

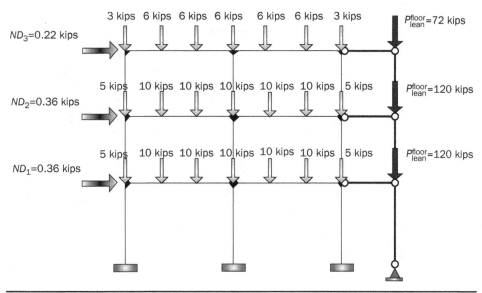

Figure 7.42   Notional dead loads.

second-order drifts to first-order drifts is less than 1.7 for all combinations, it is not required to include the notional loads in the combinations with other lateral loads. Still, to simplify the design process, the notional loads can be conservatively included in all combinations without considering this step.

The frame is subjected to the factored dead and notional dead loads. The first- and second-order lateral displacements obtained from the analysis under $1.2D + 1.0W + L + 0.5L_r$ combination are illustrated in Fig. 7.43.

Example drift calculation for Combination#4 ($1.2D + 1.0W + L + 0.5L_r$) is shown below.

$$\text{First-order}\begin{cases}\text{First-Story Drift} \rightarrow \Delta_1 = \delta_1 = 1.25'' \\ \text{Second-Story Drift} \rightarrow \Delta_2 = \delta_2 - \delta_1 = 2.19 - 1.25 = 0.94'' \\ \text{Third-Story Drift} \rightarrow \Delta_3 = \delta_3 - \delta_2 = 2.79 - 2.19 = 0.60''\end{cases}$$

$$\text{Second-order}\begin{cases}\text{First-Story Drift} \rightarrow \Delta_1 = \delta_1 = 1.30'' \\ \text{Second-Story Drift} \rightarrow \Delta_2 = \delta_2 - \delta_1 = 2.29 - 1.30 = 0.99'' \\ \text{Third-Story Drift} \rightarrow \Delta_3 = \delta_3 - \delta_2 = 2.91 - 2.29 = 0.62''\end{cases}$$

As given in Table 7.3, second-to-first-order drift ratios are less than 1.7 for all combinations. Therefore, the notional loads are not included in combinations 4 through 6.

Step V: Check if $\tau_b = 1.0$
Since τ_b values were assumed to be 1 in Step 2, the actual τ_b values need to be examined. To determine the actual τ_b values, the peak axial force demands plotted in Figs. 7.44 and 7.45 are required.

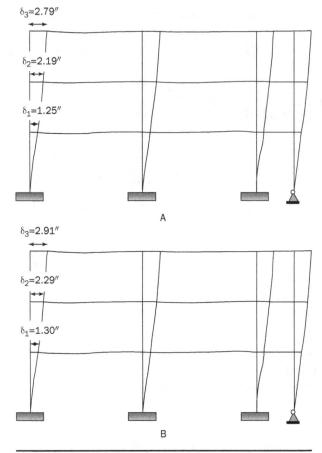

$\delta_3 = 2.79''$

$\delta_2 = 2.19''$

$\delta_1 = 1.25''$

A

$\delta_3 = 2.91''$

$\delta_2 = 2.29''$

$\delta_1 = 1.30''$

B

Figure 7.43 Comparison of (a) first- and (b) second-order lateral displacements under $1.2D + 1.0W + L + 0.5L_r$.

Example τ_b calculation for the second-story interior column is shown below (W14×109: $A_g = 32$ in.²).

$$P_{r,\max} = 96^{\text{kips}} \qquad \frac{\alpha P_{r,\max}}{P_y} = \frac{1.0 \times 96}{1600} = 0.06 < 0.50$$
$$P_y = A_g F_y = 32 \times 50 = 1600^{\text{kips}}$$

Thus, $\tau_b = 1.0$ assumption is correct for the column.

The computed τ_b values for the exterior and interior columns are summarized in Table 7.4.

Step VI: Determine capacities of members using K = 1.0

- Compressive strength, P_c
  W14×109: $A_g = 32$ in.²; $r_x = 6.22$ in.; $r_y = 3.73$ in.; *Compact for compression with* $F_y = 50$ ksi

$$K_x = K_y = 1.0$$

Combination	Story	First-order drift (in.)	Second-order drift (in.)	$\Delta_{2nd}/\Delta_{1st}$ ratio
1	1	0.015	0.016	1.021
	2	0.018	0.018	1.032
	3	0.014	0.015	1.059
2	1	0.015	0.015	1.027
	2	0.023	0.024	1.034
	3	0.018	0.019	1.077
3	1	0.018	0.018	1.024
	2	0.021	0.022	1.038
	3	0.017	0.018	1.070
4	1	0.600	0.627	1.045
	2	0.940	0.984	1.046
	3	1.246	1.302	1.045
5	1	1.25	1.30	1.045
	2	0.94	0.99	1.053
	3	0.60	0.62	1.033
6	1	1.188	1.221	1.028
	2	1.879	1.935	1.030
	3	2.493	2.567	1.030

TABLE 7.3 Summary of the Ratio of Second-Order Drifts to First-Order Drifts ($\Delta_{2nd}/\Delta_{1st}$)

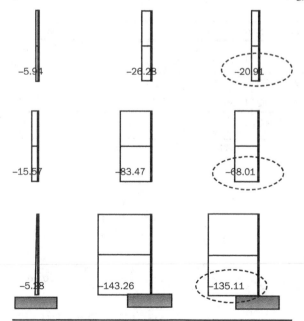

FIGURE 7.44 Peak axial force demands obtained for the exterior columns from second-order analysis.

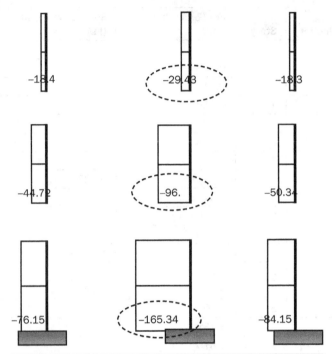

Figure 7.45   Peak axial force demands obtained for the interior columns from second-order analysis.

	Exterior column				Interior column			
Story	**Section**	P_r **(kips)**	P_r/P_y **(kips)**	τ_b	**Section**	P_r **(kips)**	P_r/P_y **(kips)**	τ_b
3	W14×53	20.9	0.027	1.0	W14×74	29.4	0.027	1.0
2	W14×82	68.0	0.057	1.0	W14×109	96.0	0.060	1.0
1	W14×82	135.1	0.113	1.0	W14×109	165.3	0.103	1.0

Table 7.4   Axial Force-to-Capacity Ratio for Columns

$$\left(\frac{L_c}{r}\right) = \left(\frac{L_c}{r}\right)_y = \frac{1.0 \times (13 \times 12)}{3.73} = 41.8$$

$$F_e = \frac{\pi^2 E}{(L_c/r)^2} = \frac{\pi^2 \times 29{,}000}{(41.8)^2} = 163.8^{\text{ksi}}$$

$$F_{cr} = (0.658^{50/163.8}) \times 50 = 44.0\ \text{ksi}$$

$$P_c = \phi_c F_{cr} A_g = 0.9 \times 44.0 \times 32 = 1{,}267\ \text{kips}$$

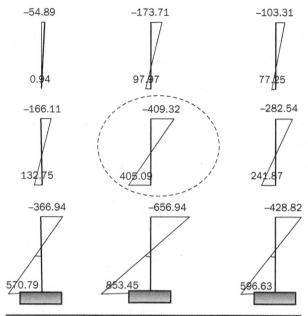

Figure 7.46 Peak bending moment demand obtained from second-order analysis.

- Flexural strength, M_c

$$L_b = 13^{ft} < L_p = 13.2^{ft} \rightarrow M_n = M_p$$

$$M_c = \phi_b M_p = \phi_b F_y Z_x = 0.9 \times 50 \times 192/12^{in/ft} = 720 \text{ kip-ft}$$

Step VII: Determine the adequacy of the second-story interior column (W14×109)
As shown in Figs. 7.45 and 7.46, the required axial force and flexural strengths are obtained from the second-order analysis as $P_r = 96$ kips and $M_r = 409.3$ kip-ft.

The axial force–bending moment interaction can be checked for beam-column members according to the Specification, AISC 360, Eqs. H1-1a and H1-1b.

$$\left.\begin{array}{l} M_r = 409.3^{kip\text{-}ft} \\ P_r = 96^{kips} \\ M_c = 720^{kip\text{-}ft} \\ P_c = 1267^{kips} \end{array}\right\} \Rightarrow \left|\begin{array}{l} \dfrac{P_r}{P_c} = \dfrac{96}{1267} = 0.076 < 0.2 \rightarrow \text{Use Eq. H1-1b} \\[2mm] \dfrac{P_r}{2P_c} + \dfrac{M_r}{M_c} = \dfrac{96}{2 \times 1267} + \dfrac{409.3}{720} = 0.61 < 1.0 \end{array}\right|$$

Typically, the demand over capacity ratio would be within 0.70 and 1.05. In this example, the preliminary design based on engineering judgment seems to be practically good (0.61 is close to 0.7, slightly conservative). If desired in real design practice, the designer may reduce the weight of all members proportionally by one or two levels, for example, replacing W14×109 with W14×99 or W14×90, and recheck them again with the same procedure illustrated in this example.

7.9   Problems

7.1   Please answer the following questions:

    a. How many failure modes (or limit states) are there for a wide-flange beam-column member?

    b. What are the differences between beam-columns and beams and columns?

    c. Why do we make moment amplification in beam-column members?

    d. What is the basis of the Direct Analysis Method in stability analyses?

7.2   Figure 7.47a shows a 35-ft-long W12×120 structural member with pinned/roller ends. The member is loaded with a lateral wind load Q at B, and possibly an axial force

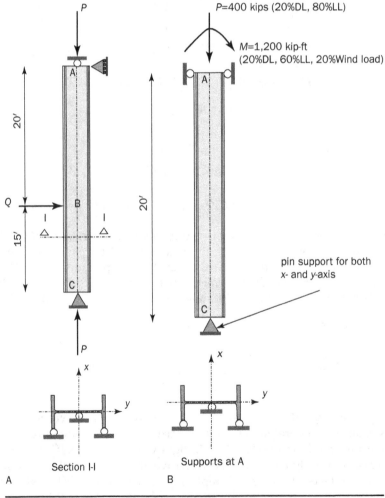

Figure 7.47

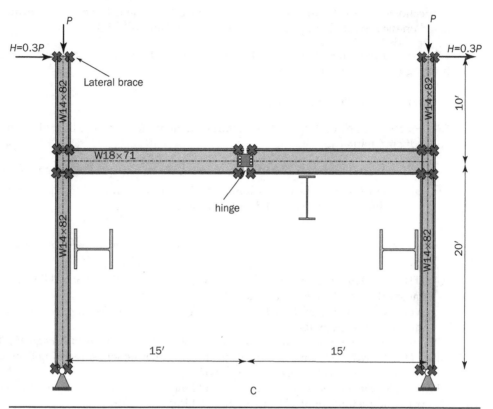

FIGURE 7.47 *(Continued)*

P, braced in both in-plane and out-of-plane of the bending at its ends (A and C). In addition, a lateral brace is provided in the weak direction at point B, 15 ft from the lower end. Neglect the self-weight. Use $F_y = 36$-ksi steel. Determine the maximum lateral wind load Q (100% wind load) for each of the following two cases:

Case 1: Axial force P is not applied ($P = 0.0$)

Case 2: $P = 0.25P_n$ is applied as a 100% dead load together with lateral wind load Q; $P_n = F_{cr}A_g$, the nominal column design strength of the member

7.3 Figure 7.47*b* shows a 20-ft-long W-shape member supporting a load P and bending moment M at its top end A. The two ends of the member have lateral supports. Use $F_y = 50$-ksi steel.

1. Select the lightest W14-section, starting with a trial section W14×82.

2. Select the lightest W24-section, starting with a trial section W24×76.

7.4 A steel frame is shown in Fig. 7.47*c*. A992 steel is used. Dead load P is applied at the top of two columns. The service wind load $H = 0.3P$. The frame is laterally braced at the tops of columns, all joints, and supports. The beam-to-column connections are rigid connections. Determine the maximum service load P. K_y and K_x need to be evaluated based on the provided information. Assume $B_1 = 1.0$ and $B_2 = 1.3$.

7.5 Write a python code that checks the adequacy of a column for a given factored bending moment and axial load.

Related to Building Project in Chapter 1:

7.6 Please design the following structural members for the building given in the appendix in Chapter 1:

    a. Determine the B1 and B2 for the second-story columns in the moment frames.

    b. Design the second-story columns in the moment frames based on moment amplification method and direct analysis method.

Bibliography

AASHTO, *LRFD Bridge Design Specification*, American Association of State Highway and Transportation Officials for Bridges, Washington, 2020.

Aghayere, A. O., and J. Vigil, *Structural Steel Design: A Practice Oriented Approach*, Pearson Education, Inc., New Jersey, 2015.

AISC, *Seismic Design Manual*, American Institute of Steel Construction, Chicago, IL, 2018.

AISC, *The Material Steel*, American Institute of Steel Construction, A Teaching Primer for Colleges of Architecture, Chicago, IL, 2007.

AISC 341, *Seismic Provisions for Structural Steel Buildings*, ANSI/AISC Standard 341-16, American Institute of Steel Construction, Chicago, IL, 2016.

AISC 358, *Prequalified Connections for Special and Intermediate Steel Moment Frames for Seismic Applications*, American Institute of Steel Construction, Chicago, IL, 2016.

AISC 360, *Specification for Structural Steel Buildings*, ANSI/AISC Standard 360-16, American Institute of Steel Construction, Chicago, IL, 2016.

AISC Manual, *Steel Construction Manual*, 15th ed., American Institute of Steel Construction, Chicago, IL, 2016.

ASCE 7, *Minimum Design Loads for Buildings and Other Structures*, ASCE/SEI 7-16, American Society of Civil Engineers, Reston, VA, 2016.

ASTM, *Annual Book of ASTM Standards*, American Society for Testing and Materials, Philadelphia, 2020.

AWS, *Structural Welding Code-Steel (AWS D1:1:2020)*, 24th ed., American Welding Society, Danvers, MA, 2020.

Salmon, C. G., J. E. Johnson, and F. A. Malhas, *Steel Structures: Design and Behavior—Emphasizing Load and Resistance Factor Design*, 5th ed., Pearson Education, Inc., New Jersey, 2009.

Shen, J., B. Akbas, O. Seker, and C. Carter, *Structural Engineering Handbook—Chapter 8: Design of Structural Steel Members*, 5th ed., McGraw-Hill, New York, 2020.

CHAPTER 8

Introduction to Connections

8.1 Introduction

Steel structures consist of structural components that have distinctive features. So far, we have discussed the behavior and associated limit states of individual structural components such as beams, columns, etc., and how to design them in accordance with the current design specifications. Nevertheless, to form any type of steel structure, either building or non-building, these structural members should be connected. This section will present a brief overview of the common connection types.

Due to their crucial role, connections in structural systems are intended to be "protected." The reason is that the limit states related to connections (e.g., bolt or weld failure) are often brittle (i.e., sudden failure). Furthermore, depending on the connection type, connection failures may endanger the stability of structural systems, let alone overall strength and stiffness loss due to the disruption to the load transfer between the connected structural members. In other words, a structure can only be as strong as its connections. The design intention is, therefore, to prevent premature connection failures. This design philosophy is traditionally implemented primarily by ordering the expected failure modes by a hierarchical chain based on the inherent ductility level of each limit state. This aspect will be further discussed in the subsequent sections after digesting the basic definitions.

8.2 Overview of Common Connection Types in Framed Structures

In framed structures, the most common connection type is beam-to-column connections. Beams and columns are connected with various elements and fasteners since the connection between these two main load–carrying structural components is of great significance to withstand and transfer loads as desired. Hence, cost, performance, and safety aspects are often taken into consideration simultaneously for the design and fabrication of beam-to-column connections. Accordingly, beam-to-column connections in the primary structural systems may differ according to the system which is designed predominantly to resist lateral forces or gravity loads. In proportion to the amount of the bending moment transferred from beam to column, these connections can be divided into three groups in general:

- *Fully Restrained Moment Connections*: Both beam flanges, through which a major portion of the bending moment is transferred, are connected to the column flange. These connections are deemed to provide sufficient restraint against the rotation of the connected beam relative to the column and, therefore, referred to as *rigid* or *fully restrained* moment connections. Rigid connections are frequently used to connect beams and columns in moment frames to resist lateral loads as well as to provide stability to the entire building consisting of the slab, moment, and gravity-only frames. Relatively speaking, rigid connections can be costly due to the additional labor and inspection cost of on-site welding, which requires qualified workers. A typical example of a rigid beam-to-column connection is depicted in Fig. 8.1, where the beam flanges are connected to the column through groove welds while the web is connected with a shear tab bolted and welded to the beam web and column flange, respectively.

- *Partially Restrained Moment Connections*: In lieu of welding the flanges directly, beam flanges are connected by means of additional elements with flexibility such as bolted angles or T-stubs. Deformation of the connectors between the two flanges provides certain flexibility to the relative movement of the beam flanges and accordingly reduces the moment transfer from beam to column as the applied deformation increases. Therefore, these connections are called *semirigid* or *partially restrained* moment connections. Even though semirigid connections require additional steel shapes, bolts, and drilling for connecting the flanges, fabricating and assembling semirigid connections are usually less labor-intensive compared with the rigid connections. Figure 8.1 demonstrates a semirigid beam-to-column connection with top and seated angles and a bolted web angle.

- *Shear Connections*: Shear connections, also known as *simple* or *pin* connections, are such that beam and column flanges are not connected at all. Beam web can be attached to the column by different shapes (e.g., single or double angles, shear tab, end plate). Figure 8.1 shows a common shear connection with double web angles. These connections are flexible, and traditionally, assumed to carry shear force only, even though a certain amount of bending moment can be transferred because of the inevitable resistance against relative rotation. The amount of the neglected bending moment may vary based on the shapes and fasteners used to connect the web. As the name suggests, simple connections are easy to make and are employed to join beams and columns that are part of gravity load–carrying frames.

Besides the above-described beam-to-column connections in moment and gravity frames (i.e., rigid, semirigid, and simple connections), beam-column-gusset connections in braced frames are also very common in steel structures. When braced frames are utilized as a lateral force–resisting system, conventionally, gusset plates are used to connect beam, column, and bracing members, as shown in Fig. 8.2. Gusset plates are responsible for distributing the axial force developed in the connected bracing to the adjacent beam and column. Therefore, the size and geometry of gusset plates are dependent on many parameters, such as the connector type (i.e., weld or bolt), expected limit

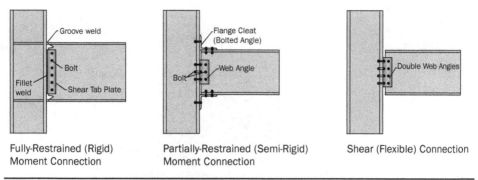

FIGURE 8.1 Typical beam-to-column connections in framed structures.

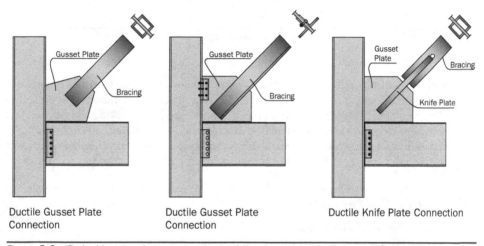

FIGURE 8.2 Typical beam-column-gusset assemblies in concentrically braced frames.

state and capacity of the bracing (e.g., tensile capacity), special detailing requirements, and so on. Besides, the design and detailing of gusset connections might as well vary according to the desired ductility level, as indicated in Fig. 8.2. Ductile connections, for instance, are detailed to accommodate in-plane or out-of-plane buckling of the bracing while nonductile (or low-ductility) detailing is not concerned with the post-buckling deformations. That is, connection design does not solely involve the selection of the elements that transfer forces from one member to another but also requires designers to consider deformation capabilities.

To comprehend connection design in accordance with the codes and specifications, first, general failure modes will be discussed. Then, we will cover each limit state through worked examples.

8.3   General Failure Modes

To learn how to design a connection in accordance with the codes and specifications, general failure modes should be understood first. In order to describe each limit state

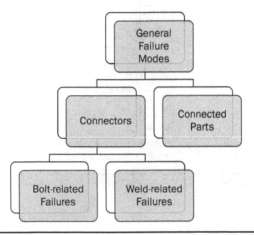

Figure 8.3   Classification of general failure modes.

within a particular group, general failure modes could be related to the connectors or the connected elements, as shown in Fig. 8.3. Since structural members are connected with either bolts or welds or a combination of both, the failure modes associated with the connectors can be categorized as bolt-related and weld-related failure modes (Fig. 8.3). Some of the failure modes, such as block shear and tensile rupture, have already been covered in Chapter 3. The rest of the limit states will be discussed in what follows.

8.4   Bolted Connections

8.4.1   General Remarks

We use bolts to assemble almost all structural member types, such as tension or compression members (Fig. 8.2), members for flexure and shear (Fig. 8.1), splices, and column bases. Figure 8.4 shows a typical structural bolted plate connection consisting of a bolt with threads, a hexagonal nut, and a washer. As shown in Fig. 8.4, bolts are used to connect two or more parts with holes by clamping them between bolt head and nut. Notice that washers are placed between the surfaces to evenly distribute the pressure.

Bolt and bolted connection types used in steel structures differ in terms of material and force transfer mechanisms. Commonly used bolt types are *common bolts* and *high-strength bolts*. Common bolts (also known as unfinished or black bolts) made of ASTM A307 are often used to connect members that are not subjected to reversed cyclic loads or vibrations (i.e., connections under static loads). The nuts on common bolts are tightened by a worker using an ordinary wrench (e.g., stubby or open-ended), which produces small friction force between the surfaces. Common bolts are only permitted for the connections where pretension is not required. The use of high-strength bolts is necessary when a pretension is specified. High-strength bolts, such as ASTM A325 (Group A) or A490 (Group B) bolts, cost more than common bolts and require less labor than installing rivets or welding. These bolts, when pretensioned, provide higher resistance to fatigue and loosening, along with rigid joints without slippage due to the frictional force induced between the connected elements. Turn-of-nut and calibrated wrench–tightening methods can be utilized for installing high-strength bolts.

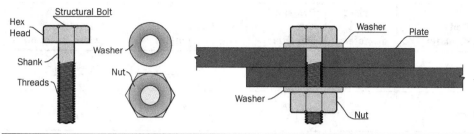

FIGURE 8.4 Typical fastener components (bolt, nut, washer) and bolted plates.

Typical bolt sizes are ranged between ½ and 1½ inches in diameter. Bolt holes can be standard, slotted, or oversized, considering the connection type as well as the manufacturing tolerances. Standard hole diameters are usually one-sixteenth of an inch larger than the bolt diameter for the bolts with a diameter of less than 1 inch whereas hole diameters are made ⅛″ larger than those of the bolts for the bolt diameters equal or greater than 1 inch. To accommodate adjustments, enlarged holes are also permitted in some cases. Limitations on the use of the three types of enlarged holes and the maximum nominal hole dimensions per *the Specification, Table J3.3* are given in Table 8.1.

The Specification specifies the spacing between the holes (*s*) and minimum and maximum edge distance (*e*) for the bolted connections to facilitate construction as well as to consider the bearing and tearout strength requirements. Table 8.2 presents a summary of these requirements for various bolt diameters. The center-to-center distance between the holes of all types (i.e., standard, slotted, oversized) should be greater than $2\frac{2}{3}$ times the nominal bolt diameter (*d*). Note that the distance between the centers of holes (*s*) is typically taken no less than 3*d* for the ease of bolt tightening. The maximum spacing varies depending on whether the elements consist of two shapes or not, according to *the Specification, Section J3.5*. The minimum edge distance, *e*, is also dependent on the nominal bolt diameter, as listed in Table 8.2, whereas the maximum edge distance from the hole center is a function of element thickness and given as the smallest of 12 times the thickness of the connected part and 6-inch distance. The limitation on the maximum edge distance intends to prevent corrosion while limiting the minimum distance aims to provide for the standard of workmanship.

Example 8.1
Check bolt spacing and edge distances for the connection shown in Fig. 8.5. The connection consists of two plates connected with ½-inch-diameter bolts.

Solution:
Per the restrictions given in *the Specification, J3.3, J3.4, and J3.5*, the limiting edge distance and fastener spacing can be summarized for the given connection:

- Min. edge distance, e_{min} = ¾″ for ½-inch bolts (*Specification, Table J3.4*)
- Max. edge distance, e_{max}= 12 times the thickness of the connected part ≥ 6 in. (*Specification, J3.5*)
- Min. spacing between holes, s_{min} = $2\frac{2}{3}$ times the nominal bolt diameter (*Specification, J3.3*)
- Max. spacing between holes, s_{max} = 24 times the thickness of the thinner part ≥ 12 in. (*Specification, J3.5a*)

Bolt diameter	Standard hole diameter	Oversized hole diameter	Short-slot (width × length)	Long-slot (width × length)
$\frac{1}{2}''$	$\frac{9}{16}''$	$\frac{5}{8}''$	$\frac{9}{16}'' \times \frac{11}{16}''$	$\frac{9}{16}'' \times 1\frac{1}{4}''$
$\frac{5}{8}''$	$\frac{11}{16}''$	$\frac{13}{16}''$	$\frac{11}{16}'' \times \frac{7}{8}''$	$\frac{11}{16}'' \times 1\frac{9}{16}''$
$\frac{3}{4}''$	$\frac{13}{16}''$	$\frac{15}{16}''$	$\frac{13}{16}'' \times 1''$	$\frac{13}{16}'' \times 1\frac{7}{8}''$
$\frac{7}{8}''$	$\frac{15}{16}''$	$1\frac{1}{16}''$	$\frac{15}{16}'' \times 1\frac{1}{8}''$	$\frac{15}{16}'' \times 2\frac{3}{16}''$
$1''$	$1\frac{1}{8}''$	$1\frac{1}{4}''$	$1\frac{1}{8}'' \times 1\frac{5}{16}''$	$1\frac{1}{8}'' \times 2\frac{1}{2}''$
$\geq 1\frac{1}{8}''$	$d_b + \frac{1}{8}''$	$d_b + \frac{5}{16}''$	$\left(d_b + \frac{1}{8}''\right) \times \left(d_b + \frac{3}{8}''\right)$	$\left(d_b + \frac{1}{8}''\right) \times 2.5d_b$
Restrictions	None	NP in bearing-type connections	Long side of the slot should be perpendicular to the loading direction in bearing-type connections	Permitted in only one of the parts being connected. Long side of the slot should be perpendicular to the loading direction in bearing-type connections
Illustration of hole type				

*d_b: Nominal bolt diameter; NP: Not permitted.

TABLE 8.1 Maximum Nominal Hole Dimensions

$$\text{Longitudinal direction} \rightarrow \begin{cases} e_{min} = 3/4'' < e = 2'' < e_{max} = 12 \times 1/2 = 6'' \\ s_{min} = 2\frac{2}{3} \times \frac{1}{2} = 1.34'' < s = 3'' < s_{max} = 24 \times 1/2 = 12'' \end{cases}$$

$$\text{Transverse direction} \rightarrow \begin{cases} e_{min} = 3/4'' < e = 1\frac{3}{4}'' < e_{max} = 12 \times 1/2 = 6'' \\ s_{min} = 2\frac{2}{3} \times \frac{1}{2} = 1.34'' < s = 2\frac{1}{2}'' < s_{max} = 24 \times 1/2 = 12'' \end{cases}$$

Thus, bolt and edge spacing are adequate.

Bolt diameter	Min. edge distance (e)	Max. edge distance (e)**	Spacing (s)*
$\frac{1}{2}''$	$\frac{3}{4}''$		
$\frac{5}{8}''$	$\frac{7}{8}''$		
$\frac{3}{4}''$	$1''$		
$\frac{7}{8}''$	$1\frac{1}{8}''$	$\begin{Bmatrix} 12t \\ 6 \text{ in.} \end{Bmatrix}_{min}$	$2\frac{2}{3}d \leq s$
$1''$	$1\frac{1}{4}''$		
$1\frac{1}{8}''$	$1\frac{1}{2}''$		
$1\frac{1}{4}''$	$1\frac{5}{8}''$		
$>1\frac{1}{4}''$	$1.25d_b$		

*Typically 3d; **t is the thickness of the thinner part connected.

TABLE 8.2 Spacing and Edge Distance for Bolts

8.4.2 Bolted Joint Types

In general, we can classify a joint that is loaded with shear or tension, or a combination of both either as non-pretensioned or pretensioned. When pretension is not required, a group of bolts in a joint can be tightened enough with an ordinary spud or impact wrench so that the plies are brought into firm contact. We refer to this non-pretensioned connection type as a snug-tightened joint. In this case, the load is resisted by bearing stress-induced against the plies and shear stress of the non-pretensioned bolts used in the shearing connection. There is no upper or lower tension requirement for snug-tight bolts.

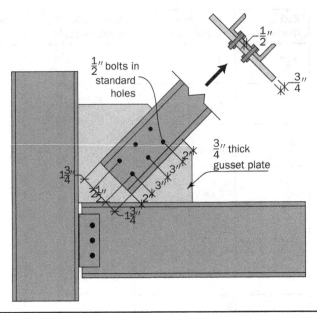

FIGURE 8.5   Example 8.1.

Comparatively speaking, snug-tightened joints are easy to design, install, and inspect and can be used in most structures where slippage is allowed, and pretension is not required. Snug-tightened joints involving A325 bolts in direct tension are also permitted for statically loaded applications, while neither bolts under nonstatic tension load nor A490 bolts under any type of tension load is permitted.

When pretension is necessary, but slippage is not a design consideration, high-strength pretensioned bolts can be used. Per *Specification for Structural Joints Using High-Strength Bolts* (RCSC, 2014), besides code-specified requirements, pretension is required when

- Load reversal is significant.
- Fatigue load exists (without reversal of loading direction).
- A325 bolts that are subject to tensile fatigue are used.
- A490 bolts that are subject to tension or combined tension and shear are used.

In addition, according to *the Specification*, for reasons other than slip-resistance, pretension is also required for several connection types, such as buildings with cranes over 5-ton capacity, members that provide bracing to columns, etc.

Slip-critical connections, where a slip of a bolted joint under service loads is undesirable, would be required in certain cases. For instance, joints that are subject to fatigue load should be designed as slip-critical, given that slippage increases the possibility of fatigue failure. Slip-critical joints are also required when oversized or slotted holes are utilized; potential slip between the faying surfaces has an influence on structural performance or at the joints where welds and bolts are sharing the load. These connections can be considerably expensive compared with other joint types due to the additional cost of faying surface preparation as well as rigorous installation and inspection requirements. A summary of the joint types is presented in Table 8.3.

Joint type*	Remarks	Load transfer
Snug-tightened	• Plies are expected to bring into firm contact. • Bolts are tightened using a spud or impact wrench. • No pretension is applied. • Both common and high-strength bolts can be used. • Easy to design, install, and inspect.	• Load is resisted by shear/bearing. • Provide resistance to shear load only, static tension load only, or combined shear and static tension load.
Pretensioned	• Bolt pretension is required when joints are subject to significant load reversal, fatigue load, and so on. • Slippage is not of concern. • Only high-strength bolts are used.	• Load is resisted by shear/bearing. • Provide resistance to shear load only, tension load only, or combined shear and tension load.
Slip-critical	• Bolts are fully pretensioned to cause a clamping force on the faying surfaces between the connected components. • Slippage is not permitted. • Required when reversed fatigue load exists. • Only high-strength bolts are used. • Preferred when joint slippage is detrimental to the serviceability of the structure. • More expensive.	• Load is transferred by the friction between the bolted components. • Provide resistance to slippage, shear, or combined shear and tension.

* Design, installation, and inspection requirements can be found in RCSC (2014) Sections 4.1 through 4.3.

TABLE 8.3   Summary of Bolted Joint Types (RCSC, 2014)

8.4.3   Limit States

Besides the previously discussed limit states, which are block shear rupture and tensile rupture in the net section, the possible failure modes possible in several bolted connection types are listed as follows:

- Bolt failure under shear
- Bolt tension fracture
- Bolt failure under combined shear and tension
- Bearing/tearout at bolt holes
- Shear yielding or rupture of connected parts

Before attending to specific bolted connection types, we need to examine and understand the basics of the above-listed limit states. Note that some of the possible limit states will not be covered in detail, such as fatigue-related limit states and the limit states under concentrated forces.

8.4.3.1 Bolt Shear

Shear failure of a fastener that connects two or more elements, such as plates as demonstrated in Fig. 8.6, physically occurs when a fastener is fractured from the plane along the contact surface between the connected plies. Although a failure due to pure shear is an idealization, the internal force developed in each failure plane of an individual fastener would be estimated as the product of gross area and the average shear stress. Using the free-body-diagram shown in Fig. 8.6 (left), the shear strength of a bolt in a lap-joint can be expressed by

$$P = F_v A_b \tag{8.1}$$

$$A_b = \frac{\pi d_b^2}{4} \tag{8.2}$$

where P is the load per fastener, F_v is the average shear stress over the failure plane, A_b is the gross cross-sectional area of an individual bolt, and d_b is the nominal bolt diameter. This case, where shear failure could only occur in a single plane, is named as "single shear." In some cases, a fastener may clamp more than two plates creating two planes of shear failure, as depicted in Fig. 8.6 (right). Considering the free-body-diagrams of a failed bolt shown in Fig. 8.6 (right), the applied load can be related to the average shear stress as

$$P / 2 = F_v A_b \tag{8.3}$$

In this double-lap-joint case, the total applied load is shared by force induced in two shear planes, and therefore this loading case is referred to as "double shear."

For design purposes, shear failure limit state is assumed to take place when the average shear stress, F_v uniformly distributed over the cross-section attains the nominal shear strength, F_{nv}. Thus, the nominal shear strength per bolt per shear plane will be

$$r_n = F_{nv} A_b \tag{8.4}$$

The nominal shear strength of a connection becomes

$$R_n = r_n \times \{\text{number of shear planes}\} \times \{\text{number of bolts}\} \tag{8.5}$$

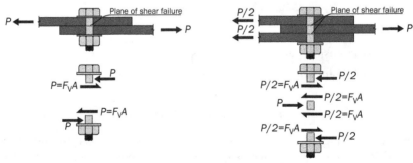

FBD of a bolt subjected to single shear   FBD of a bolt subjected to double shear

Figure 8.6 Bolted shear connection: single shear (left) and double shear (right).

Bolt description	Nominal shear strength, F_{nv} (ksi)
A307 bolts	27
A325 bolts (threads are not excluded from shear planes)	54
A325 bolts (threads are excluded from shear planes)	68
A490 bolts (threads are not excluded from shear planes)	68
A490 bolts (threads are excluded from shear planes)	84

*$F_{nv} = 0.563F_u$ when threads are excluded.
**$F_{nv} = 0.45F_u$ when threads are not excluded.

TABLE 8.4 Nominal Shear Strength of Various Fasteners in Bearing-Type Connections

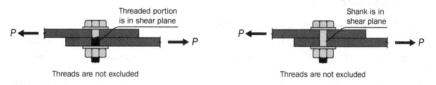

FIGURE 8.7 Bolted connection: threads are not excluded (left), and threads are excluded (right).

Nominal shear strength, F_{nv} values adopted from *the Specification, Table J3.2* are summarized in Table 8.4 for common and high-strength bolts. The nominal shear strength for the same high-strength bolt type (e.g., A325 or A490 bolts) varies depending on whether the threads are in the shear plane or not. Given the threaded portion has a smaller area than that of the shank (i.e., unthreaded part), the reduction in the bolt diameter is accounted for by decreasing the shear stress by about 20%. These connections that the threads are included in and excluded from the shear plane are illustrated in Fig. 8.7, respectively. Note that it is better to be on the safe side by assuming that threads are included in the plane if the thread length is not known when designing a bolted connection for shear.

Example 8.2
A double-shear bolted connection consisting of three rectangular plates and six ⅝″ A325 bolts (threads are not in the shear planes is shown in Fig. 8.8). Find the nominal strength of the connection, considering the shear failure of the bolts only.

Solution:
Since threads are excluded from the shear planes, $F_{nv} = 68$ ksi (*Specification, Table J3.2*)

$$\text{Nominal bolt area, } A_b = \frac{\pi d_b^2}{4} = \frac{\pi \times \left(\frac{5}{8}\right)^2}{4} = 0.307 \text{ in.}^2$$

Number of bolts $= 6$
Number of shear planes $= 2$
Nominal shear strength per bolt per shear plane, $r_n = F_{nv}A_b = 68^{\text{kips/in.}^2} \times 0.307^{\text{in.}^2} = 20.86$ kips
Nominal shear strength of the connection, $R_n = r_n \times \{\text{number of shear planes}\} \times \{\text{number of bolts}\}$

$$R_n = 20.86^{\text{kips}} \times 2 \times 6 = 250.4 \text{ kips} \quad \text{Ans.}$$

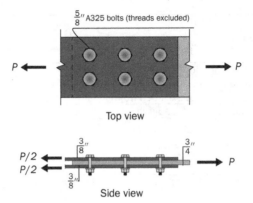

Top view

Side view

FIGURE 8.8 Example 8.2.

8.4.3.2 Bearing/Tearout at Bolt Holes

Stress developed at bolt holes due to the contact pressure between the bolt, and the connected part is called bearing stress. To better visualize this, the bearing of a connecting plate at a bolt hole is demonstrated by the simulated bolted lap-joint shown in Movie 8.1 and Fig. 8.9. The joint is subjected to ever-increasing axial load, P (Fig. 8.9a). As demonstrated by the excessive bearing deformation at the individual bolt hole shown in Fig. 8.9c, bearing limit state is associated with the connected parts and therefore influenced by the geometric parameters (e.g., plate thickness or bolt diameter) along with the strength of the connected material. Note that even though bending deformation due to the eccentricity between the two plates also seems dominant in this case, the interaction between shear and tensile deformations will not be discussed here (Fig. 8.9b).

The nominal bearing strength for a joint at each bolt hole can be expressed as the product of the bearing stress and bearing area. Given that estimating the actual stress distribution around the hole as well as the area in contact can be quite complicated, we tend to simplify the nominal bearing strength calculations. We usually use a uniform bearing stress distribution based on the average stress acting on the projected bolt area, as illustrated in Fig. 8.10. Thus, the bearing area is expressed as

$$\text{Bearing Area} = \text{thickness of the connected material} \times \text{bolt diameter} = d_b \times t \quad (8.6)$$

The nominal bearing strength against each bolt hole will be

$$r_n = C \times \left(d_b \times t_{\min}\right) \times F_u \quad (8.7)$$

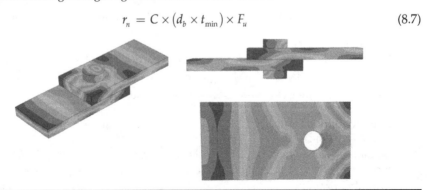

MOVIE 8.1 Bearing of a bolted plate. (To view the entire movie, go to www.mhprofessional.com/design-of-steel-structures.)

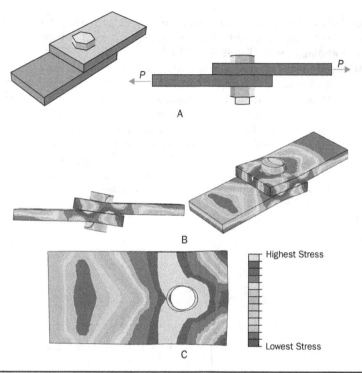

FIGURE 8.9   Illustration of bearing deformation at a hole: (a) Typical bolted lap-joint. (b) Deformed shape. (c) Bearing deformation at a hole.

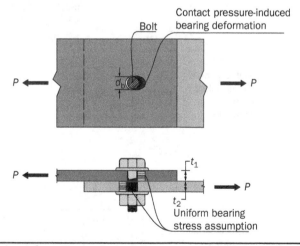

FIGURE 8.10   Bearing deformation against a hole.

in which C is a constant having 2.4 or 3.0 depending on whether the local bearing deformation at service load is a concern or not, and F_u is the ultimate stress of the connecting element. The governing bearing stress has been experimentally determined. Extensive experimental research (Kulak et al., 1987) pointed out that a hole could attain a bearing strength of around 2.4 times the product of bearing area and ultimate stress of the connected material when a reasonable bearing deformation of ¼″ is expected. A bearing constant of 3.0, on the other hand, corresponds to a total standard hole elongation equal to the bolt diameter (Kulak et al., 1987).

Bolts might thrust through the edge or adjacent bolt hole, forcing the connected material to rupture, if bearing deformation at a hole is significant. This type of shear failure is called "tearout" failure. Similar to the assumptions in the bearing strength formulation, the shear failure plane, in this case, is idealized to be straight lines for the sake of simplicity. As shown in Fig. 8.11, the nominal tearout strength at each bolt hole can be computed by considering the clear distance from hole to the edge, L_e for exterior bolts or hole-to-hole clear distance, L_c for interior bolts. Clear distances, indicated in Fig. 8.11a, for tearout strength calculations, can be determined by subtracting the nominal bolt diameter, d_h from center-to-center distances as follows:

For exterior bolts → $L_e = e - d_h/2$

For interior bolts → $L_c = s - d_h$

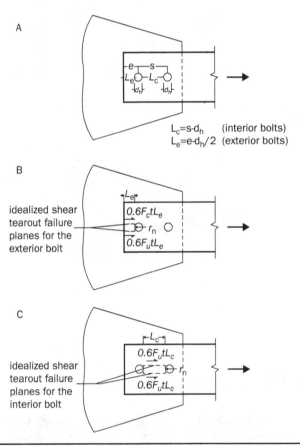

Figure 8.11 Dimensions and strength formulation for interior and exterior bolts for tearout limit state: (a) Dimensions. (b) Failure plane for edge bolt. (c) Failure plane for interior bolt.

Using the idealized shear rupture paths shown in Fig. 8.11b and c, tearout strength for each exterior and interior bolt can be determined as total shear area times shear fracture stress, $0.6F_u$:

For exterior bolts $\rightarrow r_n = 2 \times (L_e \times t) \times (0.6F_u) = 1.2F_u t L_e$ (Fig. 8.11b)

For interior bolts $\rightarrow r_n = 2 \times (L_c \times t) \times (0.6F_u) = 1.2F_u t L_c$ (Fig. 8.11c)

For standard holes, if we substitute C values back into Eq. (8.7) and consider both tearout and bearing strengths, we get

   (1) When the deformation of the bolt hole at service load is a design consideration:

$$r_n = 2.4 d_b t_{\min} F_u \geq 1.2 L_c t F_u$$

   (2) When the deformation of the bolt hole at service load is not a design consideration:

$$r_n = 3.0 d_b t_{\min} F_u \geq 1.5 L_c t F_u$$

The nominal bearing strength for a joint is the sum of the bearing (or tearout strength, whichever is smaller) of individual bolts, and can be calculated as

$$R_n = r_n \times \{\text{number of bolts}\} \tag{8.8}$$

It is important to notice that according to *the Specification*, Eq. (8.8) assumes that each bolt resists the applied force equally regardless of the fastener shear strength or presence of threads.

Example 8.3

Determine the nominal strength of the joint shown in Fig. 8.12 considering shear, bearing, and tearout limit states. Assume that bearing deformation is a design consideration.

Solution:

Nominal shear strength:
Threads are included in the shear planes, $F_{nv} = 68$ ksi

Nominal bolt area, $A_b = \dfrac{\pi d_b^2}{4} = \dfrac{\pi \times 3/4^2}{4} = 0.442$ in.^{2}

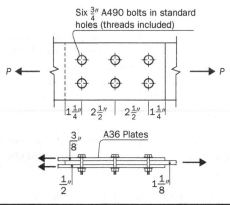

Figure 8.12   Example 8.3.

Number of shear planes = 2 (three plies)

Number of bolts = 6

Nominal shear strength per bolt, $r_n = F_{nv}A_b = 2 \times \left(68^{\text{kips/in.}^2} \times 0.442^{\text{in.}^2}\right) = 60.1$ kips/bolt

Nominal shear strength of the connection, $R_{n1} = 60.1 \times 6 = 360.6$ kips

Nominal bearing/tearout strength:

- Bearing limit state:

Nominal bolt diameter, $d_b = \dfrac{3}{4}$ in.

Ultimate strength of A36 plates, $F_u = 58$ ksi

Thickness for projected bearing area, $t_{min} = \left\{\begin{array}{l}\left[\left(\dfrac{3}{8}+\dfrac{1}{2}\right)\right] \\ \text{or} \\ 1\dfrac{1}{8}\end{array}\right\}_{min} = \dfrac{7}{8}$ in.

Nominal bearing strength per bolt hole, $r_n = 2.4d_b t_{min}F_u = 2.4 \times \dfrac{3}{4} \times \dfrac{7}{8} \times 58 = 91.4$ kips

- Tearout limit state:

Clear distance for exterior bolt $\rightarrow L_e = e - d_h/2 = 1\dfrac{1}{4} - \dfrac{1}{2} \times \left(\dfrac{3}{4} + \dfrac{1}{16}\right) = 0.84375$ in.

Clear distance for interior bolts $\rightarrow L_c = s - d_h = 2\dfrac{1}{2} - \left(\dfrac{3}{4} + \dfrac{1}{16}\right) = 1.6875$ in.

Nominal tearout strength per exterior bolt

$\rightarrow r_n = 1.2L_e tF_u = 1.2 \times 0.84375 \times 7/8 \times 58 = 51.4$ kips/bolt

Nominal tearout strength per interior bolt

$\rightarrow r_n = 1.2L_c tF_u = 1.2 \times 1.6875 \times 7/8 \times 58 = 102.8$ kips/bolt

Bearing strength per exterior bolt is greater than its tearout counterpart, whereas the tearout strength for the interior bolts is larger than their bearing strength. Therefore, the bearing limit state governs for the two exterior bolts while the available strength of the four interior bolt holes is controlled by bearing limit state. Thus, the nominal bearing/tearout strength of the connection will be

$R_{n2} = (2 \times 51.4) + (4 \times 91.4) = 468.4$ kips

Based on the limit states under consideration, the nominal strength of the connection is the smaller of the nominal capacities obtained from shear and bearing/tearout limit states. Therefore, bolt shear failure governs

$R_n = \{R_{n1}; R_{n2}\}_{min} = \{360.6^{\text{kips}}; 468.4^{\text{kips}}\}_{min} = 360.6$ kips   Ans.

Bolt description	Nominal tensile strength, F_{nt} (ksi)*
A307 bolts	45
A325 bolts	90
A490 bolts	113

*$F_{nt} = 0.75F_u$ to roughly account for the ratio of the effective tension area of the threaded portion to the area of the shank.

TABLE 8.5 Nominal Tensile Strength of Fasteners

8.4.3.3   Bolt Tension Fracture

Tensile strength of a bolt is a function of its ultimate tensile strength and the cross-sectional area of the threaded portion. The area of the threaded portion, which is dependent on the thread size, is typically 75% of the nominal bolt area. We account for this reduction in the area by using 75% of the ultimate tensile strength per unit area. Table 8.5 summarizes the nominal tensile strength, F_{nt} for a common, and the two conventional high-strength bolt types.

According to *the Specification*, the nominal tensile strength of a bolt is

$$R_n = F_{nt}A_b \tag{8.9}$$

where A_b is the gross cross-sectional area of the shank. It is essential to note that the tensile capacity of a bolt is independent of the joint type. Therefore, the tensile capacity remains virtually unchanged whether or not pretension is applied. On the other hand, depending on the connection type, the additional tension force resulting from a potential prying action should be included in the tensile strength as per *the Specification*, J3.6. If the legs of the connecting element had sufficient stiffness, the tension force in the bolts would be equal to the applied load. Otherwise, prying action causes a compression force near the tip of the outstanding leg, which increases the tension force in the bolt. Prying action only applies in bolted construction and for tensile bolt forces, such as bolted hanger connections with a tee-shape or angle. According to *the AISC Manual* (AISC, 2011), the thickness required to prevent prying action is

$$t_{min} = \sqrt{\frac{4Tb'}{\phi p F_u}} \tag{8.10}$$

where

F_u = specified minimum tensile strength of the connecting element (tee or angle)
T = required strength per bolt

$$b' = b - \frac{d_b}{2}$$

b = for a tee-type connecting element, the distance from bolt centerline to the face of the tee stem; for an angle-type connecting element, the distance from bolt centerline to centerline of angle leg
p = tributary length (maximum $2b$ but should be less than s)
s = bolt spacing
$\phi = 0.9$

If the fitting thickness is equal to or greater than t_{min}, the additional force in bolt becomes negligible. Hence, a prying action check is not necessary.

8.4.3.4 Tension and Shear Interaction in Bearing-Type Connections

At the ultimate limit state, failure of a fastener in a bearing-type connection would take place by shear or tensile failure if the bolt is loaded only with shear or tension. In many cases, however, shear and tension forces are simultaneously experienced. Based on the experimental tests (Kulak et al., 1987), the interaction between shear and tension has been defined by an elliptical relationship as follows:

$$\left(\frac{f_t}{\phi F_{nt}}\right)^2 + \left(\frac{f_v}{\phi F_{nv}}\right)^2 = 1.0 \tag{8.11}$$

where f_t and f_v are the required tensile and shear stresses, respectively. The elliptical representation can be replaced with a similar relationship consisting of three straight lines as follows:

when $\dfrac{f_v}{\phi F_{nv}} \leq 0.3 \rightarrow f_t = \phi F_{nt}$ $\qquad\qquad$ (8.12a)

when $\begin{cases} 0.3 \leq \dfrac{f_v}{\phi F_{nv}} \leq 1.0 \\ \text{and} \\ 0.3 \leq \dfrac{f_t}{\phi F_{nt}} \leq 1.0 \end{cases} \rightarrow \left(\dfrac{f_t}{\phi F_{nt}}\right) + \left(\dfrac{f_v}{\phi F_{nv}}\right) = 1.3$ $\qquad$ (8.12b)

when $\dfrac{f_t}{\phi F_{nt}} \leq 0.3 \rightarrow f_v = \phi F_{nv}$ $\qquad\qquad$ (8.12c)

The straight-line representation given in Eq. (8.12) not only accurately approximates the original elliptical interaction curve in Eq. (8.11) but also neglects the adverse impact of combined effects when the required stress is equal to or less than 30% of the corresponding available design stress.

According to *the Specification, Section J3.7*, the nominal tensile strength of a bolt subjected to combined effects of shear and tension is

$$R_n = F'_{nt} A_b \tag{8.13a}$$

$$F'_{nt} = 1.3 F_{nt} - \frac{F_{nt}}{\phi F_{nv}} f_{rv} \leq F_{nt} \tag{8.13b}$$

where
$\quad F'_{nt}$ = nominal tensile stress modified to include the effects of shear stress
$\quad F_{nt}$ = nominal tensile stress
$\quad F_{nv}$ = nominal shear stress
$\quad f_{rv}$ = required shear stress

Example 8.4

A column-to-bracing connection is shown in Fig. 8.13. Diagonal bracing made with double-channels welded to a tapered gusset plate. The gusset plate is also welded to an end plate while the column and the gusset assembly are bolted together with eight ½" A490 bolts. Factored axial load P_u is estimated as 75 kips. Threads are included in the planes, and the bolts are snug-tightened. Determine the nominal tensile strength of each bolt.

Solution:
The bolts are subjected to combined tension and shear. Since the longitudinal axis of bracing coincides with the center of gravity of the bolt group, each bolt will share the tension and shear load equally.

For ½" A490 bolts:

Nominal tensile stress, $F_{nt} = 113$ ksi

Nominal shear stress, $F_{nv} = 68$ ksi

Nominal cross-sectional area of each bolt, $A_b = \dfrac{\pi d_b^2}{4} = \dfrac{\pi \times 1/2^2}{4} = 0.196$ in.²

Required shear stress for each bolt is

$$f_{rv} = \frac{P_u \sin 45^0}{n A_b} = \frac{75 \times \sin 45^0}{8 \times 196} = 33.8 \text{ ksi}$$

The modified tensile stress becomes

$$F_{nt}' = 1.3 \times 113 - \frac{113}{0.75 \times 68} \times 33.8 = 72 \text{ ksi} \le F_{nt} = 113 \text{ ksi}$$

Thus, the nominal tensile strength of an individual bolt will be

$$r_n = F_{nt}' A_b = 72 \times 0.196 = 14.1 \text{ kips}$$

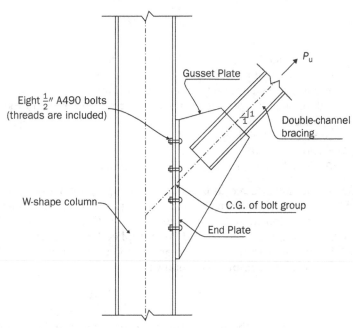

FIGURE 8.13 Example 8.4.

8.4.3.5 *Slip Prevention in Slip-Critical Connections*

Slip-critical connections are designed considering the strength limit states applicable to bearing-type connections as well as to prevent slip at working loads. To satisfy the serviceability requirement, it is assumed that the load is transmitted solely through the friction-induced between the faying surfaces under service loads. In other words, if properly pretensioned, slip-critical connections are unlikely to subject to bearing or shear, although they are designed to have sufficient shear and bearing strength.

For design purposes, the foregoing slip-resistance is provided by the friction force resulting from the applied bolt pretension-induced contact pressure. Figure 8.14 illustrates how the load is transmitted from one member to another. To determine the slip-resistance limit on the applied force on the simple slip-critical joint shown in Fig. 8.14a, we should recognize the force transfer as well as the interaction between bolt assembly and plates. Subsequent to applied pretension, T_b the clamping force develops washer-to-plate and plate-to-plate contact (Fig. 8.14a). As demonstrated by the free-body-diagram of the pretensioned bolt shown in Fig. 8.14b, the total contact force, N, produced between the washer below the bolt head and the top plate has to balance the internal bolt tension force, T_b. If we transfer the total pressure, N to the top plate, as shown in Fig. 8.14c, we can easily associate the pretension force, T_b to the applied force, P. Satisfying the vertical and horizontal force equilibrium of each plate using the free-body-diagram given in Fig. 8.14c indicates that the force on the connection, P should not exceed the following to avoid slippage in theory.

$$P \leq F = \mu N = \mu T_b \tag{8.14}$$

Per *the Specification, Section J3.8*, the nominal slip-resistance per bolt can be determined as follows:

$$r_n = \mu D_u h_f T_b n_s \tag{8.15}$$

where
  D_u = a multiplier to account for the mean installed bolt pretension to the specified minimum bolt pretension ratio, can be taken as 1.13
  T_b = minimum fastener pretension listed in Table 8.6
  h_f = 1.0 for one filler between connected parts, 0.85 for two or more fillers between connected parts
  n_s = number of slip planes
  μ = mean slip coefficient, 0.3 for Class A surfaces (e.g., unpainted clean steel) and 0.5 for Class B surfaces (e.g., unpainted blast-cleaned steel surfaces)

The Specification also recognizes that the clamping force due to applied pretension would be reduced when a tensile load is acting on a connection. Reduction in the net clamping force is considered by introducing an additional factor, k_{sc}. The nominal slip resistance per bolt, therefore, should be multiplied by k_{sc} factor, which is a function of the ratio of the factored tension force on the connection and the modified pretension force. k_{sc} factor can be computed as follows:

$$k_{sc} = 1 - \frac{T_u}{D_u T_b n_b} \geq 0 \tag{8.16}$$

in which T_u and n_b are the required tension force and number of bolts carrying the applied tension, respectively.

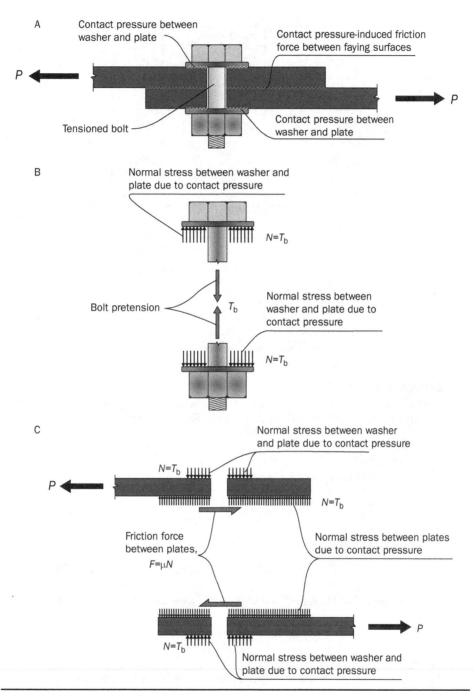

A — Contact pressure between washer and plate

Contact pressure-induced friction force between faying surfaces

P

P

Tensioned bolt

Contact pressure between washer and plate

B — Normal stress between washer and plate due to contact pressure

$N = T_b$

Bolt pretension

T_b

Normal stress between washer and plate due to contact pressure

$N = T_b$

C — Normal stress between washer and plate due to contact pressure

$N = T_b$

P

$N = T_b$

Friction force between plates,

$F = \mu N$

Normal stress between plates due to contact pressure

$N = T_b$

P

$N = T_b$

Normal stress between washer and plate due to contact pressure

FIGURE 8.14 Load transfer in slip-critical connections: (*a*) Slip-critical joint. (*b*) FBD of bolt assembly. (*c*) FBD of plates.

Bolt diameter	A325 Bolts	A490 Bolts
$\frac{1}{2}''$	12	15
$\frac{5}{8}''$	19	24
$\frac{3}{4}''$	28	35
$\frac{7}{8}''$	39	49
$1''$	51	64
$1\frac{1}{8}''$	64	80
$1\frac{1}{4}''$	81	102
$1\frac{3}{8}''$	97	121
$1\frac{1}{2}''$	118	148

TABLE 8.6 Minimum Bolt Pretension (kips)

Example 8.5
Rework Example 8.3 by assuming that the connection is slip-critical and find the nominal slip-resistance of the connection given in Fig. 8.12. Surfaces are unpainted mill scale.

Solution:
$$D_u = 1.13$$
$$T_b = 35 \text{ kips for } \frac{3}{4}'' \text{ A490 bolts}$$
$$h_f = 1.0$$
$$n_s = 2$$
$$\mu = 0.3$$

Each bolt can resist

$$r_n = \mu D_u h_f T_b n_s = 0.3 \times 1.13 \times 1.0 \times 35 \times 2 = 23.73 \text{ kips/bolt}$$

There are six bolts. Therefore, nominal slip-resistance of the connection is

$$R_n = r_n \times \text{number of bolts} = 23.73 \times 6 = 142.4 \text{ kips} \qquad \text{Ans.}$$

Example 8.6
Rework Example 8.4 by assuming that the connection is slip-critical and find the nominal slip-resistance of each fastener given in Fig. 8.13. Surfaces are unpainted blast-cleaned.

Solution:
Given the slip-critical connection is subjected to shear combined with tension, the reduction in the clamping force should be considered by introducing the reduction factor, k_{sc}

given in Eq. (8.16). Required tension force, T_u will be the horizontal component of the total factored axial load, P_u:

$$T_u = P_u \cos 45^0 = 75 \times \cos 45^0 = 53.03 \text{ kips}$$

The rest of the parameters needed for k_{sc} calculation are
   $D_u = 1.13$
   $T_b = 15$ kips for ½″ A490 bolts
   $n_b = 8$

Thus, the reduction will be

$$k_{sc} = 1 - \frac{T_u}{D_u T_b n_b} = 1 - \frac{53.03^{\text{kips}}}{1.13 \times 15 \times 8} = 0.609$$

Since $h_f = 1.0$, $n_s = 1$, $\mu = 0.5$ (Class B surface), the nominal slip-resistance of each bolt is

$$k_{sc} r_n = k_{sc} \mu D_u h_f T_b n_s = 0.609 \times (0.5 \times 1.13 \times 1.0 \times 15 \times 1) = 5.16 \text{ kips/bolt} \quad \text{Ans.}$$

8.5   Welded Connections

8.5.1   General Remarks

Structural welding, in general, is a fabrication process during which two or more metal structural parts, such as steel gusset plate and beam or column, are melted together to join with the addition of molten filler material between them using heat treatment. Welding has a wide range of applications and can be preferred in many structural applications, such as industrial or multistory buildings, for their advantages over bolting:

(1) Welded connections are more rigid because of the absence of additional flexible connector elements often needed for bolted connections.

(2) Likewise, the direct connection between the welded parts requires no or fewer connector elements (e.g., angles, plates, etc.) and therefore welded connections reduces the overall steel weight, as well as the labor cost of hole drilling and bolt installment, not to mention complete exclusion of bolt cost.

(3) Elimination of the area reduction and associated limit states due to the drilled bolt holes.

(4) Welding offers the possibility of joining a much wide range of shapes, including hollow shape-to-hollow shape connections.

Still, there exist certain difficulties in making welded connections, primary of which are as follows:

(1) Welded connections require skilled workers that are certified.

(2) Minor flaws may result in serious problems, such as cracking under load reversals. Therefore, quality control is very important but usually difficult and would necessitate inspection techniques other than visual inspection, such as ultrasonic testing or radiographic procedures.

(3) Employment of qualified welders and inspectors would bring in a significant additional cost.

(4) Heat treatment during the welding process generates residual stresses (we discussed the potential issues associated with residual stresses in Chapter 4) along with distortion, and alters the mechanical properties or microstructure of the material in the vicinity of the welded base metal.

Prequalified welding methods that are recognized by the American Welding Society (AWS) are shielded metal arc welding (SMAW), submerged arc welding (SAW), gas metal arc welding (GMAW), gas tungsten arc welding (GTAW), and flux-cored arc welding (FCAW) (AWS, 2000). SMAW and SAW, however, are more common than the other prequalified welding processes. SMAW process, also known as "stick welding," is common for both on-field and shop applications. This process uses an electrical circuit to form an arc between a covered electrode and the base metal. The electrode core, which acts as a sole filler material and fills the gap between the base metal and electrode, is covered with shielding. This protective coating is used as "shielding" to prevent (or reduce) the welded area from oxidation and contamination (AWS, 2000). Due to its simplicity, SMAW process has been one of the most popular welding processes in steel construction (Weman, 2003). Similar to SMAW, SAW is an arc welding process that uses electricity to create enough heat to melt metal. However, the SAW is often an automated shop-welding process rather than manual. Instead of a shielded electrode, the molten weld metal and the arc are shielded by a blanket of granular flux (AWS, 2000). Although it requires relatively complex equipment compared with the SMAW process, the SAW provides deeper weld penetration, faster, and ductile and corrosion-resistant welds.

8.5.2   Weld Types and Symbols

We usually divide the weld types into two broad categories as fillet welds and groove welds. However, despite that they are not as common as fillet and groove welds, plug and slot welds are also used in certain cases when it is not adequate to use fillet or groove welds. These four weld types and their symbols are illustrated in Figs. 8.15 and 8.16, respectively.

Selection of proper joint-weld type may not always be an easy task because of the quality, and constructive considerations need to be taken into account in advance during design and fabrication. One of such considerations that we recognize besides the strength requirement when detailing is lamellar tearing where the anisotropy of the material due to the rolling process results in separation of base metal. Therefore, as shown in Fig. 8.15, depending on the position of the adjoining members with respect to each other, different welded connections can be used in several joint types. Still, most welded connections are made with fillet welds, which have, in theory, right-angled triangular cross-sections, as demonstrated in the typical lap-joint shown Fig. 8.15a. The fillets of these triangular cross-sections can be either concave or convex and are defined by their leg size, which determines the throat size and the strength of the connection consequently (Fig. 8.15a). If the angle between the legs is less or greater than 90 degrees, the welds are called skewed fillets. It is permitted for legs of a fillet weld to intersect at angles within 60 and 120 degrees. The welds are considered to be groove welds if the angular limits are exceeded.

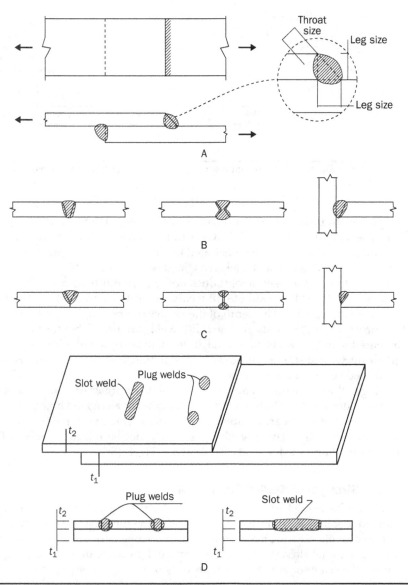

FIGURE 8.15 Classification of structural welds: (*a*) fillet-welded lap-joint; (*b*) complete-joint-penetration (CJP) groove welds; (*c*) partial-joint-penetration (PJP) groove welds; (*d*) plug and slot welds.

Groove welds are the welds made in the groove between the pieces to be joined. This welding type can be, therefore, more suitable in butt or tee joints, as illustrated in Figs. 8.15*b* and *c*. Butt joints can be either single- or double-beveled with various configurations, such as V, U, or J grooves. For detailed information on how a joint with groove welds should be detailed, readers should refer to AWS D1.1 (AWS, 2000). Depending on the penetration required, groove welds can be further categorized as

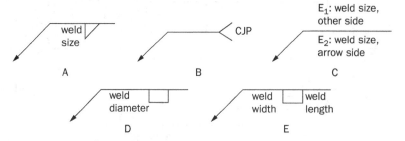

FIGURE 8.16 Weld symbols: (a) fillet weld; (b) CJP groove weld; (c) PJP groove weld; (d) plug weld; (e) slot weld.

complete-joint-penetration (CJP) or partial-joint-penetration (PJP) groove welds. A groove weld is named as CJP when full penetration throughout the depth of the workpieces is achieved, as demonstrated in Fig. 8.15b. Thus, the thickness of the thinner of the connected pieces can be taken as the throat size of a CJP weld for strength calculations. But, in some cases, a joint does not require full penetration strength. In those cases we employ PJP instead of CJP welds. Typical examples of joints with PJP welds are shown in Fig. 8.15c. The depth of the groove can be roughly considered as the effective throat when PJP welds are used. PJP welds are preferable for their cost-efficiency compared with CJP welds. In general, unless CJP welds are required by design codes, or PJP (or fillet) weld strength is not sufficient to transfer loads, CJP groove welds should be avoided because of their high-cost.

Plug welds are made in circular holes drilled or punched in the adjoining member, whereas slot welds are similar but made in slotted or elongated holes. The holes formed in plug and slot welds are completely or partially filled with the filler material deposited during welding. These weld types should not be mixed up with a fillet-welded hole or slot. Plug and slot welds are shown in Fig. 8.15d.

8.5.3 Strength of Welded Connections

Welds can be loaded in any direction, but are weakest in shear and are therefore assumed to fail in shear. Hence, we determine the nominal weld strength by considering the shear stresses on an effective weld area. Considering the limit state of weld rupture, the nominal strength, R_n, for a linear weld group with a uniform leg size, loaded through the center of gravity is as follows as per *the Specification, J2.4a*:

$$R_n = F_{nw} A_{nw} \qquad (8.17a)$$

where A_{we} is the effective area of weld, which is the product of the effective length and the effective throat. Thus, considering the throat size of the right triangle shown in Fig. 8.17a, the effective area of a fillet weld is

$$A_{nw} = t_{eff} \times \text{weld length} = 0.707 t_w \times \text{weld length} \qquad (8.17b)$$

As previously mentioned, the effective throat size of groove welds can be taken as the thickness of the thinner workpiece. Therefore, the effective area of groove welds can be computed in the same manner with fillet welds as the length of the weld multiplied by the effective throat. It should be noted that the effective throat of a PJP weld might be dependent on the process used and the weld position (AISC, 2016).

According to *the Specification, J2.4b(1)*, for a linear weld with uniform leg size group loaded in the plane of the weld, loaded through the center of gravity, the nominal weld strength in shear is

$$F_{nw} = 0.60F_{EXX}\left(1.0 + 0.50\sin^{1.5}\theta\right) \tag{8.17c}$$

where
F_{EXX} = electrode (filler metal) strength in ksi
θ = angle of loading measured from the weld longitudinal axis

As depicted in Fig. 8.17b, the angles for the weld groups with respect to the direction of loading are 0 and 90 degrees, respectively, for the transversely and longitudinally oriented fillet welds. According to *the Specification, J2.4b(2)*, when concentrically loaded fillet weld groups with a uniform leg size are oriented both longitudinally and transversely to the direction of applied load, the combined strength, R_n, of the fillet weld group should be determined based on strain compatibility. Therefore, the combined strength, R_n, can be determined as the greater of Eqs. (8.18a) and (8.18b):

$$R_n = R_{nwl} + R_{nwt} \tag{8.18a}$$

$$R_n = 0.85R_{nwl} + 1.5R_{nwt} \tag{8.18b}$$

where
R_{nwl} = total nominal strength of longitudinally loaded fillet welds
R_{nwt} = total nominal strength of transversely loaded fillet welds

A welded connection can be as strong as its weld material and base metal. Thus, the nominal strength of a welded joint is determined as the lower of the weld and base metal strengths. The base material strength can be determined according to the limit states of tensile rupture and shear rupture as follows:

$$R_n = F_{nBM}A_{BM} \tag{8.19}$$

where F_{nBM} and A_{BM} are the nominal stress of the base metal and cross-sectional area of the ruptured base metal, respectively.

In addition to the strength criteria, the limitation on the minimum fillet weld size is summarized in Table 8.7. The maximum fillet weld size cannot be greater than the thickness of the material if the material is thinner than ¼″. If the material thickness is equal or greater than ¼″, a clearance of $^1/_{16}$″ should be left from the edge. Likewise, the weld length should not be less than four times the weld size, t_w. Even though there is no upper limit on the weld length, the actual length is reduced to an effective length by multiplying the actual length by β when the weld length exceeds 100 times the weld size (t_w). The reduction factor β is defined as a linear function of weld length and size:

$$\beta = 1.2 - 0.002\left(L_w / t_w\right) \leq 1.0 \tag{8.20}$$

When the length of the weld exceeds 300 times the leg size, t_w, the effective length is taken as $180t_w$.

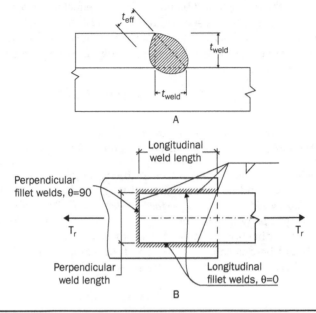

Figure 8.17   Fillet weld strength: (a) effective weld thickness; (b) weld orientation.

Material thickness of thinner part joined	Minimum fillet weld size
$t_{min} \leq \frac{1}{4}''$	$\frac{1}{8}''$
$\frac{1}{4}'' < t_{min} \leq \frac{1}{2}''$	$\frac{3}{16}''$
$\frac{1}{2}'' < t_{min} \leq \frac{3}{4}''$	$\frac{1}{4}''$
$\frac{3}{4}'' < t_{min}$	$\frac{5}{16}''$

Table 8.7   Minimum Fillet Weld Size

Example 8.7

Design the welds to develop the full strength of the angles L7 × 4 × 1/2 and minimize eccentricity for the connection shown in Fig. 8.18. A36 steel and E70 (F_{EXX} = 70 ksi) electrodes are used. The strength of the connection $T_r = \phi_t A_g F_y$ of the tension member. Gusset plate is designed to be stronger than the weld or the angle. Thus, all possible limit states come from the weld and angle. A_g = 5.25 in.² for L7 × 4 × 1/2.

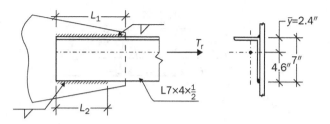

FIGURE 8.18 Example 8.7.

Solution:

Design strength of the connection is based on the gross section yielding limit state as follows:

$$T_r = \phi_t F_y A_g = 0.9 \times 36 \times 5.25 = 170 \text{ kips}$$

Since the minimum thickness is greater than ¼″, the fillet weld size, t_w, can be determined as

$$t_w = t_{min} - \frac{1}{16}'' = \frac{1}{2} - \frac{1}{16} = \frac{7}{16} \text{ in.}$$

Nominal shear rupture strength of the $\frac{7}{16}$-in.-thick weld with $F_{EXX} = 70$ ksi per unit length is

$$r_n = (0.6 F_{EXX}) t_{eff} = (0.6 \times 70)\left(0.707 \times \frac{7}{16}\right) = 13 \text{ kips/inch}$$

There are two lines of longitudinal fillet welds with lengths of L_1 and L_2. Therefore, we can eliminate the eccentricity due to the applied tensile load by selecting two different lengths considering the load distribution among them. By satisfying the longitudinal force and the moment equilibrium, we can get L_1 and L_2 as follows:

$$\left.\begin{array}{l} \sum F_{long} = 0 \rightarrow \phi_t r_n L_1 + \phi_t r_n L_2 = T_r \\ \sum M = 0 \rightarrow (\phi_t r_n L_2) \times 7 - T_r \times 2.4 = 0 \end{array}\right\} \Rightarrow L_1 = 11.5 \text{ in. and } L_2 = 6 \text{ in.}$$

After calculating the weld lengths to eliminate eccentricity, the limit state of base metal rupture needs to be checked to make sure that the yielding in the gross section governs.

$$\phi_t R_n = \phi_t 0.6 F_u A_{nw} = 0.75 \times 0.6 \times 58 \times \left[\frac{1}{2} \times (11.5 + 6)\right] = 228 \text{ kips} > T_r = 170 \text{ kips}$$

Example 8.8

Determine the design strength, T_r, of the fillet-welded connection shown in Fig. 8.19. E70 electrode is used. Plates are made of A36 steel.

Solution:

The design strength of the connection will be controlled by the limit state with the smallest strength. Therefore, each limit state should be investigated one by one.

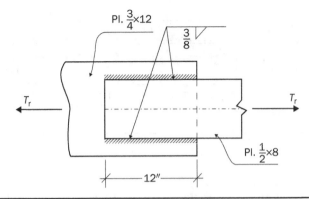

Figure 8.19   Example 8.8.

The design strength of the connection based on the gross section yielding limit state is

$$\phi_t T_n = \phi_t F_y A_g = 0.9 \times 36 \times \left(\frac{1}{2} \times 8\right) = 129.6 \text{ kips}$$

The minimum weld size for $\frac{1}{2}$"-thick plate is $\frac{3}{16}$". The minimum thickness is greater than $\frac{1}{4}$"; the maximum fillet weld size, t_w, can be determined as

$$t_{w,min} = \frac{3}{16} \text{ in.} < t_w = \frac{3}{8} \text{ in.} < t_{w,max} = t_{min} - \frac{1}{16}^{''} = \frac{1}{2} - \frac{1}{16} = \frac{7}{16} \text{ in.}$$

Thus, the fillet weld size is within limits. Note that there is no need to reduce the strength of the fillet weld since the length does not exceed 100 times the weld size ($L_w = 12'' < 100 \times 3/8 = 37.5''$).

The design strength of the $\frac{7}{16}$-in.-thick weld with $F_{EXX} = 70$ ksi is

$$\phi_t R_n = \phi_t F_{nw} A_{nw} = 0.75 \times 70 \times \left[2 \times \left(0.707 \times \frac{3}{8}\right) \times 12\right] = 334 \text{ kips}$$

The design strength based on the limit state of base metal rupture will be

$$\phi_t R_n = \phi_t 0.6 F_u A_{nw} = 0.75 \times (0.6 \times 58) \times \left(2 \times \frac{1}{2} \times 12\right) = 313 \text{ kips}$$

The limit state of gross section yielding governs. Thus, the design strength of the connection is

$$\phi_t R_n = \{129.6; \ 334; \ 313\}_{min} = 129.6 \text{ kips}$$

8.6   Design Procedures for Common Beam-to-Column Connections

Design of a moment connection is based on the types of lateral loads acting on the moment frame, as follows:

- For wind load or low-seismicity demand, the frame is expected to remain elastic, and the moment connection only needs to meet requirements in Chapter J, AISC 360.

- For high-seismicity demand, the frame is expected to sustain large inelastic deformation, and the moment connection needs to meet requirements in both AISC 360 and the seismic provisions in AISC 341 and AISC 358.

Typical fully restrained beam-to-column connection details for wind or seismic applications are shown in Fig. 8.20*a* and *b*. In the welded-flange bolted web moment connection, beam flanges are directly connected to column flange through CJP groove welds, as indicated in Fig. 8.20*a*. Web of the girder, however, bolted to a web plate, which is shop-welded to column flange (Fig. 8.20*a*). This moment connection is rigid and relatively easier to make but requires on-site welding, which would not be preferable to ensure the quality of groove welds at flanges. With the bolted flange plate (BFP) connection, on the other hand, field welding can be avoided. Typical BFP connection, shown in Fig. 8.20*b*, consists of flange plates shop-welded to the column, bolted to beam flanges and a single plate for web connection. Two identical plates are bolted to top and bottom flanges, respectively, as "Flange Connections." A shear tab plate is bolted to the beam web and fillet welded to the column flange as "Web Connection." This connection is more flexible and can be employed for lateral force–resisting systems that resist wind, low- and high-seismicity. However, the necessity of incorporating additional components (i.e., top and bottom flange plates) might not enhance the attractiveness of BFP connections.

Simple connections given in Fig. 8.20*c* and *d* are typical examples of column-to-beam and girder-to-beam shear connections, respectively. In the all-bolted, double-angle column-to-beam connection shown in Fig. 8.20*c*, web angles are shop-bolted to the column and field-bolted to the girder. All-bolted double-angle connections are quite flexible to accommodate end rotations but require more components than other options, such as single-angle or shear tab connections. Shear tab or single-plate connections, either column-to-beam or beam-to-beam, are simpler, stiffer, and require fewer components compared with their double-angle counterparts. Two examples of simple beam-to-beam connections with the shear tab are demonstrated in Fig. 8.20*d*. These shear tabs are shop-welded to the supporting girder and field-bolted to the supported secondary beam. Since top flanges of both members connected (i.e., girder and beam) have to be at the same floor level, the supported beam should be either single- or double-coped depending on its depth, as presented in Fig. 8.20*d*.

Basic design assumption for a fully restrained (rigid) connection is that shear force is resisted solely by the web connection while the bending moment is resisted solely by the flange connection. Likewise, a shear connection, unless it is coped, is designed considering only the limit states associated with shear force. Design procedures for the commonly employed fully restrained and shear connections will be presented in detail in what follows along with the worked examples.

8.6.1   Bolted Flange Plate Connections

Typically, ASTM A992 steel is used for beam and column sections and A36 for plates. Fully pretensioned high-strength A325 or A490 bolts designed for bearing are used to connect the web plate and beam web, whereas the A36 web plate is fillet-welded to the column flange usually using 70-ksi electrodes. Consistent with the preceding design assumptions, the design of the fully restrained BFP moment connection, illustrated in Fig. 8.21*b*, can be classified as a two-phase procedure: (1) design for bending; and

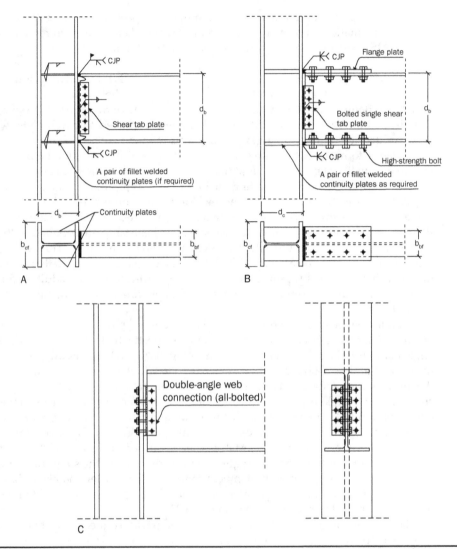

Figure 8.20 Common connection types: (a) welded-flange bolted-web (FR); (b) bolted flange plate (FR); (c) all-bolted double-angle shear connection; (d) coped beam-to-beam connection.

(2) design for shear. Therefore, the limit states can also be divided into two categories based on the associated demand-type as follows:

(1) Design for bending:

  a. Tensile rupture of tension flange

  b. Gross-section yielding in flange plate

  c. Net section fracture in flange plate

  d. Block shear in flange plate

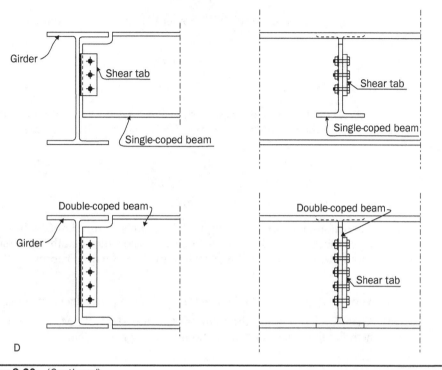

FIGURE 8.20   (Continued)

    e. Bearing/tearout at flange plate

    f. Weld failure at flange plates

    g. Block shear in beam flange

    h. Bearing/tearout at beam flange

    i. Shear failure in flange bolts

    j. Limit states associated with concentrated forces

(2) Design for shear:

    a. Shear tab plate yielding in shear

    b. Shear tab plate fracture in shear

    c. Block shear of shear tab

    d. Shear tab bearing/tearout

    e. Weld failure at shear tab

    f. Shear failure in web bolts

    g. Beam web bearing

Step-by-step design procedure for BFP moment connections is summarized in what follows.

A. Design of flange plates based on bending moment-associated limit states

Step 1. Determine design load

Design load can be determined by converting the required bending moment, M_r, to a force couple using the free-body-diagram given in Fig. 8.21. Thus, the design load is

$$F_r = \frac{M_r}{d_b} \tag{8.21}$$

Step 2. Check tensile rupture of the tension flange

For BFP connections, the flexural strength of the beam is reduced due to the presence of holes at the connection. *The Specification, Section F13* specifies that the nominal flexural strength should be limited according to the limit state of tensile rupture of the tension flange:

(a) When $F_u A_{fn} \geq Y_t F_y A_{fg}$, the limit state of tensile rupture does not apply.

(b) When $F_u A_{fn} \geq Y_t F_y A_{fg}$, the nominal flexural strength, M_n, at the location of the holes in the tension flange should not be greater than

$$M_n = \frac{F_u A_{fn}}{A_{fg}} S_x \tag{8.22}$$

where
  A_{fg} = gross tension flange area
  A_{fn} = net tension flange area
  Y_t = 1.0 for steel with $F_y / F_u \leq 0.8$, 1.1 otherwise

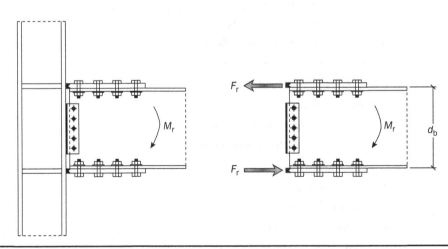

FIGURE 8.21 Concentrated design load on the flange plates.

Step 3. Determine number and diameter of bolts

Since the design load F_r has already been determined, if F_{nv} and bolt diameter, d_b (with common sizes of ½", ¾", ⅞", and 1") are known, the number of bolts, N can also be calculated considering bolt shear failure as follows:

$$F_r \le \phi_t \left[N(F_{nv} A_b) \right] \tag{8.23a}$$

$$N \ge \frac{F_r}{\phi_t (F_{nv} A_b)} \tag{8.23b}$$

Note that two bolts per row (one on each side of the beam web) should be used. Therefore, N has to be an even whole number.

Step 4. Determine the plate dimensions, b_{PL} and t_{PL}, based on the limit state of yielding in gross area

After selecting the flange plate thickness, t_{PL}, which can be slightly larger than beam flange thickness, to ensure a proper yielding of gross area strength, the plate width can be selected using the following inequalities:

$$F_r \le \phi_t F_{y,\text{plate}} A_{g,\text{plate}} = \phi_t F_{y,\text{plate}} \left(b_{PL} t_{PL} \right) \tag{8.24a}$$

$$b_{PL} \ge \frac{F_r}{\phi_t F_{y,\text{plate}} t_{PL}} \tag{8.24b}$$

Note that the flange plate width, b_{PL}, should be equal or less than the column flange width, b_{cf} (Fig. 8.22).

Step 5. Check fracture of effective net area in the tension plate

If design strength based on the limit state of net section rupture is smaller than the design load, F_r, fracture in the net flange plate will govern, and thus the design should be altered.

$$F_r \le \phi_t F_u A_e \tag{8.25a}$$

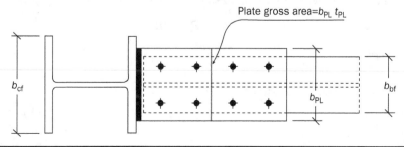

FIGURE 8.22   Gross area of the flange plate.

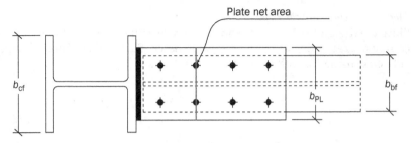

Figure 8.23 Net area of the flange plate.

Plate net area, A_n (shown in Fig. 8.23), which is equal to the effective net area, A_e, in this case, can be computed as follows:

$$A_e = A_n = A_g - 2t_{PL}\left(d_b + \frac{1''}{8}\right) \tag{8.25b}$$

Step 6. Check block shear for flange plate and beam flange

Under the tensile force F_r, two-block shear failure cases might occur in the plate (indicated in Fig. 8.24a). One case of block shear failure might occur in the beam flange (Fig. 8.24b) since the beam web would not allow the block within the bolt rows to separate.

Step 7. Check flange plates and beam flange against bearing/tearout

Bearing at flange plate and beam flange should be checked considering the upper limit based on the limit state of bolt tearout. Clear distance between the interior bolt holes, L_c, and clear edge distance, L_e, are demonstrated in Fig. 8.25a and b, respectively, for flange plate and beam flange.

Step 8. Check weld fracture for the top flange plate

There are two options for the welded connection between the column flange and the top and bottom flange plates. These connections can be made either with two-sided fillet welds, as shown in Fig. 8.26a, or using CJP groove welds (Fig. 8.26b), even though using CJP groove welds is more preferable. Thus, adequacy of the welds can be examined using the following depending on the weld type:
For CJP groove welded option shown in Fig. 8.26a:

$$F_r \le \phi R_n \tag{8.26a}$$

$$R_n = 0.6F_{EXX}t_{PL}b_{PL} \tag{8.26b}$$

For fillet-welded option shown in Fig. 8.26b:

$$F_r \le \phi R_n \tag{8.27a}$$

$$R_n = 2 \times \left[0.6F_{EXX}\left(0.707t_{weld}\right)b_{PL}\right] \tag{8.27b}$$

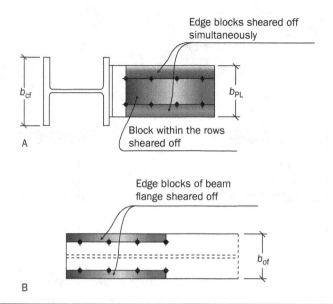

Edge blocks sheared off simultaneously

b_{cf}

b_{PL}

Block within the rows sheared off

A

Edge blocks of beam flange sheared off

b_{of}

B

FIGURE 8.24 Possible block shear failure modes: (*a*) top flange plate, (*b*) beam flange.

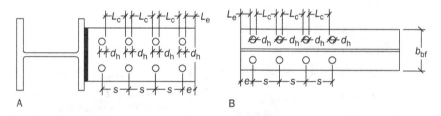

L_c L_c L_c L_e L_e L_c L_c L_c

d_h d_h d_h d_h d_h d_h d_h d_h

b_{bf}

s s s e e s s s

A   B

FIGURE 8.25 Clear hole-to-hole distances: (*a*) top flange plate; (*b*) beam flange.

B. Design of web plate based on shear force–associated limit states

Step 1. Determine number and diameter of bolts

With nominal shear strength, F_{nv}, and bolt diameter, d_b, the number of bolts, N, required to transfer the shear force, V_r, can be determined considering bolt shear failure as follows:

$$V_r \leq \phi\left[N\left(F_{nv}A_b\right)\right] \tag{8.28a}$$

$$N \geq \frac{V_r}{\phi\left(F_{nv}A_b\right)} \tag{8.28b}$$

Web-plate geometry is illustrated in Fig. 8.27. The typical 3-in. spacing between bolts can be used for all bolted plates, and edge spacing is taken as 1-½ in. unless specified otherwise. Note that the web-plate length, L_s, should be smaller than the T distance,

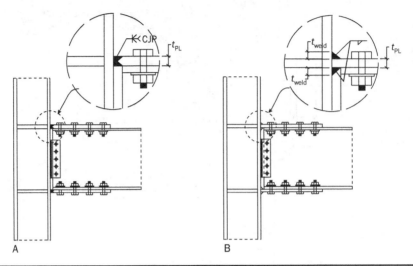

FIGURE 8.26 Welded-flange plate-to-column flange connection options: (a) CJP groove welds; (b) two-sided fillet welds.

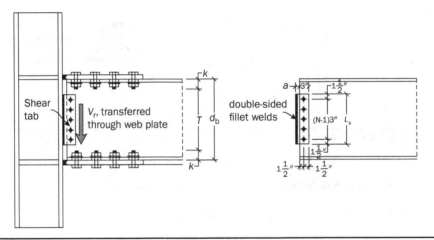

FIGURE 8.27 Web-plate geometry.

which can be computed by subtracting two time the fillet size, k, from the beam depth, d_b (Fig. 8.27) so that the plate can be installed.

Step 2. Yielding of gross web-plate area in shear

Design strength based on the limit state of shear yielding in the gross web-plate area (Fig. 8.28) should be greater than the required shear force, V_r.

$$V_r \leq \phi_t(0.6F_{y,\text{plate}})A_{g,\text{plate}} = \phi_t(0.6F_{y,\text{plate}})(L_s t_{PL}) \qquad (8.29)$$

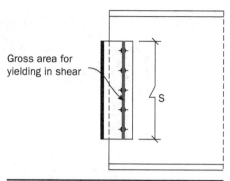

FIGURE 8.28 Gross area of shear tab.

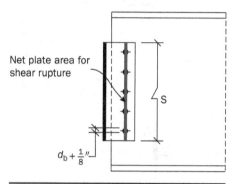

FIGURE 8.29 Net area of shear tab.

Step 3. Fracture of net area in shear

Design strength based on the limit state of fracture in the net web-plate area under shear (Fig. 8.29) should be greater than the required shear force, V_r:

$$V_r \leq \phi_t 0.6 F_u A_e \qquad (8.30a)$$

$$A_e = A_n = A_g - N t_{pl}\left(d_b + \frac{1''}{8}\right) \qquad (8.30b)$$

Step 4. Check block shear

Block shear failure could only take place in the web plate under the shear force V_r. The shaded portion to be sheared off is indicated in Fig. 8.30.

Step 5. Check bearing/tearout

Limit state of bearing/tearout at shear tab and beam web should be checked. Clear distance between the interior bolt holes, L_c, and clear edge distance, L_e, are demonstrated in Fig. 8.31 for the shear tab.

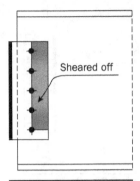

FIGURE 8.30   Block shear under V_r.

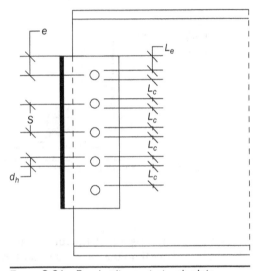

FIGURE 8.31   Bearing/tearout at web plate.

Step 6. Fillet weld rupture in web plate

Based on the limit state of weld shear rupture, the two-sided fillet welds that connect shear tab to the column flange should satisfy the following:

$$V_r \leq \phi R_n = 2 \times \phi \left[0.6 F_{EXX} (0.707 t_{weld}) L_s \right] \qquad (8.31)$$

Example 8.9

Design of a fully restrained BFP moment connection to support the factored bending moment of $1000^{kip\text{-}ft}$ and factored shear force of 80^{kips} due to wind and gravity loads. Use 3-in. spacing between the bolts, and 1-½ in. edge spacing. The steel grade is A992 for the W24 × 162 beam and W14 × 132 column and A36 for the steel plate. Use $F_{EXX} = 70$ ksi electrodes and $^7/_8''$ A490 bolts (threads excluded) for the flange plate ($F_{nv} = 84$ ksi), ¾" A325 bolts (threads included) for the shear tab ($F_{nv} = 54$ ksi).

Solution:

Geometric and material properties of the beam, column, and plates are summarized below.

Section and geometric properties:
  Beam: W24×162 ($Z_x = 468$ in.^{3}, $b_{bf} = 13''$, $d_b = 25''$, $t_{bf} = 1.22''$, $t_{bw} = 0.705''$, $k = 1.72''$)
  Column: W14×132 ($b_{cf} = 14.7''$)

Material properties:
  Beam: W24×162 (ASTM A992, $F_y = 50$ ksi and $F_u = 65$ ksi)
  Column: W14×132 (ASTM A992, $F_y = 50$ ksi and $F_u = 65$ ksi)
  Plates: ASTM A36, $F_y = 36$ ksi and $F_u = 58$ ksi

Design of flange plates

Step 1. Determine the design load

$$F_r = \frac{M_r}{d_b} = \frac{1000 \times 12}{25} = 480 \text{ kips}$$

Step 2. Check tensile rupture of the tension flange

$$A_{fg} = t_{bf} b_{bf} = 1.22 \times 13 = 15.86 \text{ in.}^2$$

$$A_{fn} = A_{fg} - \sum A_{holes} = 15.86 - 2 \times \left[1.22 \times \left(\frac{7}{8} + \frac{1}{8}\right)\right] = 13.42 \text{ in.}^2$$

$$Y_t = 1.0 \text{ for A992 Grade 50 Steel}$$

$$\left. \begin{array}{l} F_u A_{fn} = 65 \times 13.42 = 872.3 \text{ kips} \\ Y_t F_y A_{fg} = 1.0 \times 50 \times 15.86 = 793 \text{ kips} \end{array} \right\} F_u A_{fn} = 872.3 \text{ kips} \geq Y_t F_y A_{fg} = 793 \text{ kips}$$

Thus, the limit state of tensile rupture does not apply.

$$\phi_b M_n = \phi_b F_y Z_x = 0.9 \times 50 \times 468 / 12 = 1{,}755 \text{ kip-ft.} > M_r = 1{,}000 \text{ kip-ft.} \quad \text{Adequate.}$$

Step 3. Determine the number of flange bolts

$$\left. \begin{array}{l} A_b = \dfrac{\pi d_b^2}{4} = \dfrac{\pi \times 7/8^2}{4} = 0.601 \text{ in.}^2 \\ F_{nv} = 84 \text{ ksi} \\ \phi_t = 0.75 \end{array} \right\} N \geq \dfrac{480}{0.75 \times (84 \times 0.601)} = 12.7$$

Use $N = 14$.

Step 4. Determine the plate dimensions, b_{PL} and t_{PL}, based on the limit state of yielding in gross area.

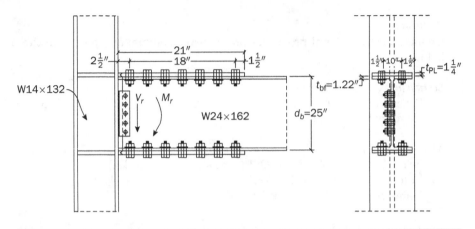

FIGURE 8.32   Flange plate dimensions and connection configuration.

Assume $t_{PL} = 1.25''$

$$
\left.
\begin{array}{l}
F_{y,plate} = 36 \text{ ksi} \\
t_{PL} = 1.25'' \\
\phi_t = 0.9
\end{array}
\right\}
b_{PL} \geq \frac{F_r}{\phi_t F_{y,plate} t_{PL}} = \frac{480}{0.9 \times 36 \times 1.25} = 11.85''
$$

The plate width, b_{PL}, should be equal or less than the column flange width $b_f = 14.7''$. Thus,

Use Pl. 13" × 1.25" for flanges (Fig. 8.32).

Step 5. Check fracture of effective net area in the tension plate

$$
\begin{aligned}
F_u &= 58 \text{ ksi} \\
A_g &= t_{PL} b_{PL} = 1.25 \times 13 = 16.25 \text{ in.}^2 \\
A_n &= A_g - 2t_{PL}\left(d_b + \frac{1''}{8}\right) = 16.25 - 2 \times 1.25 \times \left(\frac{7}{8} + \frac{1}{8}\right) = 13.75 \text{ in.}^2 \\
U &= 1.0 \\
A_e &= U A_n = 1.0 \times 13.75 = 13.75 \text{ in.}^2
\end{aligned}
$$

$$
F_r = 480 \text{ kips} < \phi_t F_u A_e = 0.75 \times 58 \times 13.75 = 598 \text{ kips}   \text{ Adequate.}
$$

Step 6. Check block shear

Since the shear area in failure mode#1 (Fig. 8.33a) is smaller than that of failure mode#2 (Fig. 8.33a), failure mode#1 is critical for the flange plate.

Referring to Fig. 8.33a,

$$
\begin{aligned}
A_{gv} &= 2L_{v,PL} t_{PL} = 2 \times 19.5 \times 1.25 = 48.75 \text{ in.}^2 \\
A_{nv} &= A_{gv} - (N-1) t_{PL}\left(d_b + \frac{1''}{8}\right) = 48.75 - 13 \times 1.25 \times \left(\frac{7}{8} + \frac{1}{8}\right) = 32.5 \text{ in.}^2 \\
A_{nt} &= t_{PL}\left[2L_{t,PL} - \left(d_b + \frac{1''}{8}\right)\right] = 1.25 \times \left[2 \times 1\frac{1}{2} - \left(\frac{7}{8} + \frac{1}{8}\right)\right] = 2.5 \text{ in.}^2
\end{aligned}
$$

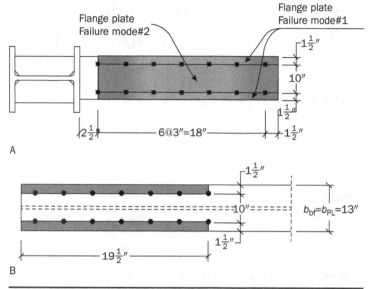

FIGURE 8.33 Block shear failure modes possible in (a) flange plate and (b) beam flange.

$$R_n = \left\{0.6F_u A_{nv}; 0.6F_y A_{gv}\right\}_{min} + F_u A_{nt}$$

$$R_n = \begin{cases} 0.6F_u A_{nv} = 0.6 \times 58 \times 32.5 = 1{,}131^{kips} \\ 0.6F_y A_{gv} = 0.6 \times 36 \times 48.75 = 1{,}053^{kips} \end{cases}_{min} + 58 \times 2.5 = 1{,}198 \text{ kips}$$

Referring to Fig. 8.33b,

$$A_{gv} = 2 \times 19.5 \times 1.22 = 47.58 \text{ in.}^2$$

$$A_{nv} = 47.58 - 13 \times 1.22 \times \left(\frac{7}{8} + \frac{1}{8}\right) = 31.72 \text{ in.}^2$$

$$A_{nt} = 1.22 \times \left[2 \times 1\frac{1}{2} - \left(\frac{7}{8} + \frac{1}{8}\right)\right] = 2.44 \text{ in.}^2$$

$$R_n = \begin{cases} 0.6F_u A_{nv} = 0.6 \times 65 \times 31.72 = 1{,}237^{kips} \\ 0.6F_y A_{gv} = 0.6 \times 50 \times 47.58 = 1{,}427^{kips} \end{cases}_{min} + 65 \times 2.44 = 1{,}396 \text{ kips}$$

$$F_r = 480 \text{ kips} < \phi_t R_n = 0.75 \times 1{,}198 = 898.5 \text{ kips} \qquad \text{OK.}$$

Step 7. Check bearing/tearout for top flange plate and beam flange

- Top flange plate:

  Interior bolts $\rightarrow L_c = s - d_h = 3 - \left(\frac{7}{8} + \frac{1}{16}\right) = 2\frac{1}{16}$ in.

  Exterior bolt $\rightarrow L_e = e - d_h/2 = 1\frac{1}{2} - \frac{1}{2} \times \left(\frac{7}{8} + \frac{1}{16}\right) = 1.03125$ in.

Nominal bearing/tearout strength per exterior bolt hole,

$$r_n = \begin{cases} 2.4 d_b t_{PL} F_u = 2.4 \times \dfrac{7}{8} \times 1.25 \times 58 = 152.3^{\text{kips/bolt}} \\ 1.2 L_e t_{PL} F_u = 1.2 \times 1.03125 \times 1.25 \times 58 = 89.7^{\text{kips/bolt}} \end{cases}_{\text{min}} = 89.7 \text{ kips/bolt}$$

Nominal bearing/tearout strength per interior bolt hole,

$$r_n = \begin{cases} 2.4 d_b t_{PL} F_u = 2.4 \times \dfrac{7}{8} \times 1.25 \times 58 = 152.3^{\text{kips/bolt}} \\ 1.2 L_c t_{PL} F_u = 1.2 \times 2\dfrac{1}{16} \times 1.25 \times 58 = 179.4^{\text{kips/bolt}} \end{cases}_{\text{min}} = 152.3 \text{ kips/bolt}$$

The design strength is

$$\phi R_n = 0.75 \times (2 \times 89.7 + 12 \times 152.3) = 1{,}505 \text{ kips} > F_r = 480 \text{ kips} \quad \text{Adequate.}$$

- Beam flange:

Nominal bearing/tearout strength per exterior bolt hole,

$$r_n = \begin{cases} 2.4 d_b t_{bf} F_u = 2.4 \times \dfrac{7}{8} \times 1.22 \times 65 = 166.5^{\text{kips/bolt}} \\ 1.2 L_e t_{bf} F_u = 1.2 \times 1.03125 \times 1.22 \times 65 = 98.1^{\text{kips/bolt}} \end{cases}_{\text{min}} = 98.1 \text{ kips/bolt}$$

Nominal bearing/tearout strength per interior bolt hole,

$$r_n = \begin{cases} 2.4 d_b t_{bf} F_u = 2.4 \times \dfrac{7}{8} \times 1.22 \times 65 = 166.5^{\text{kips/bolt}} \\ 1.2 L_c t_{bf} F_u = 1.2 \times 2\dfrac{1}{16} \times 1.22 \times 65 = 196.3^{\text{kips/bolt}} \end{cases}_{\text{min}} = 166.5 \text{ kips/bolt}$$

The design strength is

$$\phi R_n = 0.75 \times (2 \times 98.1 + 12 \times 166.5) = 1{,}646 \text{ kips} > F_r = 480 \text{ kips} \quad \text{OK.}$$

Step 8. Check CJP groove weld at top flange plate

$$F_r = 480 \text{ kips} \le \phi R_n = \phi(0.6 F_{EXX} t_{PL} b_{PL}) = 0.75 \times (0.6 \times 70 \times 1.25 \times 13) = 512 \text{ kips} \quad \text{OK.}$$

Design of beam web-to-column connection

Step 1. Determine number of bolts

$$\left. \begin{array}{l} A_b = \dfrac{\pi d_b^2}{4} = \dfrac{\pi \times 3/4^2}{4} = 0.442^{\text{in.}^2} \\[2mm] F_{nv} = 54^{\text{ksi}} \\[2mm] \phi_t = 0.75 \end{array} \right\} N \ge \dfrac{V_r}{\phi(F_{nv} A_b)} = \dfrac{80}{0.75 \times (54 \times 0.442)} = 4.47$$

Use $N = 5$.

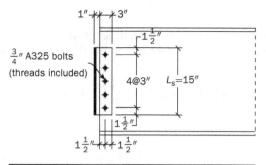

Figure 8.34   Shear tab dimensions.

Step 2. Shear yielding in web plate

Plate dimensions are determined based on bolt spacing as shown in Fig. 8.34. Selected plate length should also be compared with T value of beam to ensure that beam is sufficiently deep for the selected plate.

$$T = d_b - 2k = 25 - 2 \times 1.72 = 21.56 \text{ in.} > L_s = 15 \text{ in.} \qquad \text{OK.}$$

Design strength based on the limit state of shear yielding in the gross web-plate area (Fig. 8.34) should be greater than the required shear force, V_r.

$$V_r = 80 \text{ kips} \leq \phi_t (0.6 F_{y,\text{plate}})(L_s t_{PL})$$

$$t_{PL} \geq \frac{V_r}{\phi_t (0.6 F_{y,\text{plate}}) L_s} = \frac{80}{0.9 \times (0.6 \times 36)(15)} = 0.274 \text{ in.}$$

Try $t_{PL} = 0.5''$.

Step 3. Fracture of net area in shear

$$A_e = U A_n = A_g - N t_{PL}\left(d_b + \frac{1''}{8}\right) = (15 \times 0.5) - 5 \times 0.5 \times \left(\frac{3}{4} + \frac{1}{8}\right) = 5.3125 \text{ in.}^2$$

$$V_r = 80 \text{ kips} < \phi_t 0.6 F_u A_e = 0.75 \times 0.6 \times 58 \times 5.3125 = 138.7 \text{ kips} \qquad \text{OK.}$$

Step 4. Check block shear

$$A_{gv} = L_{v,PL} t_{PL} = 13.5 \times 0.5 = 6.75 \text{ in.}^2$$

$$A_{nv} = A_{gv} - (N - 0.5) t_{PL}\left(d_b + \frac{1''}{8}\right) = 6.75 - 4.5 \times 0.5 \times \left(\frac{3}{4} + \frac{1}{8}\right) = 4.78 \text{ in.}^2$$

$$A_{nt} = t_{PL}\left[1\frac{1}{2} - \frac{1}{2}\left(d_b + \frac{1''}{8}\right)\right] = 0.5 \times \left[1\frac{1}{2} - 0.5 \times \left(\frac{3}{4} + \frac{1}{8}\right)\right] = 0.53125 \text{ in.}^2$$

$$R_n = \{0.6 F_u A_{nv}; 0.6 F_y A_{gv}\}_{\min} + F_u A_{nt}$$

$$R_n = \left. \begin{cases} 0.6 F_u A_{nv} = 0.6 \times 58 \times 4.78 = 166.3^{\text{kips}} \\ 0.6 F_y A_{gv} = 0.6 \times 36 \times 6.75 = 145.8^{\text{kips}} \end{cases} \right\}_{\min} + 58 \times 0.53125 = 176.6 \text{ kips}$$

$$V_r = 80 \text{ kips} < \phi_t R_n = 0.75 \times 176.6 = 132.5 \text{ kips} \qquad \text{OK.}$$

Step 5. Check bearing/tearout

Both beam web thickness and the ultimate beam strength are greater than the web-plate thickness and strength, respectively. Therefore, the web plate will control.

$$\text{Interior bolts} \rightarrow L_c = s - d_h = 3 - \left(\frac{3}{4} + \frac{1}{16}\right) = 2\frac{3}{16} \text{ in.}$$

$$\text{Exterior bolt} \rightarrow L_e = e - d_h/2 = 1\frac{1}{2} - 0.5 \times \left(\frac{3}{4} + \frac{1}{16}\right) = 1.09375 \text{ in.}$$

Nominal bearing/tearout strength per exterior bolt hole,

$$r_n = \begin{cases} 2.4 d_b t_{PL} F_u = 2.4 \times \dfrac{3}{4} \times 0.5 \times 58 = 52.2^{\text{kips/bolt}} \\ 1.2 L_e t_{PL} F_u = 1.2 \times 1.09375 \times 0.5 \times 58 = 38.1^{\text{kips/bolt}} \end{cases}_{\min} = 38.1 \text{ kips/bolt}$$

Nominal bearing/tearout strength per interior bolt hole,

$$r_n = \begin{cases} 2.4 d_b t_{PL} F_u = 2.4 \times \dfrac{3}{4} \times 0.5 \times 58 = 52.2^{\text{kips/bolt}} \\ 1.2 L_c t_{PL} F_u = 1.2 \times 2\dfrac{3}{16} \times 0.5 \times 58 = 76.1^{\text{kips/bolt}} \end{cases}_{\min} = 52.2 \text{ kips/bolt}$$

Thus, the design strength will be

$$\phi R_n = 0.75 \times (38.1 + 4 \times 52.2) = 185.2 \text{ kips} > V_r = 80 \text{ kips} \quad \text{OK.}$$

Step 6. Check fillet welds

$$V_r \le \phi R_n = 2 \times \phi[0.6 F_{\text{EXX}}(0.707 t_{\text{weld}}) L_s] \rightarrow t_{\text{weld}} \ge \frac{V_r}{2 \times \phi[0.6 F_{\text{EXX}}(0.707) L_s]}$$

$$t_{\text{weld}} \ge \frac{80}{2 \times 0.75 \times 0.6 \times 70 \times 0.707 \times 15} = 0.12 \text{ in.}$$

$t_{\min} = {}^3/_{16}''$ as per *Table J2.4, the Specification.* Thus,

$$t_{\text{weld}} = \left\{ 0.12^{\text{in.}}; \frac{3}{16}^{\text{in.}} \right\}_{\min} = \frac{3}{16} \text{ in.}$$

8.6.2   All-Bolted Double-Angle Connection

Limit states associated with all-bolted double-angle connections are as follows:

    a. Block shear rupture

    b. Bolt bearing

    c. Bolt shear

    d. Shear yielding

    e. Shear rupture

    f. Flexural strength

The nominal governing strength of the connection will be the smallest among all.

Example 8.10

Design an all-bolted double-angle connection between a W24×146 beam and a W14×90 column to support the following beam end reactions due to dead and live loads: $V_D = 37.5$ kips and $V_L = 112.5$ kips. Use ¾-in.-diameter ASTM A325 bolts (threads included) in standard holes.

Material properties:

Beam: W24×146 (ASTM A992, $F_y = 50$ ksi and $F_u = 65$ ksi)
Column: W14×90 (ASTM A992, $F_y = 50$ ksi and $F_u = 65$ ksi)
Angles: 2L5×3½ SLBB (ASTM A36, $F_y = 36$ ksi and $F_u = 58$ ksi)

Section properties:
W24×146: $d_b = 24.7$ in., $t_{bw} = 0.65$ in., $k = 1.59$ in.
W14×90: $t_{cf} = 0.710$ in.

Solution:

Step 1. Determine the required strength

$$V_r = 1.2V_D + 1.6V_L = 1.2 \times 37.5 + 1.6 \times 112.5 = 225 \text{ kips}$$

Step 2. Find number of bolts based on bolt shear

$$V_r \leq \phi R_n = \phi N r_n$$

ASTM A325 bolts (threads are included in the shear planes), $F_{nv} = 54$ ksi

Nominal bolt area, $A_b = \dfrac{\pi d_b^2}{4} = \dfrac{\pi \times {3/4}^2}{4} = 0.442$ in.^{2}

Number of shear planes $= 2$
Nominal shear strength per bolt, $r_n = F_{nv}A_b = 2 \times (54^{\text{kips/in.}^2} \times 0.442^{\text{in.}^2}) = 47.7$ kips/bolt

Required shear strength of the connection, $V_r = 225$ kips

$$N \geq \frac{V_r}{\phi(F_{nv}A_b)} = \frac{225}{0.75 \times 47.7} = 6.3$$

Use seven ¾-in.-diameter ASTM A325 bolts (Fig. 8.35).
The length of the angle should be less than T,

$$T = d_b - 2k = 24.7 - 2 \times 1.59 = 21.52 \text{ in.} > L_s = 21 \text{ in.} \quad \text{OK.}$$

Step 3. Check bearin g/tearout strength
Since the total leg thickness of the connected angles ($2 \times {}^5/_{16}''$) is smaller than the web thickness of the beam ($t_{bw} = 0.65''$), the angles will control the bearing strength.

$$\text{Interior bolts} \rightarrow L_c = s - d_h = 3 - \left(\frac{3}{4} + \frac{1}{16}\right) = 2\frac{3}{16} \text{ in.}$$

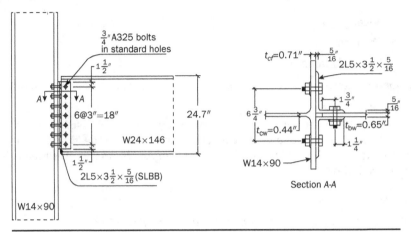

FIGURE 8.35   Example 8.10.

Exterior bolt $\rightarrow L_e = e - d_h/2 = 1\frac{1}{2} - 0.5 \times \left(\frac{3}{4} + \frac{1}{16}\right) = 1.09375$ in.

Nominal bearing/tearout strength per exterior bolt hole,

$$r_n = \begin{cases} 2.4d_b t_{\text{angle}} F_u = 2.4 \times \dfrac{3}{4} \times \left(2 \times \dfrac{5}{16}\right) \times 58 = 65.2^{\text{kips/bolt}} \\[2mm] 1.2L_e t_{\text{angle}} F_u = 1.2 \times 1.09375 \times \left(2 \times \dfrac{5}{16}\right) \times 58 = 47.6^{\text{kips/bolt}} \end{cases}_{\min} = 47.6 \text{ kips/bolt}$$

Nominal bearing/tearout strength per interior bolt hole,

$$r_n = \begin{cases} 2.4d_b t_{\text{angle}} F_u = 2.4 \times \dfrac{3}{4} \times \left(2 \times \dfrac{5}{16}\right) \times 58 = 65.2^{\text{kips/bolt}} \\[2mm] 1.2L_c t_{\text{angle}} F_u = 1.2 \times 2\dfrac{3}{16} \times \left(2 \times \dfrac{5}{16}\right) \times 58 = 95.2^{\text{kips/bolt}} \end{cases}_{\min} = 65.2 \text{ kips/bolt}$$

Thus, the design strength will be

$$\phi R_n = 0.75 \times (47.6 + 6 \times 65.2) = 329.1 \text{ kips} > V_r = 225 \text{ kips} \quad \text{OK.}$$

Step 4. Check shear yielding
Shear yielding does not apply to an uncoped beam. Hence, it should be checked for the angles only based on the yielding line shown in Fig. 8.36.

$$\phi R_n = 0.90 \times \left[2 \times (0.60 \times 36) \times \left(21 \times \frac{5}{16}\right)\right] = 255 \text{ kips} > V_r = 225 \text{ kips} \quad \text{OK.}$$

Step 5. Check net section fracture in shear
Using the net area shown in Fig. 8.36, the effective net area can be computed as

$$A_e = UA_n = A_g - Nt_{\text{angle}}\left(d_b + \frac{1}{8}''\right) = 2 \times \left[\left(21 \times \frac{5}{16}\right) - 7 \times \frac{5}{16} \times \left(\frac{3}{4} + \frac{1}{8}\right)\right] = 9.3 \text{ in.}^2$$

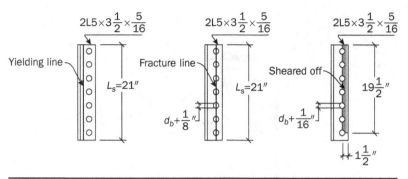

FIGURE 8.36   Shear yielding, fracture, and block shear for angles.

$$V_r = 225 \text{ kips} < \phi_t 0.6 F_u A_e = 0.75 \times 0.6 \times 58 \times 9.3 = 242.7 \text{ kips} \quad \text{OK.}$$

Step 6. Check block shear

$$A_{gv} = 2L_v t_{angle} = 2 \times 19.5 \times \frac{5}{16} = 12.19 \text{ in.}^2$$

$$A_{nv} = A_{gv} - 2 \times (N - 0.5) t_{angle} \left(d_b + \frac{1''}{8} \right) = 12.19 - 2 \times \left[6.5 \times \frac{5}{16} \times \left(\frac{3}{4} + \frac{1}{8} \right) \right] = 8.63 \text{ in.}^2$$

$$A_{nt} = 2 t_{angle} \left[1\frac{1}{2} - \frac{1}{2} \left(d_b + \frac{1''}{8} \right) \right] = 2 \times \frac{5}{16} \times \left[1\frac{1}{2} - 0.5 \times \left(\frac{3}{4} + \frac{1}{8} \right) \right] = 0.66 \text{ in.}^2$$

$$R_n = \begin{cases} 0.6 F_u A_{nv} = 0.6 \times 58 \times 8.63 = 300^{kips} \\ 0.6 F_y A_{gv} = 0.6 \times 36 \times 12.19 = 263^{kips} \end{cases}_{min} + 58 \times 0.66 = 301.8 \text{ kips}$$

$$V_r = 225 \text{ kips} < \phi_t R_n = 0.75 \times 301.8 = 226.3 \text{ kips} \quad \text{OK.}$$

Design summary: Use seven ¾-in. A325 (threads included) bolts and two 21-in.-long L5×3½×⁵/₁₆ SLBB.

8.7   Problems

8.1   For the connection shown in Fig. 8.37, check whether bolt and edge spacing is adequate.

8.2   For the connection shown in Fig. 8.37, find the maximum required axial load, P_r, that can be applied to the angles safely. Assume that connection is bearing type and ignore the limit states associated with the gusset plate.

8.3   Rework 8.2 by assuming a slip-critical connection with Class A surface.

8.4   For the connection shown in Fig. 8.38, check bolt and edge spacing.

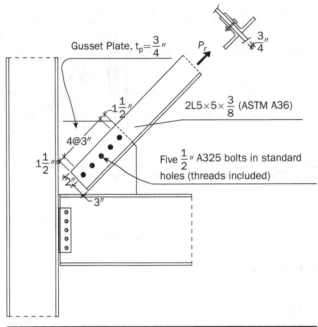

Gusset Plate, $t_p = \frac{3}{4}''$

P_r

$*\frac{3}{4}''$

$1\frac{1}{2}''$

$2L5 \times 5 \times \frac{3}{8}$ (ASTM A36)

4@3"

Five $\frac{1}{2}''$ A325 bolts in standard holes (threads included)

$1\frac{1}{2}''$

2"

3"

Figure 8.37   Problems 8.1 through 8.3.

8.5   For the connection shown in Fig. 8.38, find the design strength considering all possible limit states for bolts and plates. Assume that the connection is bearing-type, and thread are excluded from the shear planes.

8.6   Rework 8.5 by assuming that the connection is slip-critical and thread are not excluded from the shear planes.

8.7   If the connection in Fig. 8.38 is subjected to a dead load of 80 kips and a live load of 50 kips, would you expect slippage between the plates under service loads? Connection is slip-critical, and surfaces are unpainted blast-cleaned.

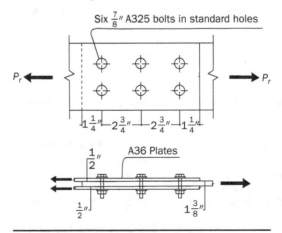

Six $\frac{7}{8}''$ A325 bolts in standard holes

P_r

P_r

$1\frac{1}{4}''$ — $2\frac{3}{4}''$ — $2\frac{3}{4}''$ — $1\frac{1}{4}''$

$\frac{1}{2}''$ A36 Plates

$\frac{1}{2}''$

$1\frac{3}{8}''$

Figure 8.38   Problems 8.4 through 8.7.

8.8   Find the design strength of the bearing-type connection shown in Fig. 8.39 based on the limit states related to bolts and channels. Channels are made of A36 steel. Threads are not excluded from the shear planes.

8.9   For the bearing-type connection shown in Fig. 8.39, what would your strategy be to ensure that yielding in the gross channel section will occur?

8.10   A bracing made with a square hollow section, shown in Fig. 8.40, is subjected to $P_D = 50$ kips, $P_L = 30$ kips, and $P_{wind} = 150$ kips. HSS6 × 6 × 3/8 bracing is made of ASTM A500 Gr. C steel and fillet welded to the gusset plate. Using the minimum fillet weld size, find the required weld length, l_w. 70-ksi electrode is used.

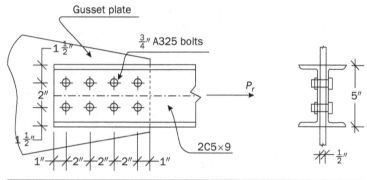

FIGURE 8.39   Problems 8.8 and 8.9.

8.11   Determine the design load that can be supported by the bracing shown in Fig. 8.40. Consider only net section fracture and gross-section yielding limit states for the bracing.

8.12   Determine the design strength of the connection shown in Fig. 8.40 based on the base metal rupture limit state for bracing only. HSS6 × 6 × $^1/_2$ bracing is made of ASTM A500 Gr. B steel. Fillet weld length, l_w, is 10 in.

8.13   Design the welds to develop the full strength of the angles 2L4 × 4 × $^1/_4$ and minimize eccentricity for the connection shown in Fig. 8.41. A36 steel and E70 ($F_{EXX} = 70$ ksi) electrodes are used. The strength of the connection $P_r = \phi_t A_g F_y$ of the tension member.

8.14   Determine the design strength of the ½-in. E70 fillet-welded connection shown in Fig. 8.42.

8.15   Rework 8.14 by removing the welds perpendicular to loading.

8.16   Check the adequacy of the connection shown in Fig. 8.42 considering all possible limit states for welds and plates. P_r is composed of dead, live, snow, and wind loads. $P_{dead} = 100$ kips, $P_{live} = 50$ kips, $P_{snow} = 50$ kips, and $P_{wind} = 150$ kips. Plates are made of 50-ksi-steel and E60 electrodes are used for the ½-in. fillet welds.

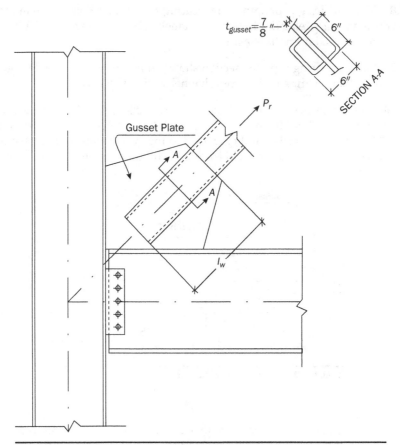

$t_{gusset} = \frac{7}{8}''$

6″

6″

SECTION A-A

P_r

Gusset Plate

A

A

l_w

Figure 8.40   Problems 8.10 through 8.12.

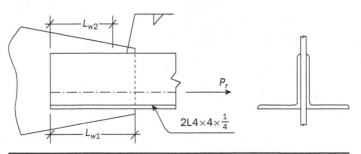

L_{w2}

P_r

$2L4 \times 4 \times \frac{1}{4}$

L_{w1}

Figure 8.41   Problem 8.13.

8.17   Design a single-plate shear connection between a W24×94 beam and a W14×109 column to support the following beam end reactions due to dead and live loads: $V_D =$ 75 kips and $V_L = 50$ kips. Use ½-in.-diameter ASTM A325 bolts (threads excluded) in standard holes. Shear tab is bolted to the beam web and fillet-welded to the column flange using E70 electrodes.

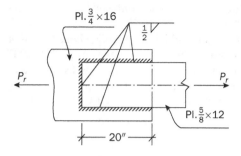

PI. $\frac{3}{4}$ ×16

PI. $\frac{5}{8}$ ×12

P_r

P_r

20"

FIGURE 8.42   Problems 8.14 through 8.16.

8.18   Design the same simple beam-to-column connection given in 8.17 using all-bolted double-angles instead of shear tab.

8.19   Design of a rigid BFP connection to support the factored bending moment of 600$^{kip\text{-}ft}$ and factored shear force of 80^{kips} due to wind and gravity loads. A992 steel is used for both W27´102 beam and W14 159 column, and A36 for the steel plate. Use $F_{EXX} =$ 70 ksi electrodes and ¾" A490 bolts (threads included) for the flange plate and ½" A325 bolts (threads included) for the shear tab.

8.20   Develop a python code that can compute the nominal strength of a

  a. bolted connection considering shear, bearing, and tearout limit states.

  b. welded connection.

Related to Building Project in Chapter 1:

8.21   Please design connections for the building given in the appendix in Chapter 1:

  a. a typical shear connection for the secondary beams connecting to a main girder of the SMF.

  b. a typical beam-to-column connection in the gravity-only frames.

  c. a BFP connection for the exterior beam in the second-story of SMF.

  d. a gusset plate connection for the braces connecting to the beam and column in the second-story of SCBF.

Bibliography

AISC, *Specification for Structural Steel Buildings*, ANSI/AISC Standard 360-16, American Institute of Steel Construction, Chicago, IL, 2016.

AISC, *Steel Construction Manual*, 15th ed., American Institute of Steel Construction, Chicago, IL, 2016.

American Welding Society, *Structural Welding Code—Steel: AWS D1.1:2000*, 17th ed., Miami, FL, 2000.

Kulak, G. L., J. W. Fisher, and J. H. A. Struik, *Guide to Design Criteria for Bolted and Riveted Joints*, 2nd ed., John Wiley & Sons, New York, NY, 1987.

Research Council on Structural Connections (RCSC), *Specification for Structural Joints Using High-Strength Bolts*, August 1, 2014.

Weman, K., *Welding Processes Handbook*, CRC Press LLC, New York, NY, 2003.

CHAPTER 9

Plate Girders

9.1 Introduction

A plate girder (PG) is a special type of beam that is subjected to bending moment and shear force. The major difference in appearance between the plate girder and a hot-rolled W-section is the overall depth.

The plate girder has a depth often larger than 60 in., which is the reason that the plate girder is sometimes referred to as a "deep beam" (Fig. 9.1). A typical plate girder would have a large depth-to-thickness ratio, as compared to typical rolled steel W-shapes. Such ratio in plate girders is often as follows:

$$\frac{h}{t_w} > \lambda_r = 5.70\sqrt{\frac{E}{F_y}} \tag{9.1}$$

Major features of a PG compared with typical hot-rolled W-section can be listed as follows:

1. PG includes all limit states that a hot-rolled steel beam has. However, the governing limit state in a PG is often dramatically different due to its large h/t_w ratio.

2. The bending strength and shear strength are significantly related to the unstable behavior of the web.

Plate girders are predominantly used in highway bridges and sometimes are also seen in building structures as well as other structures where hot-rolled shapes are not available for required strength and/or stiffness. Some examples of plate girder structures are shown in Fig. 9.2.

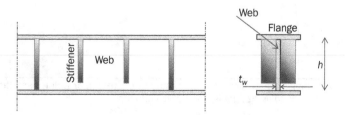

FIGURE 9.1 Plate girders.

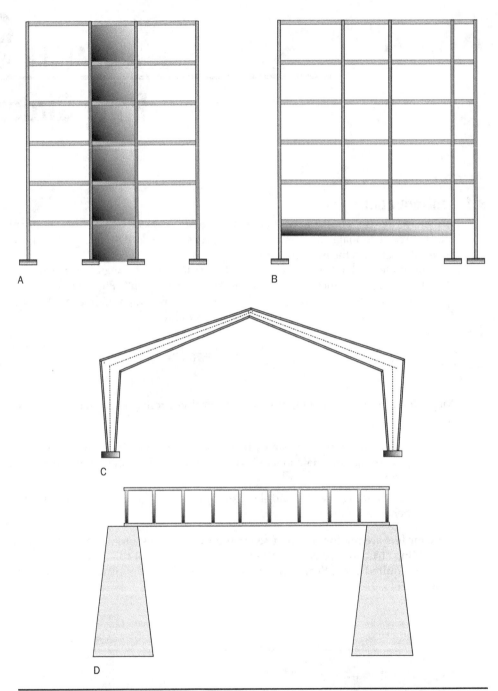

Figure 9.2   Application of plate girders in structures: (*a*) shear wall in a multistory frame, (*b*) deep beam in a building, (*c*) gable frames, (*d*) bridges.

9.2 Behavior and Design of Web under Shear

A plate girder is subjected to bending moment and shear force. All limit states (failure modes) in the plate girder thus are related to either bending moment, shear force, or the combination of them. The plate girder under shear force action is discussed first. Special attention is paid to local buckling of the web, mainly due to its large depth–to–thickness ratio (i.e., large h/t_w). A typical plate girder under shear force is demonstrated by a tested cantilever in Fig. 9.3 beam under a cyclic load V going up and down (simulating a seismic force).

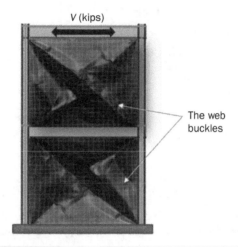

FIGURE 9.3 Behavior of a plate girder under cyclic loads.

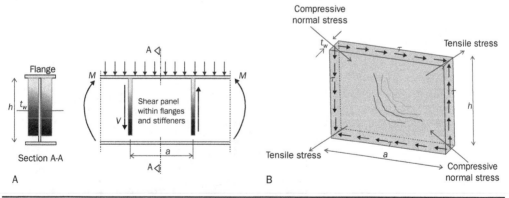

FIGURE 9.4 A plate girder (*a*) subject to bending moment, *M*, and shear force, *V*, (*b*) compressive and tensile stresses developing in the shear panel.

When tension field action (postbuckling stage) is not considered:

A typical plate girder consists of top and bottom flange plates, a web plate, and stiffeners with spacing "*a*" along the plate girder (Fig. 9.4*a*). When h/t_w ratio is high, the web panel (Fig. 9.4*b*) will buckle under the shear-induced compressive normal stress. Note that the illustrative buckling shape is only one of many possible shapes, depending on the *a/h* ratio.

a. *Elastic web buckling under shear*

A plate girder would most likely suffer an elastic web buckling under shear force. With the assumptions of an ideal web plate, i.e.,

- No initial out-of-straightness

- No residual stress

- The web plate remains elastic

Timoshenko described, in his *Theory of Plates and Shells* (1959) book, critical buckling shear stress as follows:

$$\tau_{cr} = k_v \frac{\pi^2 E}{12(1 - \mu^2)(h / t_w)^2} \tag{9.2}$$

where

$$k_v = \text{buckling coefficient, function of } a/h \text{ (Fig. 9.5)}$$

It can be observed in Fig. 9.5 that k_v is strongly related to *a/h* as well as buckling mode shapes. An approximation was proposed for k_v for design purpose with the lower bound taken as

$$k_v = 4 + \frac{5.34}{\left(\frac{a}{h}\right)^2} \text{ for } \frac{a}{h} \le 1.0 \tag{9.3a}$$

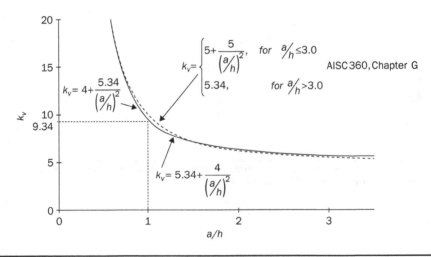

FIGURE 9.5 Approximate buckling coefficient, k_v, for design purposes.

$$k_v = 5.34 + \frac{4}{\left(\frac{a}{h}\right)^2} \quad \text{for } \frac{a}{h} > 1.0 \tag{9.3b}$$

Using the k_v approximation, and taking $\mu = 0.3$, $E = 29{,}000$ ksi and $\tau_y = 0.6F_y$, one can find the AISC design formula as

$$\tau_{cr} = k_v \frac{\pi^2 29{,}000}{12(1 - 0.3^2)(h / t_w)^2} = \frac{26{,}200k_v}{(h / t_w)^2} \tag{9.4}$$

or

$$C_v = \frac{\tau_{cr}}{\tau_y} = \frac{\tau_{cr}}{0.6F_y} \cong \frac{44{,}000k_v}{(h / t_w)^2 F_y} = \frac{1.51Ek_v}{F_y(h / t_w)^2} \text{ (Fig. 9.6)} \tag{9.5}$$

where

$$k_v = \begin{cases} 5 + \dfrac{5}{\left(\frac{a}{h}\right)^2}, & \text{for } \frac{a}{h} \leq 3.0 \\[4mm] 5.34, & \text{for } \frac{a}{h} > 3.0 \end{cases} \quad \text{AISC 360, Chapter G}$$

b. Inelastic buckling capacity

Elastic buckling result is valid where the web plate remains elastic but is less likely for a smaller h/t_w ratio (i.e., in a shallow or thick web). Also, residual stress and out-of-straightness have a more significant effect on the buckling of the web plate. An empirical approach was used for design purpose for inelastic buckling stress:

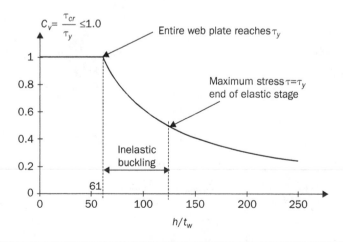

Figure 9.6 Web shear strength coefficient, C_v for $F_y = 50$ ksi and $k_v = 5.34$.

$$\tau_{cr}\big|_{\text{inelastic}} = C(\tau_{cr})^n_{\text{elastic}} \tag{9.6}$$

From curve fitting of the test data $C = \sqrt{0.8\tau_y}$ and $n=1/2$, we have

$$\tau_{cr} = \sqrt{0.8\tau_y \, \tau_{cr}\big|_{\text{elastic}}} \tag{9.7}$$

$$C_v\big|_{\text{inelastic}} = \frac{\tau_{cr}\big|_{\text{inelastic}}}{\tau_y} \tag{9.8}$$

$$= \frac{187}{\left(h/t_w\right)}\sqrt{\frac{k_v}{F_y}} \quad\text{or}\quad \frac{1.10}{\left(h/t_w\right)}\sqrt{\frac{k_v E}{F_y}}$$

The maximum value: $C_v\big|_{\text{inelastic}} = 1.0$ when $\tau_{cr}\big|_{\text{inelastic}} = \tau_y$

When $\dfrac{187}{\left(h/t_w\right)}\sqrt{\dfrac{k_v}{F_y}} = 1.0$,

$$h/t_w = 187\sqrt{\frac{k_v}{F_y}} \tag{9.9}$$

When $h/t_w < 187\sqrt{\dfrac{k_v}{F_y}}$, the plate will yield before any buckling.

c. *Yielding under shear*

Yielding under shear will occur when (Fig. 9.7)

$$\tau_{cr} = \tau_y \tag{9.10}$$

or

$$C_v = 1.0$$

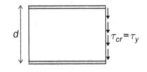

FIGURE 9.7   Yielding under shear.

9.3   Tension Field Action (Postbuckling Stage)

The sequence of the progress deformation pattern in a typical plate girder is

    i. web buckles along its diagonal in the direction of this compressive stress (as discussed above),

    ii. the tension in the other diagonal portion remains functional, providing a post-buckling shear strength; and

    iii. finally, the plate girder reaches its full shear capacity when either the tension-field yields or stiffeners fail (Fig. 9.8).

Angle, ϕ:
The portion of the web plate that might be involved in the tension field action is not well defined and is heavily dependent on the panel dimensions, *a* and *h*. An approximate determination of angle, ϕ is often used to define the tension field as described below (Fig. 9.8):

    i. The partial force, ΔV_t, developed in one stiffener by the tension force, $\sigma_t b_e t_w$, can be expressed as

$$\Delta V_t = \sigma_t b_e t_w \sin\phi \qquad (9.11)$$

Since $b_e = h\cos\phi - a\sin\phi$ and $\sin\phi\cos\phi = \dfrac{\sin 2\phi}{2}$, we have

$$\Delta V_t = \sigma_t t_w \left(\frac{h}{2}\sin 2\phi - a\sin^2\phi\right) \qquad (9.12)$$

The angle would eventually reach the value that enables the maximum shear component to develop from tension field, i.e.,

$$\frac{\Delta V_t}{d\phi} = \sigma_t t_w (h\cos 2\phi - a\sin 2\phi) = 0 \qquad (9.13)$$

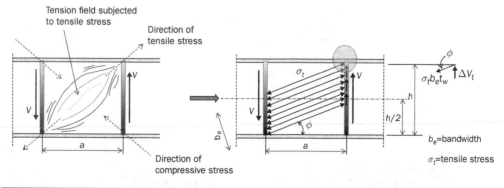

FIGURE 9.8   Tension field action.

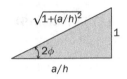

Figure 9.9 A right triangle to construct trigonometric functions for 2ϕ.

Trigonometric functions for 2ϕ can be written as (Fig. 9.9):

$$\tan 2\phi = \frac{h}{a} = \frac{1}{\left(a/h\right)} \tag{9.14a}$$

$$\sin 2\phi = \frac{1}{\sqrt{1 + \left(a/h\right)^2}} \tag{9.14b}$$

$$\cos 2\phi = \frac{a/h}{\sqrt{1 + \left(a/h\right)^2}} \tag{9.14c}$$

$$\sin^2 \phi = \frac{1 - \cos 2\phi}{2} = \frac{1}{2}\left[1 - \frac{a/h}{\sqrt{1 + \left(a/h\right)^2}}\right] \tag{9.14d}$$

Shear strength from tension field action:
From the free-body-diagram in Fig. 9.10, we have (let $\Delta F_w = 0$ since F_w is small)

$$\sum F_x = 0 \rightarrow \Delta F_f = (\sigma_t t_w a \sin\phi)\cos\phi = \frac{\sigma_t}{2} t_w a \sin 2\phi \tag{9.15}$$

$$\sum M_o = 0 \rightarrow \left[\Delta F_f \frac{h}{2}\right] - \left[\frac{V_t}{2} a\right] = 0 \tag{9.16}$$

From the above two equations and $\sin 2\phi$ value, one can find

$$V_t = \sigma_t \frac{t_w h}{2} \frac{1}{\sqrt{1 + (a/h)^2}} \tag{9.17}$$

where
  F_w = normal stress in the web due to bending moment, ksi
  F_f = normal stress in the flanges due to bending moment, ksi
  V_t = additional shear strength due to tension field action, kips

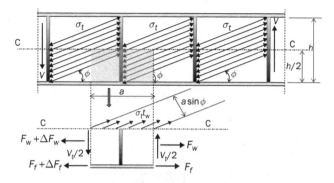

FIGURE 9.10   Free-body-diagram from tension field action.

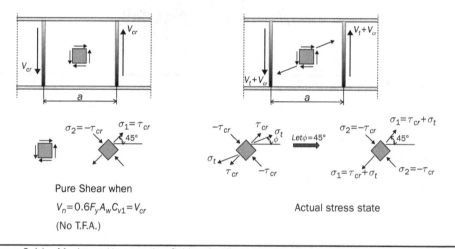

FIGURE 9.11   Maximum V_t at tension field action limit state.

Maximum V_t at tension field action limit state:
V_t reaches its maximum value when the actual stress state reaches the failure (limit state) condition (somewhere on the yielding surface) (Fig. 9.11).

The energy of distortion yielding theory is used as shown in Fig. 9.12, in which the yielding surface is described as

$$\sigma_1^2 + \sigma_2^2 - \sigma_1\sigma_2 = F_y^2 \tag{9.18}$$

In Fig. 9.12, point A represents a pure shear failure state, whereas point C represents a pure tension failure state. Thus, the failure point of the actual stress state, Fig. 9.11, should be somewhere on the curve between A and C. Using a straight line to replace curve AC approximately, that is,

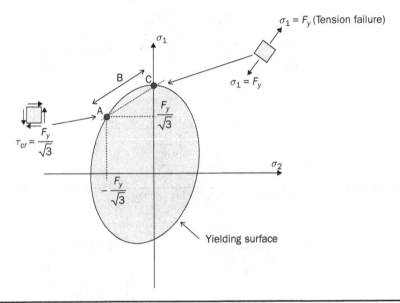

Figure 9.12 Yielding surface.

$$\sigma_1 = F_y + \sigma_2(\sqrt{3} - 1) \tag{9.19}$$

Note that $\sigma_1 = \tau_{cr} + \sigma_t$ and $\sigma_2 = -\tau_{cr}$

$$\frac{\sigma_t}{F_y} = 1 - \frac{\tau_{cr}}{F_y / \sqrt{3}} = 1 - C_v \tag{9.20}$$

where $C_v = \dfrac{\tau_{cr}}{\tau_y}, \tau_y = \dfrac{F_y}{\sqrt{3}}$

$$V_t = F_y(1 - C_v)\frac{t_w h}{2}\frac{1}{\sqrt{1 + \left(\frac{a}{h}\right)^2}} = \frac{t_w h}{2}F_y\frac{1 - C_v}{\sqrt{1 + \left(\frac{a}{h}\right)^2}} \tag{9.21}$$

Thus, the total nominal shear capacity with tension field action becomes

$$V_n = V_{cr} + V_t$$

$$= 0.6F_y A_w C_v + \frac{t_w h}{2}F_y\frac{1 - C_v}{\sqrt{1 + \left(\frac{a}{h}\right)^2}}$$

$$= 0.6F_y A_w \left[\underbrace{C_v}_{\text{from web buckling}} + \underbrace{\frac{1 - C_v}{1.15\sqrt{1 + \left(\frac{a}{h}\right)^2}}}_{\text{from T.F.A. after that}} \right] \tag{9.22}$$

In Chapter G in AISC 360, nominal shear strength with tension field action is defined as

a. For $h/t_w \leq 1.10\sqrt{\dfrac{k_v E}{F_y}}$, no buckling

$$V_n = 0.6F_y A_w \tag{9.23}$$

b. For $h/t_w > 1.10\sqrt{\dfrac{k_v E}{F_y}}$, inelastic or elastic buckling

$$V_n = 0.6F_y A_w \left[C_v + \frac{1 - C_v}{1.15\sqrt{1 + \left(a/h\right)^2}} \right] \tag{9.24}$$

Limits on use of tension field action:
Tension field action is possible when the web plate is adequately supported on all its four sides by flanges and stiffeners. In the following cases, tension field action is not permitted to be used:

a. End panels

Use of tension field action is not allowed at end panels (Fig. 9.13).

b. $a/h > 3.0$ or $a/h > \left(\dfrac{260}{h/t_w}\right)^2$

c. $\dfrac{A_w}{\dfrac{A_{fc} + A_{ft}}{2}} > 2.5$ flanges are not "rigid" enough for the given web

A_w =web area, in.^{2}
A_{fc} = area of top flange, in.^{2}
A_{ft} = area of bottom flange, in.^{2}

d. $\dfrac{h}{b_{fc}} > 6.0$ or $\dfrac{h}{b_{ft}} > 6.0$ (Fig. 9.14)

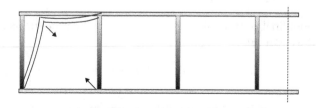

FIGURE 9.13 Tension field action at end panels (not allowed).

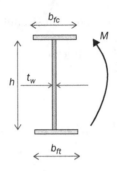

FIGURE 9.14 Plate girder section under bending moment, M.

9.4 Behavior and Design of Web under Bending Moment

9.4.1 General Discussions

Some useful definitions:

   i. Related to flanges (Fig. 9.15):

$$\lambda_{fc} = \frac{b_{fc}}{2t_{fc}} \quad \text{the width-to-thickness ratio for compression flange} \tag{9.25a}$$

$$\lambda_{pf} = 0.38\sqrt{\frac{E}{F_y}} \quad \text{limiting width-to-thickness ratio for compact flange} \tag{9.25b}$$

$$\lambda_{rf} = 0.95\sqrt{\frac{k_c E}{F_L}} \quad \text{limiting width-to-thickness ratio for noncompact flange} \tag{9.25c}$$

k_c = coefficient for slender unstiffened elements

$$0.35 \le k_c = \frac{4}{\sqrt{h/t_w}} \le 0.76 \tag{9.25d}$$

$$F_L = \begin{cases} 0.7F_y, & \text{when slender web } \lambda > \lambda_{rw} \text{ is used; or} \\ & \text{nonslender web with } S_{xt}/S_{xc} < 0.7 \\ F_y\,S_{xt}/S_{xc} \ge 0.5F_y, & \text{nonslender web with } S_{xt}/S_{xc} < 0.7 \end{cases} \tag{9.25e}$$

F_L = nominal compressive strength above which the inelastic buckling limit states apply (ksi) \qquad (9.25f)

$$S_{xt} = \frac{I_x}{y_t}, \text{ elastic section modulus for tension flange (in.}^3) \tag{9.25g}$$

$$S_{xc} = \frac{I_x}{y_c}, \text{ elastic section modulus for compression flange (in.}^3) \tag{9.25h}$$

ii. Related to web (Fig. 9.16):

$$\lambda = \frac{h_c}{t_w} \text{ the width-to-thickness ratio for compression flange} \qquad (9.25i)$$

$h_c = h$ for equal flanges (doubly symmetric sections) (in.)
     $= 2$ times compression web depth for unequal flanges (in.) $\qquad$ (9.25j)

$$\lambda_{rw} = 5.70\sqrt{\frac{E}{F_y}} \text{ limiting width-to-thickness ratio for noncompact web} \qquad (9.25k)$$

iii. Related to plate girder member (Fig. 9.17):

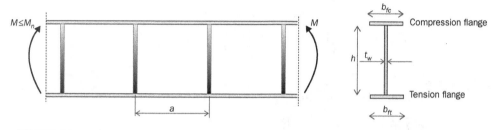

FIGURE 9.15 Plate girder under bending moment, *M* and girder section.

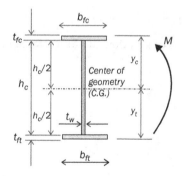

FIGURE 9.16 Doubly symmetric plate girder section under bending moment, *M*.

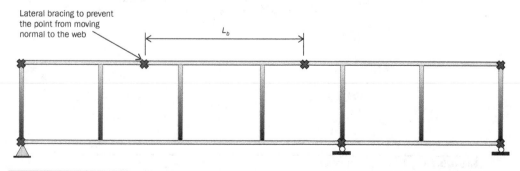

FIGURE 9.17 Lateral bracing in plate girders.

$L_p = 1.10r_t\sqrt{\dfrac{E}{F_y}}$, limiting laterally unbraced length for the limit state of

yielding (in.)

(9.25l)

$L_r = 1.95r_t\dfrac{E}{F_L}\sqrt{\dfrac{J}{S_{xc}h_o} + \sqrt{\left(\dfrac{J}{S_{xc}h_o}\right)^2 + 6.76\left(\dfrac{F_L}{E}\right)^2}}$, limiting unbraced length for

the limit state of inelastic lateral-torsional buckling (in.)

(9.25m)

or

$L_r = 1.95r_t\dfrac{E}{F_L}\sqrt{\dfrac{J}{S_{xc}h_o}}\sqrt{1 + \sqrt{1 + 6.76\left(\dfrac{F_L}{E}\dfrac{S_{xc}h_o}{J}\right)^2}}$

where

J = torsional constant, in.^{4}

$\quad = \sum\dfrac{1}{3}b_i t_i^3 = \dfrac{1}{3}\left(b_{fc}t_{fc}^3 + b_{ft}t_{ft}^3 + ht_w^3\right)$

h = section height measured between centroids of the flanges, in.

$\quad = \sum\dfrac{1}{3}b_i t_i^3 = \dfrac{1}{3}\left(b_{fc}t_{fc}^3 + b_{ft}t_{ft}^3 + ht_w^3\right)$

r_t = effective radius of gyration lateral-torsional buckling, in.

$r_t = \dfrac{b_{fc}}{\sqrt{12\left(1 + \dfrac{1}{6}a_w\right)}} = \sqrt{\dfrac{I_y^*}{A_T^*}}$ (Fig. 9.18)

$a_w = \dfrac{h_c t_w}{b_{fc}t_{fc}}$, but $a_w \le 10$

$$C_b = \dfrac{12.5M_{max}}{2.5M_{max} + 3M_A + 4M_B + 3M_C}R_m \le 3.0 \qquad (9.25n)$$

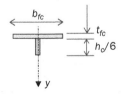

FIGURE 9.18 "*T*" for r_t calculation.

where

M_{max} = maximum value of bending moment over the unbraced length (Fig. 9.19), kip-in.

R_m = 1.0, doubly symmetric for any type of bending

= 1.0, singly symmetric under single curvature bending

= $0.5 + 2\left(\dfrac{I_{yc}}{I_y}\right)^2$, singly symmetric under reverse curvature

I_y = moment of inertia about the principal y-axis, in.^{4}

I_{yc} = moment of inertia about y-axis referred to the compression flange, or if reverse curvature bending, referred to the smaller flange, in.^{4}

9.4.2 Limit States Related to the Flange

Limit states related to the flange (Fig. 9.20) are listed as follows:

i. For $\lambda_{fc} > \lambda_{rf}$, the compression flange is a slender flange, and elastic local buckling occurs under local stress less than F_y.

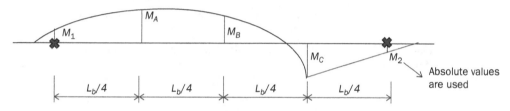

FIGURE 9.19 Moment diagram between lateral bracing points.

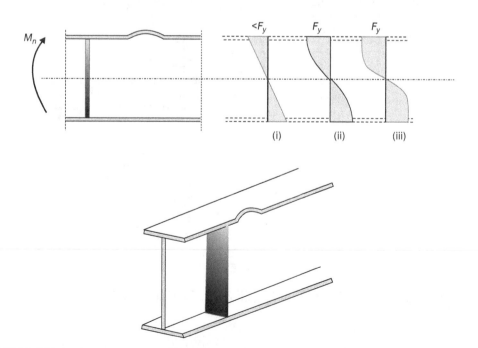

FIGURE 9.20 Normal stress distribution for flange-related limit states.

ii. For $\lambda_{pf} < \lambda_{fc} \leq \lambda_{rf}$ noncompact compression flange, and inelastic local buckling occurs. Some part of the flange might have reached yield stress, F_y, but not entire flange. Bending moment strength would be higher than that in the elastic buckling case.

iii. For $\lambda_{fc} \leq \lambda_{pf}$ compact compression flange. The compression flange can reach yield stress before any buckling occurs in the flange. The section has the potential to reach M_p as far as the flange is concerned.

9.4.3 Limit States Related to the Web

Limit states related to the web (Fig. 9.21) are:

i. For $\lambda = \dfrac{h_c}{t_w} > \lambda_{wr} = 5.70\sqrt{\dfrac{E}{F_y}}$, the section has a slender web, and elastic web

buckling would occur before any yielding. In practical applications, most plate girders are in this category.

ii. For $\lambda = \dfrac{h_c}{t_w} \leq \lambda_{wr} = 5.70\sqrt{\dfrac{E}{F_y}}$, the section might have noncompact web, or

even compact web. This case is unlikely to be encountered in "deep beams" or "plate girders."

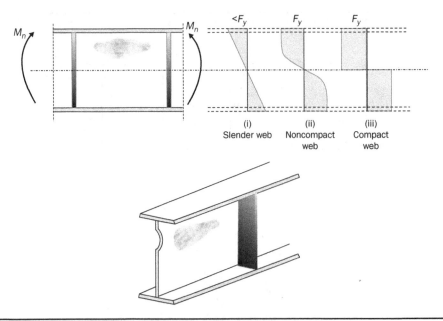

FIGURE 9.21 Normal stress distribution for web-related limit states.

M_n based on limit states of compression flange (Fig. 9.22):

$$M_n = R_{pg}F_{cr}S_{xc}$$ (9.26)

where
  R_{pg} = bending strength reduction factor

$$R_{pg} = 1 - \frac{a_w}{1,200 + 300a_w}\left(\frac{h_c}{t_w} - 5.70\sqrt{\frac{E}{F_y}}\right)$$

F_{cr} = critical stress in the compression flange as determined by λ_{fc} values, ksi

a. When $\lambda_{fc} < \lambda_{pf}$

$$F_{cr} = F_y \text{ (plastic behavior)}$$ (9.27)

b. When $\lambda_{pf} < \lambda_{fc} \leq \lambda_{rf}$

$$F_{cr} = F_y - (0.3F_y)\frac{\lambda_{fc} - \lambda_{pf}}{\lambda_{rf} - \lambda_{pf}} < F_y \text{ (inelastic buckling)}$$ (9.28)

c. When $\lambda_{fc} > \lambda_{rf}$

$$F_{cr} = \frac{0.9Ek_c}{\lambda_{fc}^2} \text{ (elastic buckling)}$$ (9.29)

M_n based on lateral-torsional buckling (LTB) when $L_b > L_p$:

$$M_n = R_{pg}F_{cr}S_{xc}$$ (9.30)

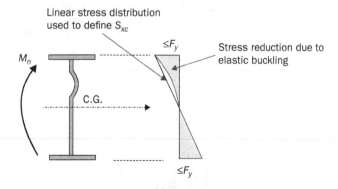

FIGURE 9.22   Stress distribution for noncompact webs in plate girders.

a. When $L_p < L_b < L_r$

$$F_{cr} = C_b \left[F_y - (0.3F_y)\frac{L_b - L_p}{L_r - L_p} \right] \leq F_y \qquad (9.31)$$

b. When $L_b > L_r$

$$F_{cr} = C_b \left[\frac{\pi^2 E}{\left(L_b / r_t \right)^2} \right] \leq F_y \qquad (9.32)$$

where

$$L_r = \pi r_t \sqrt{\frac{E}{0.7F_y}}$$

$$L_p = 1.10 r_t \sqrt{\frac{E}{F_y}}$$

9.4.4 Limit State of Lateral-Torsional Buckling (LTB)

LTB is an issue related to the member stability or global buckling (as opposed to the local buckling of web or flange) (Fig. 9.23).

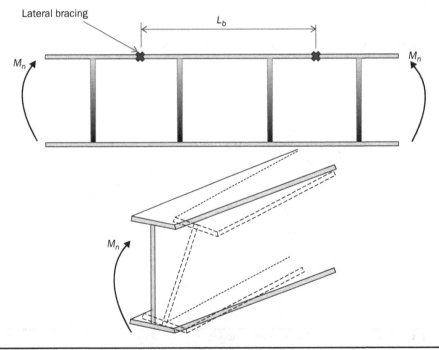

FIGURE 9.23 Lateral torsional buckling (LTB) of a plate girder member.

i. For $L_b > L_r$

The plate girder would have elastic lateral-torsional buckling under a single curvature moment over L_b segment. For a reverse moment, effective L_b might be smaller than actual unbraced length, of which the effect on LTB is accountable for by C_b.

ii. For $L_p < L_b \leq L_r$

Inelastic lateral-torsional buckling might occur, which results in a higher bending strength than elastic buckling case.

iii. For $L_b < L_p$

The girders would reach the maximum bending strength that the section can provide, depending on how flanges behave.

9.5   Intermediate Transverse Stiffeners (ITS)

Force in stiffener:
From Fig. 9.24,

$$P_s = (\sigma_t t_w a \sin\phi)\sin\phi$$

$$= \sigma_t t_w a \sin^2\phi$$

$$= \sigma_t t_w a \frac{1 - \cos^2\phi}{2} \tag{9.33}$$

$$= \sigma_t \frac{a t_w}{2}\left[1 - \frac{a/h}{\sqrt{1 + (a/h)^2}}\right]$$

Use $\dfrac{\sigma_t}{F_y} = 1 - C_v$, and assume $a/h \approx 1.0$

$$P_s = 0.15F_y\left(1 - C_v\right)ht_w \tag{9.34}$$

Let A_{st} = required stiffener area

$$A_{st}F_{yst} + (18t_w)(t_w)F_y \geq P_s$$

$$\tag{9.35}$$

$$A_{st} \geq \frac{F_y}{F_{yst}}\left[0.15ht_w(1 - C_v) - 18t_w^2\right]$$

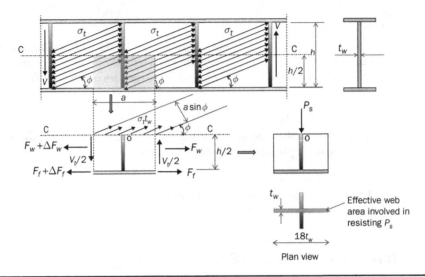

FIGURE 9.24   Intermediate transverse stiffeners (ITS).

a. No ITS is needed when $\dfrac{h}{t_w} \leq 2.46 \sqrt{\dfrac{E}{F_y}}$

$$\phi_v V_n \geq V_u \tag{9.36}$$

$\phi_v V_n \geq V_u$

where

$V_n = 0.6 F_y A_w C_v$

$\phi_v = 0.90$

$C_v = C_{v|k_v = 5}$

b. ITS design for a given $\dfrac{h}{t_w}$

  i. Spacing, a

  For end panel, estimate from:

$$C_v = \frac{V_r}{0.6 F_y A_w} = \begin{cases} \dfrac{1.51 E k_v}{F_y (h / t_w)^2}, & \text{elastic buckling} \\[3mm] \dfrac{1.10}{h / t_w} \sqrt{\dfrac{k_v E}{F_y}}, & \text{inelastic buckling} \end{cases} \tag{9.37}$$

where

$$k_v = 5 + \frac{5}{\left(\frac{a}{h}\right)^2} \rightarrow \frac{a}{h} = \sqrt{\frac{5}{k_v - 5}}$$

For the intermediate panel, suggest to use:

$$\frac{a}{h} \leq \left(\frac{260}{\frac{h}{t_w}}\right)^2 < 3.0 \tag{9.38}$$

ii. Stiffness requirement

$$I_{st} = I_{y-y} \geq jat_w^3 \tag{9.39}$$

and

$$\frac{b_{st}}{t_{st}} \leq 0.56\sqrt{\frac{E}{F}} \text{ to prevent local buckling of stiffeners (Fig. 9.25)} \tag{9.40}$$

where

$$j = \frac{2.5}{\left(\frac{a}{h}\right)^2} - 2 \geq 0.5$$

F_{yst} = yield stress of stiffeners, ksi

c. Strength requirement

$$A_{st} \geq \frac{F_{yw}}{F_{yst}}\left[0.15D_sA_w(1 - C_v)\frac{V_r}{V_c} - 18t_w^2\right] \geq 0 \tag{9.41}$$

where
$V_r = V_u$ = required shear strength at stiffener location, kips
$V_c = \phi_v V_n \geq V_u$
$V_n = V_{cr} + V_t$
D_s values and spacing requirements for the stiffener-to-web are given in Fig. 9.26.

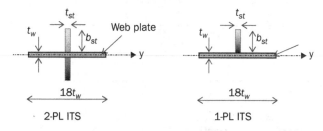

Figure 9.25 2-PL and 1-PL ITS.

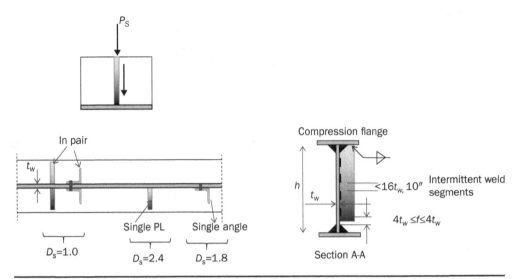

FIGURE 9.26 D_s values and spacing requirements for the stiffener.

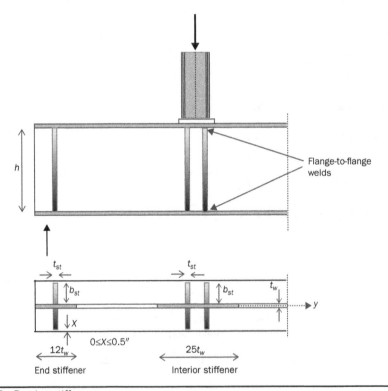

FIGURE 9.27 Bearing stiffeners.

9.6 Bearing Stiffeners (BS)

Bearing stiffener (BS) is needed when one of the failure modes due to concentrated loads occur (Fig. 9.27).

Design BS as a column member:

   i. Global stability

$$P_u \le \phi_c P_n \qquad (9.42)$$

where

$P_n = F_{cr} A_g$

$\phi_c = 0.90$

A_g = shadowed area in Fig. 9.27

$$F_{cr} = \begin{cases} \left(0.658^{F_y/F_e}\right)F_y, & \text{for } F_e \ge 0.44F_y \\ 0.877F_e, & \text{for } F_e < 0.44F_y \end{cases}$$

$$F_e = \frac{\pi^2 E}{\left(KL/r_y\right)^2}$$

$K = 0.75$

$$r_y = \sqrt{\frac{I_y}{A}}$$

I_y = moment of inertia about y-y of shadowed area, in.[4]

   ii. Local buckling

$$\frac{b_{st}}{t_{st}} \le 0.56\sqrt{\frac{E}{F}} \qquad (9.43)$$

   iii. Bearing strength

$$\phi R_n \ge P_u \qquad (9.44)$$

where

$\phi = 0.75$

$R_n = 1.8F_y(b_{st} - a)(t_{st})(m)$ (Fig. 9.28)

m = number of plates

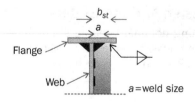

FIGURE 9.28 Effective width for bearing strength.

9.7 Design of Bearing Stiffeners

Stiffeners for concentrated forces in beams are designed as short columns to reinforce the web of a beam along the length or the web of a beam at an end reaction. For columns, stiffener plates are required when the applied forces are greater than the design strength for each failure mode (flange local bending, web local yielding, web crippling, compression buckling). The stiffeners should extend the full depth of the column when there are applied forces on both sides of the column. For columns, stiffener plates are welded to both the web and the flange. The weld to the flange is designed for the difference between the required strength and the design strength of the controlling limit state (flange local bending, web local yielding, web crippling, compression buckling) (Fig. 9.29).

The bearing strength, R_n, is determined for the limit state of bearing as

$$R_n = 1.8F_y A_{pb} \tag{9.45}$$

And the design strength for web local yielding, ϕR_n, is

$$P_u \leq \phi_{pb} R_n \tag{9.46}$$

where
$\phi_{pb} = 0.75$
P_u = ultimate reaction or concentrated force, kips
A_{pb} = projected area in bearing, in.^{2}

The available strength of the stiffener is determined for the limit states of yielding and buckling for the connecting elements as

a. For $KL/r \leq 25$

$$P_n = F_y A_g \tag{9.47}$$

b. For $KL/r > 25$

Determine the available strength based on elements in compression (Chapter 4). And the design compressive strength of the stiffener, $\phi_c P_n$, is

$$P_u \leq \phi_c P_n \tag{9.48}$$

where
$\phi_c = 0.90$

9.8 Summary for Design of Plate Girders

Summary for calculating nominal bending strength, M_n, nominal shear strength, V_n, design of intermediate transverse stiffeners and bearing stiffeners for plate girders are given in Tables 9.1, 9.2, 9.3, and 9.4, respectively.

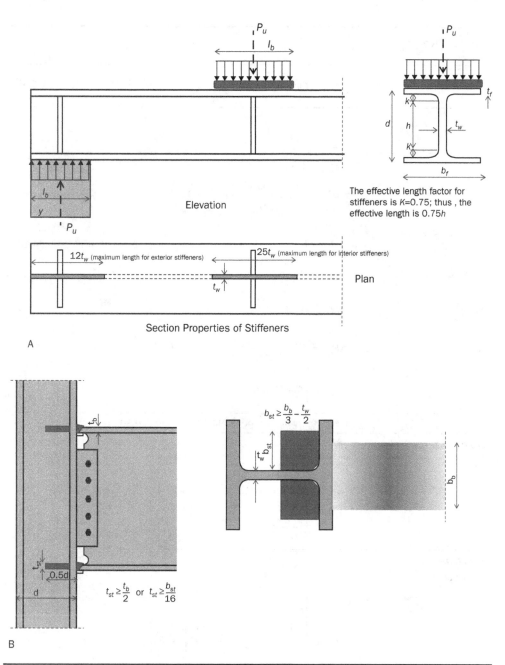

The effective length factor for stiffeners is $K=0.75$; thus, the effective length is $0.75h$

FIGURE 9.29 Bearing stiffeners: (a) for beams, (b) for columns.

For $\dfrac{h}{t_w} > 5.7\sqrt{E\big/F_y}$

$$M_n = \min\begin{cases} R_{pg}F_{cr}S_{xc} \\ F_yS_{xt} \end{cases} \qquad \text{AISC 360 (F5-7) and (F5-10)}$$

For $\dfrac{h}{t_w} \leq 5.7\sqrt{E\big/F_y}$

$\lambda_{fc} \leq \lambda_{pf} \rightarrow$

$$M_n = R_{pc}M_{yc} \qquad\qquad \text{AISC 360 (F4-1)}$$

$$R_{pc} = \frac{M_p}{M_{yc}} - \left(\frac{M_p}{M_{yc}} - 1\right)\left(\frac{\lambda - \lambda_{pw}}{\lambda_{rw} - \lambda_{pw}}\right) \leq \frac{M_p}{M_{yc}} \qquad \text{AISC 360 (F4-9b)}$$

$\lambda_{pf} < \lambda_{fc} \leq \lambda_{rf} \rightarrow$

$$M_n = R_{pc}M_{yc} - \left(R_{pc}M_{yc} - F_LS_{xc}\right)\left(\frac{\lambda - \lambda_{pf}}{\lambda_{rf} - \lambda_{pf}}\right) \qquad \text{AISC 360 (F4-13)}$$

$\lambda_{fc} > \lambda_{rf} \rightarrow$

$$M_n = \frac{0.9Ek_cS_{xc}}{\lambda_{fc}^2} \qquad\qquad \text{AISC 360 (F4-14)}$$

$S_{xt} < S_{xc} \rightarrow$

$$M_n = R_{pt}M_{yt} \qquad\qquad \text{AISC 360 (F4-15)}$$

$$R_{pg} = 1 - \frac{a_w}{1{,}200 + 300a_w}\left(\frac{h_c}{t_w} - 5.70\sqrt{\frac{E}{F_y}}\right)$$

$$a_w = \frac{h_ct_w}{b_{fc}t_{fc}} \text{ , but } a_w \leq 10$$

F_{cr} = the lowest value in the compression flange (ksi)

---

1) Local Flange Buckling

$$F_{cr} = \begin{cases} F_y & ,\lambda_{fc} \leq \lambda_{pf} \\[2mm] F_y - \left(0.3F_y\right)\dfrac{\lambda_{fc} - \lambda_{pf}}{\lambda_{rf} - \lambda_{pf}} < F_y & ,\lambda_{pf} < \lambda_{fc} \leq \lambda_{rf} \\[2mm] \dfrac{0.9Ek_c}{\lambda_{fc}^2} & ,\lambda_{fc} > \lambda_{rf} \end{cases}$$

---

2) Lateral-Torsional Buckling

$$F_{cr} = \begin{cases} F_y & ,L_b \leq L_p \\[2mm] C_b\left[F_y - \left(0.3F_y\right)\left(\dfrac{L_b - L_p}{L_r - L_p}\right)\right] \leq F_y & ,L_p < L_b \leq L_r \\[3mm] C_b\left[\dfrac{\pi^2E}{\left(L_b\big/r_t\right)^2}\right] \leq F_y & ,L_b > L_r \end{cases}$$

---

*Equation numbers in parentheses refer to the AISC equation number.

TABLE 9.1 Nominal Bending Strength, M_n^*

$$V_n = 0.6F_y A_w \left[C_v + C_v^* \right]$$

$$
C_v = \begin{cases}
1.0 & \text{, when } \dfrac{h}{t_w} \leq 1.10 \sqrt{\dfrac{k_v E}{F_y}} & \text{(G2-3)} \\[2ex]
\dfrac{1.10}{\frac{h}{t_w}} \sqrt{\dfrac{k_v E}{F_y}}, & \text{when } 1.10\sqrt{\dfrac{k_v E}{F_y}} < \dfrac{h}{t_w} \leq 1.37\sqrt{\dfrac{k_v E}{F_y}} & \begin{array}{l}\text{(G2-4)}\\ \text{Inelastic buckling}\end{array} \\[3ex]
\dfrac{1.51 E k_v}{F_y \left(\frac{h}{t_w}\right)^2}, & \text{when } \dfrac{h}{t_w} > 1.37\sqrt{\dfrac{k_v E}{F_y}} & \begin{array}{l}\text{(G2-5)}\\ \text{Elastic building}\\ \text{limit state}\end{array}
\end{cases}
$$

$$
C_v = \begin{cases}
1.0 & \text{, when } \dfrac{h}{t_w} \leq 1.10 \sqrt{\dfrac{k_v E}{F_y}} & \text{(G2-3)} \\[2ex]
\dfrac{1.10}{\frac{h}{t_w}} \sqrt{\dfrac{k_v E}{F_y}}, & \text{when } 1.10\sqrt{\dfrac{k_v E}{F_y}} < \dfrac{h}{t_w} \leq 1.37\sqrt{\dfrac{k_v E}{F_y}} & \begin{array}{l}\text{(G2-4)}\\ \text{Inelastic buckling}\end{array} \\[3ex]
\dfrac{1.51 E k_v}{F_y \left(\frac{h}{t_w}\right)^2}, & \text{when } \dfrac{h}{t_w} > 1.37\sqrt{\dfrac{k_v E}{F_y}} & \begin{array}{l}\text{(G2-5)}\\ \text{Elastic building}\\ \text{limit state}\end{array}
\end{cases}
$$

$$
k_v = \begin{cases}
5 + \dfrac{5}{\left(\frac{a}{h}\right)^2}, & \text{for } \dfrac{a}{h} \leq 3.0 \\[3ex]
5, & \text{for } \dfrac{a}{h} > 3.0 \text{ or } \dfrac{a}{h} > \left(\dfrac{260}{\frac{h}{t_w}}\right)^2
\end{cases}
$$

*Equation numbers in parentheses refer to the AISC equation number.

TABLE 9.2   Nominal Shear Strength, V_n*

Example 9.1

For the simply supported beam in Fig. 9.30, dead loads of $P_D = 20$ kips and a series of live loads of $P_D = 10$ kips with 7′6″ spacing are given. The beam is fully laterally supported by floor system. Steel grade A992 Grade 50 ($F_y = 50$ ksi). Determine the adequacy of the bearing stiffeners of $4 \times 1/2$ at the end point of the W24×68 beam (Fig. 9.31).

Solution:

a. Structural analysis (from Example 5.4 in Chapter 5):

Factored required design strengths (Fig. 9.32):

1) If $\dfrac{h}{t_w} > 2.46\sqrt{E/F_y}$, ITS is required

2) Nominal spacing, a, suggested as

$$\frac{a}{h} \le \left(\frac{260}{h/t_w}\right)^2 \quad < 3.0$$

3) Stiffness requirement $I_{st} = I_{y\text{-}y} \ge j\, a(t_w)^3$

$$j = \frac{2.5}{(a/h)^2} - 2 \ge 0.5$$

$$\frac{b_{st}}{t_{st}} \le 0.56\sqrt{E/F_y}$$

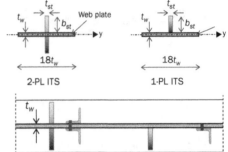

4) Strength requirement

$$A_{st} \ge \frac{F_{yw}}{F_{yst}}\left[0.15DA_w(1-C_v)\frac{V_r}{V_c} - 18t_w^2\right] \ge 0$$

TABLE 9.3 Intermediate Transverse Stiffeners (ITS)

$$M_u = 1.2M_D + 1.6M_L = 1.2(300^{\text{kip-ft}}) + 1.6(150^{\text{kip-ft}}) = 600 \text{ kip-ft}$$

$$V_u = 1.2V_D + 1.6V_L = 1.2(30^{\text{kips}}) + 1.6(12.5^{\text{kips}}) = 56 \text{ kips}$$

    b. Section properties of the area with stiffeners (Table 9.5): Try stiffeners with 0.4 in. thickness.

$$r = \sqrt{\frac{I}{A}} = \sqrt{\frac{12.12}{3.67}} = 3.30 \text{ in.}$$

    c. Determine the adequacy of the stiffener:

$$\frac{KL}{r} = \frac{0.75 \times 21.52^{\text{in.}}}{3.30} = 4.89 < 25$$

$$\phi_c P_n = \phi_c F_y A_g = 0.9 \times 50^{\text{ksi}} \times 3.67^{\text{in.}^2} = 165.2 \text{ kips} > 56 \text{ kips}$$

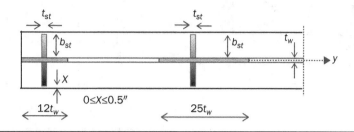

Global stability:

$$P_u \le \phi_c P_n \quad (\phi_c = 0.90)$$

$$P_n = F_y A_g, \quad \text{for } \frac{KL}{r} \le 25$$

$$P_n = F_{cr} A_g, \quad \text{for } \frac{KL}{r} > 25$$

$$A_g = \text{Shadowed area}$$

$$F_e = \frac{\pi^2 E}{\left(\dfrac{KL}{r_y}\right)^2}, \qquad K = 0.75$$

$$F_{cr} = \begin{cases} \left[0.658^{F_y/F_e}\right] F_y, & \text{for } F_e \ge 0.44F_y \quad (\text{E3-2}) \\ 0.877F_e, & \text{for } F_e < 0.44F_y \quad (\text{E3-3}) \end{cases}$$

$$r_y = \sqrt{\frac{I_y}{A}}$$

$$I_y = \text{Moment of inertia about y-y of shadowed area}$$

$$\frac{b_{st}}{t_{st}} \le 0.56\sqrt{\frac{E}{F_y}}$$

Bearing strength:

$$\phi R_n \ge P_u$$
$$\phi = 0.75$$
$$R_n = 1.8F_y\left(b_{st} - a\right)\left(t_{st}\right)(m)$$
$$m = \text{number of plates}$$
$$a = \text{weld size}$$

*Equation numbers in parentheses refer to the AISC equation number.

TABLE 9.4 Bearing Stiffeners*

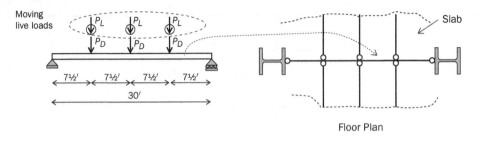

FIGURE 9.30   Simply supported beam and the floor plan.

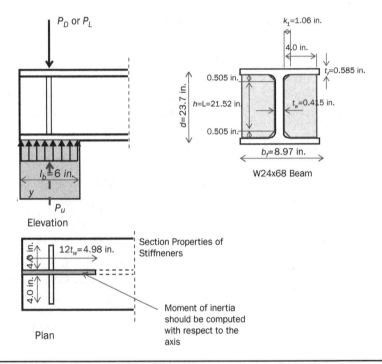

FIGURE 9.31   Bearing stiffener for W24×68.

d. Check bearing strength:

$$A_{pb} = 2 \times 0.4^{\text{in.}} \times (4^{\text{in.}} - 1.06^{\text{in.}}) = 2.35 \text{ in.}^2 \quad \text{the area at the end of the plate excluding}$$
the fillets

$$\phi_{pb} R_n = \phi_{pb} 1.8 F_y A_{pb} = 0.75 \times 1.8 \times 50^{\text{ksi}} \times 2.35^{\text{in.}^2} = 158.63 \text{ kips} > 56 \text{ kips}$$

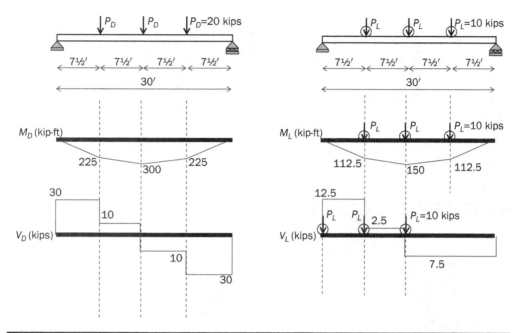

Figure 9.32 Bending moment and shear force diagrams.

Shape	Area (in.²)	I (in.⁴)	d (in.)	Ad² (in.⁴)	I+Ad² (in.⁴)
Stiffener plate (0.4×4.0 in.)	1.6	4.27	2.21	7.815	12.09
Web (0.415×4.98 in.)	2.07	0.0297	0	0	0.0297 (may be ignored)
Σ	3.67				12.12

Table 9.5 Section Properties for the Stiffener Plate and Web

Example 9.2

Figure 9.33 shows details of the plate girder in a four-girder structure simply supported and subject to three concentrated loads. $F_y = 50$ ksi steel is used. Lateral braces (shown as "x" sign in Fig. 9.33) are provided at the loading points and supports. Calculate the maximum value of P (80% dead and 20% live load). Assume all stiffeners are properly designed.

Solution:

1. Bending moment (M) and shear force diagrams (V)

   Segment (A):

   $M = 15P$

   $V = 1.5P$

   $a = 10' = 120''$

   $t_w = 1/2''$

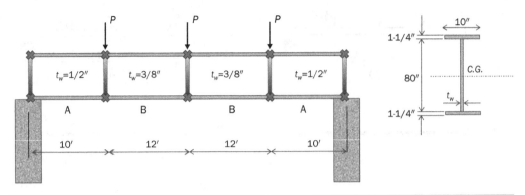

FIGURE 9.33   Plate girder in a four-girder simply supported structure.

Segment (B):
$M = 21P$
$V = 0.5P$
$a = 12' = 144''$
$t_w = 3/8''$

Factored required design strengths:
$M_u = 1.2M_D + 1.6M_L = 1.2(0.8)M + 1.6(0.2)M$
$V_u = 1.2V_D + 1.6V_L = 1.2(0.8)V + 1.6(0.2)V$

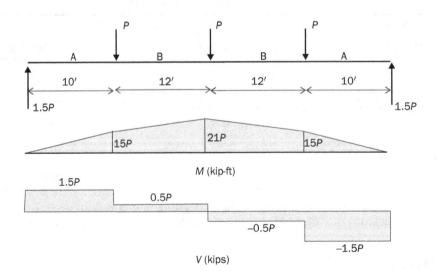

FIGURE 9.34   The bending moment and shear force diagrams.

Segment A (Fig. 9.34):
$M_u = 1.2(0.8)(15P) + 1.6(0.2)(15P) = 19.2P$ kip-ft
$V_u = 1.2(0.8)(1.5P) + 1.6(0.2)(1.5P) = 1.92P$ kips

Segment B (Fig. 9.34):
$M_u = 1.2(0.8)(21P) + 1.6(0.2)(21P) = 26.88P$ kip-ft
$V_u = 1.2(0.8)(0.5P) + 1.6(0.2)(0.5P) = 0.64P$ kips

2. Bending strength: Lateral-torsional and flange local buckling

a. Cross-section property

$$I_x = \frac{1}{12}(t_w)(h)^3 + 2(b_f t_f)\left(\frac{h + t_f}{2}\right)^2$$ (Fig. 9.35)

$$= \begin{cases} 62,590 \text{ in.}^4 \text{ (Segment A)} \\ 57,260 \text{ in.}^4 \text{ (Segment B)} \end{cases}$$

$$S_{xt} = S_{xc} = \frac{I_x}{\left(\dfrac{h + t_f}{2}\right)} = \begin{cases} 1,520 \text{ in.}^3 \text{ (Segment A)} \\ 1,390 \text{ in.}^3 \text{ (Segment B)} \end{cases}$$

$$I_T = \frac{1}{12}(t_f)(b_f)^3 = \frac{1}{12}(1.25'')(10'')^3 = 104 \text{ in.}^4$$

$$A_T = A_f + \frac{1}{6}A_w = (10'')(1.25'') + \frac{1}{6}(80'')(t_w) = \begin{cases} 19.2 \text{ in.}^2 \text{(Segment A)} \\ 17.5 \text{ in.}^2 \text{(Segment B)} \end{cases}$$

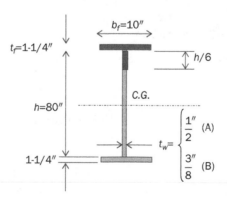

FIGURE 9.35   The plate girder section.

$$r_T = \sqrt{\frac{I_T}{A_T}} = \begin{cases} 2.33 \text{ in. (Segment A)} \\ 2.44 \text{ in. (Segment B)} \end{cases}$$

Check slenderness ratio of the web: $h_c = h$

$$\lambda = \frac{h_c}{t_w} = \begin{cases} 160 \text{ (Segment A)} > 5.70\sqrt{E / F_y} = 137 \\ \\ 213 \text{ (Segment B)} > 5.70\sqrt{E / F_y} = 137 \end{cases} \rightarrow \text{Slender web (elastic web buckling)}$$

b. F_{cr}

    i. Lateral-torsional buckling

$$L_p = 1.10r_t\sqrt{\frac{E}{F_y}} = \begin{cases} 61.7'' = 5.14' < L_{b,A} = 10' \\ 64.6'' = 5.39' < L_{b,B} = 12' \end{cases}$$

$$L_r = \pi r_t\sqrt{\frac{E}{0.7F_y}} = \begin{cases} 210.7'' = 17.6' < L_{b,A} = 10' \\ 220.7'' = 18.4' < L_{b,B} = 12' \end{cases}$$

$$C_b = \begin{cases} 1.67 \text{ (Segment A)} \\ 1.13 \text{ (Segment B)} \end{cases}$$

Note: C_b is calculated using AISC 360 (F1-1)

$$F_{cr} = C_b\left[F_y - (0.3F_y)\frac{L_b - L_p}{L_r - L_p}\right] = \begin{cases} 74 \text{ ksi} > 50 \text{ ksi} \quad \rightarrow \quad F_{cr} = 50 \text{ ksi (Segment A)} \\ 47.89 \text{ ksi} < 50 \text{ ksi} \quad \rightarrow \quad F_{cr} = 47.89 \text{ ksi (Segment B)} \end{cases}$$

    ii. Check flange local buckling

$$\lambda = \frac{b_f}{2t_f} = \frac{10''}{2(1.25'')} = 4.0 < \lambda_{pf} = 0.38\sqrt{\frac{E}{F_y}} = 9.15 \rightarrow \text{no flange local buckling}$$

    $F_{cr} = 50$ ksi   (Segments A and B)

    From (i) and (ii) the smaller value: $F_{cr} = \begin{cases} 50 \text{ ksi} \quad \text{(Segment A)} \\ 47.89 \text{ ksi} \quad \text{(Segment B)} \end{cases}$

c. $M_n = R_{pg}F_{cr}S_{xc}$

$$a_w = \frac{h_c t_w}{b_{fc}t_{fc}} = \frac{(80'')t_w}{(10'')(1.25'')} = \begin{cases} 3.2 \quad \text{(Segment A)} \\ 2.4 \quad \text{(Segment B)} \end{cases}$$

$$R_{pg} = 1 - \frac{a_w}{1,200 + 300a_w}\left(\frac{h_c}{t_w} - 5.70\sqrt{\frac{E}{F_y}}\right) = \begin{cases} 0.966 \text{ (Segment A)} \le 1.0 \\ 0.905 \text{ (Segment B)} \le 1.0 \end{cases}$$

$$M_n = R_{pg}F_{cr}S_{xc} = \begin{cases} 7,342 \text{ kip-ft} \quad \text{(Segment A)} \\ 6,024 \text{ kip-ft} \quad \text{(Segment B)} \end{cases}$$

3. Nominal shear strength: web shear buckling

Segment A:

$$\frac{a}{h} = 1.5 < \left\{\begin{array}{l} 3.0 \\ \left(\dfrac{260}{h/t_w}\right)^2 = 2.64 \end{array}\right.$$

End panel is not allowed to use tension field action, i.e., $C_v = 0$

$$k_v = 5 + \frac{5}{\left(a/h\right)^2} = 7.22$$

$$\lambda_p = 1.10\sqrt{\frac{k_v E}{F_y}} = 71 < \frac{h}{t_w} = 160$$

$$\lambda_r = 1.37\sqrt{\frac{k_v E}{F_y}} = 89 < \frac{h}{t_w} = 160 \quad \rightarrow \text{the web in Segment A has an elastic buckling}$$

limit state.

$$C_v = \frac{1.51 E k_v}{F_y\left(h/t_w\right)^2} = \frac{1.51(29,000^{\text{ksi}})7.22}{(50^{\text{ksi}})(160)^2} = 0.247$$

Segment B:

$$\frac{a}{h} = 1.8 \left\{\begin{array}{l} < 3.0 \\ > \left(\dfrac{260}{h/t_w}\right)^2 = 1.49 \end{array}\right.$$

Thus, tension field is not allowed for Segment B.

$$k_v = 5 \text{ for } \frac{a}{h} > \left(\frac{260}{h/t_w}\right)^2$$

$$\lambda_p = 1.10\sqrt{\frac{k_v E}{F_y}} = 59 < \frac{h}{t_w} = 213$$

$$\lambda_r = 1.37\sqrt{\frac{k_v E}{F_y}} = 74 < \frac{h}{t_w} = 213 \quad \rightarrow \text{the web in Segment B has an elastic buckling}$$

limit state.

$$C_v = \frac{1.51Ek_v}{F_y\left(h/t_w\right)^2} = \frac{1.51(29{,}000^{ksi})5}{(50^{ksi})(213)^2} = 0.097$$

$$V_n = \begin{cases} 0.6F_yA_wC_v, & \text{(Segment A)} \\ 0.6F_yA_wC_v, & \text{(Segment B)} \end{cases}$$

$$= 0.6(50^{ksi})(80'')\begin{bmatrix} 1/2'' \\ 3/8'' \end{bmatrix}\begin{bmatrix} 0.247 \\ 0.097 \end{bmatrix} = \begin{cases} 296 \text{ kips} & \text{(Segment A)} \\ 87 \text{ kips} & \text{(Segment B)} \end{cases}$$

4. Design P from bending and shear

   Segment A:

   From bending:

   $$M_u \le \phi M_n$$

   $19.2P \le 0.9(7{,}324^{kip\text{-}ft})$
   $P \le 344$ kips

   From shear:

   $$V_u \le \phi V_n$$

   $1.92P \le 0.9(296^{kips})$
   $P \le 138$ kips

   Segment B:

   From bending:

   $$M_u \le \phi M_n$$

   $26.88P \le 0.9(6{,}024^{kip\text{-}ft})$
   $P \le 344$ kips

   From shear:

   $$V_u \le \phi V_n$$

   $0.64P \le 0.9(87^{kips})$
   $P \le 122$ kips

Thus, maximum service load $P = 122$ kips (controlled by elastic shear buckling of the web in the panel in Segment B).

Example 9.3

A 60-ft-long plate girder supports column loads and its own weight. Design was done with "ITS," shown in Fig. 9.36. A36 steel is used. Lateral bracings are provided at the top and bottom flanges at the ends, and at the top flange under the two loads. Determine maximum service load P (the self-weight included in P approximately), and bearing stiffeners if needed.

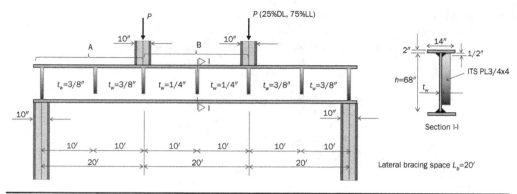

FIGURE 9.36   Plate girder supporting column loads.

Solution:

1. Bending moment (*M*) and shear force diagrams (*V*)

Segment (A) (Fig. 9.37):
$M = 20P$
$V = P$
$a = 10' = 120''$
$t_w = 3/8''$

Segment (B) (Fig. 9.37):
$M = 20P$
$V = 0$
$a = 10' = 120''$
$t_w = 1/4''$

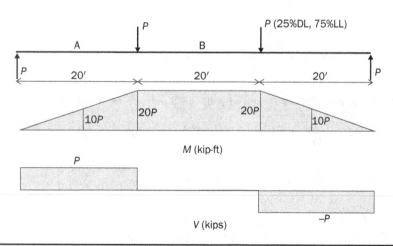

FIGURE 9.37   The bending moment and shear force diagrams.

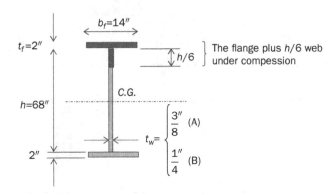

Figure 9.38   The plate girder section.

Factored required design strengths:

$$M_u = 1.2M_D + 1.6M_L = (1.2\times25\% + 1.6\times75\%)M = 1.5M$$
$$V_u = 1.2V_D + 1.6V_L = (1.2\times25\% + 1.6\times75\%)V = 1.5V$$

2. Bending strength: Lateral-torsional and flange local buckling

a. Cross-section property

$$I_x = \frac{1}{12}(t_w)(h)^3 + 2(b_f t_f)\left(\frac{h + t_f}{2}\right)^2 \quad \text{(Fig. 9.38)}$$

$$= \begin{cases} 78,400 \text{ in.}^4 \text{ (Segment A)} \\ 75,150 \text{ in.}^4 \text{ (Segment B)} \end{cases}$$

$$S_{xt} = S_{xc} = \frac{I_x}{\left(\dfrac{h + t_f}{2}\right)} = \begin{cases} 2,180 \text{ in.}^3 \text{ (Segment A)} \\ 2,090 \text{ in.}^3 \text{ (Segment B)} \end{cases}$$

The properties of the flange plus $h/6$ web under compression:

$$I_y^* = \frac{1}{12}(t_f)(b_f)^3 = \frac{1}{12}(2'')(14'')^3 = 457 \text{ in.}^4$$

$$A_T^* = A_f + \frac{1}{6}A_w = (14'')(2'') + \frac{1}{6}(68'')(t_w) = \begin{cases} 32.0 \text{ in.}^2 \text{(Segment A)} \\ 31.0 \text{ in.}^2 \text{(Segment B)} \end{cases}$$

$$r_t = \sqrt{\frac{I_y^*}{A_T^*}} = \begin{cases} 3.76 \text{ in. (Segment A)} \\ 3.85 \text{ in. (Segment B)} \end{cases}$$

Check slenderness ratio of the web: $h_c = h$

$$\lambda = \frac{h_c}{t_w} = \begin{cases} 181 \ (\text{Segment A}) > 5.70\sqrt{E \, / \, F_y} = 162 \\ 272 \ (\text{Segment B}) > 5.70\sqrt{E \, / \, F_y} = 162 \end{cases}$$

The web is slender, indicating that the limit state of the web elastic buckling occurs.

b. F_{cr} from lateral-torsional or local flange buckling with *Slender Web*

i. Lateral-torsional buckling

$L_b = 20'$ for Segments A and B (Note: L_b is unbraced length, in which panels A and B are included)

$$L_p = 1.10r_t\sqrt{\frac{E}{F_y}} = \begin{cases} 9.79' < L_{b,A} = 20' \\ 10.02' < L_{b,B} = 20' \end{cases}$$

$$L_r = \pi r_t\sqrt{\frac{E}{0.7F_y}} = \begin{cases} 33.4' < L_{b,A} = 20' \\ 34.2' < L_{b,B} = 20' \end{cases}$$

$$C_b = \begin{cases} 1.67 \ (\text{Segment A}) \\ 1.00 \ (\text{Segment B}) \end{cases}$$

$$F_{cr} = C_b\left[F_y - (0.3F_y)\frac{L_b - L_p}{L_r - L_p}\right] = \begin{cases} 52.33 \ \text{ksi} > 36 \ \text{ksi} & \rightarrow & F_{cr} = 36 \ \text{ksi} \ (\text{Segment A}) \\ 31.54 \ \text{ksi} < 36 \ \text{ksi} & \rightarrow & F_{cr} = 31.54 \ \text{ksi} \ (\text{Segment B}) \end{cases}$$

ii. Check flange local buckling

$$\lambda_{fc} = \frac{b_{fc}}{2t_{fc}} = \frac{14''}{2(2'')} = 3.5 < \lambda_{pf} = 0.38\sqrt{\frac{E}{F_y}} = 10.8 \rightarrow \text{compact compression flange}$$

(no flange local buckling)

$F_{cr} = F_y = 36$ ksi (Segments A and B)
From (i) and (ii) the smaller value:

$$F_{cr} = \begin{cases} 36 \ \text{ksi} & (\text{Segment A}) \\ 31.54 \ \text{ksi} & (\text{Segment B}) \end{cases}$$

c. $M_n = R_{pg}F_{cr}S_{xc}$

$$a_w = \frac{h_c t_w}{b_{fc}t_{fc}} == \begin{cases} 0.911 < 10.0 & (\text{Segment A}) \\ 0.607 < 10.0 & (\text{Segment B}) \end{cases}$$

$$R_{pg} = 1 - \frac{a_w}{1,200 + 300a_w}\left(\frac{h_c}{t_w} - 5.70\sqrt{\frac{E}{F_y}}\right) = \begin{cases} 0.988 \ (\text{Segment A}) \le 1.0 \\ 0.952 \ (\text{Segment B}) \le 1.0 \end{cases}$$

$$M_n = R_{pg}F_{cr}S_{xc} = \begin{cases} 0.988(36^{ksi})(2,180^{in.^3})(\frac{1}{12}^{in./ft}) = 6,460 \text{ kip-ft} \quad \text{(Segment A)} \\ 0.952(31.54^{ksi})(2,090^{in.^3})(\frac{1}{12}^{in./ft}) = 5,230 \text{ kip-ft} \quad \text{(Segment B)} \end{cases}$$

3. Nominal shear strength: web shear buckling

Segment B is in pure bending, no shear failure mode.
Segment A:

$$\frac{a}{h} = 1.76$$

$$k_v = 5 + \frac{5}{\left(\frac{a}{h}\right)^2} = 6.61$$

$$\frac{h}{t_w} = 181 > 1.37\sqrt{\frac{k_v E}{F_y}} = 100 \quad \text{elastic buckling limit state}$$

$$C_v = \frac{1.51 E k_v}{F_y\left(\frac{h}{t_w}\right)^2} = 0.245$$

Check tension field action:

$$\frac{a}{h} = 1.76 < \begin{cases} 3.0 \\ \left(\frac{260}{\frac{h}{t_w}}\right)^2 = 2.1 \end{cases} \qquad \text{O.K.}$$

$$\frac{2A_w}{A_{fc} + A_{ft}} = \frac{2(68'')(\frac{3}{8}'')}{2(14'')(2'')} = 0.91 < 2.5 \qquad \text{O.K.}$$

End panel is not allowed to use tension field action, i.e., $C_v^* = 0$
Second panel can use tension field action:

$$C_v^* = \frac{1 - C_v}{1.15\sqrt{1 + \left(\frac{a}{h}\right)^2}} = 0.324$$

Shear strength:

$$V_n = \begin{cases} 0.6F_y A_w C_v, & \text{First panel (end)} \\ 0.6F_y A_w (C_v + C_v^*), & \text{Second panel} \end{cases}$$

$$= 0.6(36^{ksi})(68'')(\frac{3}{8}'')\begin{cases} 0.245 \\ 0.245 + 0.324 \end{cases} = \begin{cases} 135 \text{ kips} & \text{End panel} \\ 313 \text{ kips} & \text{Second panel} \end{cases}$$

4. Maximum service load P from bending and shear

Segment A:
End panel:

From bending:
$M_u \leq \phi M_n$

$1.5M \leq 0.9(6{,}459^{\text{kip-ft}})$
$1.5(10P) \leq 0.9(6{,}459^{\text{kip-ft}})$
$P \leq 388$ kips

From shear:
$V_u \leq \phi V_n$

$1.5V \leq 0.9(135^{\text{kips}})$
$1.5(P) \leq 0.9(135^{\text{kips}})$
$P \leq 81$ kips

Second panel:
From bending:
$M_u \leq \phi M_n$

$1.5M \leq 0.9(6{,}459^{\text{kip-ft}})$
$1.5(20P) \leq 0.9(6{,}459^{\text{kip-ft}})$
$P \leq 194$ kips

From shear:
$V_u \leq \phi V_n$

$1.5V \leq 0.9(313^{\text{kips}})$
$1.5(P) \leq 0.9(313^{\text{kips}})$
$P \leq 188$ kips

End panel controls (shear failure) in segment A.
$P \leq 81$ kips   (Segment A)

Segment B: (pure bending)
From bending:
$M_u \leq \phi M_n$

$1.5M \leq 0.9(5{,}230^{\text{kip-ft}})$
$1.5(20P) \leq 0.9(5{,}230^{\text{kip-ft}})$
$P \leq 157$ kips

Therefore, maximum service load $P = 81$ kips (controlled by shear failure of the end panel in Segment A).

5. Check web local yielding for $P = 81$ kips

$P_u = 1.5P = 1.5(81 \text{ kips}) = 122 \text{ kips}$ Use $\phi = 1.0$

$k = t_f + a = 2'' + \frac{1}{2}'' = 2.5''$ $N = 10''$ (the bearing length)

End:

$(1.0)[N+2.5k](t_w)F_{yw} = (1.0)[10''+2.5(2.5'')](3/8'')(36^{ksi})$
$= 207 \text{ kips} > 122 \text{ kips}$

Interior:

$(1.0)[N+5k](t_w)F_{yw} = (1.0)[10''+5(2.5'')]\left(\dfrac{3/8'' + 1/4''}{2}\right)(36^{ksi})$

$= 251 \text{ kips} > 122 \text{ kips}$

No bearing stiffener is needed for web local yielding.

6. Web crippling $\phi = 0.75$

The end panel controls; the support centerline is near the web:

$e = 5'' < \dfrac{d}{2} = 36''$, and $N/d = 10''/72'' = 0.139 < 0.2$, so we use:

$$R_n = 0.40t_w^2\left[1 + 3\left(\frac{N}{d}\right)\left(\frac{t_w}{t_f}\right)^{1.5}\right]\sqrt{\frac{EF_{yw}t_f}{t_w}} \text{ , for } y \le d \text{ and } N/d \le 0.2$$

$$= 0.40\left(\frac{3''}{8}\right)^2\left[1 + 3\left(\frac{10''}{72''}\right)\left(\frac{3/8''}{2''}\right)^{1.5}\right]\sqrt{\frac{(29{,}000^{ksi})(36^{ksi})(2'')}{\left(\frac{3}{8}''\right)}} = 137 \text{ kips}$$

$\phi R_n = 0.75(137^{kips}) = 102.8 \text{ kips} < 122 \text{ kips}$

Bearing stiffener is needed to get $P_u = 122$ kips

7. Bearing stiffeners at supports

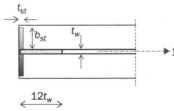

(Per J10.8, J4.4, E6.2, and/or J7 might be relevant.)

FIGURE 9.39 The stiffener and portion of the web.

The stiffener and portion of the web are treated as the cross-section of the column with $KL = 0.75h$, as shown in Fig. 9.39.

For a preliminary selection, assume $F_{cr} = 35$ ksi
$$P_u = 0.90F_{cr}A_g$$

$$A_g = \frac{P_u}{0.90F_{cr}} = \frac{122.2^{kips}}{0.90(35^{ksi})} = 4 \text{ in.}^2$$

$$A_{st} = A_g - 12t_w(t_w) = 2.3 \text{ in.}^2$$

Try 2 PLs—(3/8)×4, $A_{st} = 3$ in.^{2}, $\dfrac{b_{st}}{t_{st}} = \dfrac{4''}{3/8''} = 10.7 < \dfrac{95}{\sqrt{36^{ksi}}} = 15.8$,

the stiffener is not slender.

$$A_g = A_{st} + 12t_w(t_w) = 3^{in.^2} + 12\left(\frac{3}{8}''\right)^2 = 4.69 \text{ in.}^2$$

$$I_y = \frac{1}{12}(t_{st})(2b_{st} + t_w)^3 = \frac{1}{12}\left(\frac{3}{8}''\right)\left(8'' + \frac{3}{8}''\right)^3 = 18.36 \text{ in.}^4$$

$$r_y = \sqrt{\frac{I_y}{A_g}} = \sqrt{\frac{18.36^{in.^4}}{4.69^{in.^2}}} = 1.98 \text{ in.}$$

$$\frac{KL}{r} = \frac{0.75h}{r_y} = 25$$

Thus, J4.4 applies, use (J4.6) as

$$\phi P_n = 0.9F_yA_g = 0.9(36^{ksi})(4.69^{in.^2}) = 152 \text{ kips} > 122 \text{ kips} \quad \text{O.K.}$$

- The stiffener should not be a slender plate: $\dfrac{b_{st}}{t_{st}} = 10.7 < \dfrac{95}{\sqrt{36^{ksi}}} = 15.8$ O.K.
- Bearing strength of the stiffener:

  $1.8F_{yst}(b_{st} - D)(t_{st})(m) = 1.8(36^{ksi})(4'' - 0.5'')(3/8'')(2) = 170 \text{ kips} > 122 \text{ kips O.K.}$

8. Conclusions
   - Add bearing stiffeners at supports with 2PL3/8×4×68 and under P with 2PL1/4×4×68 (Fig. 9.40)
   - Maximum allowable service load $P = 81$ kips (Shear elastic buckling at end panel)

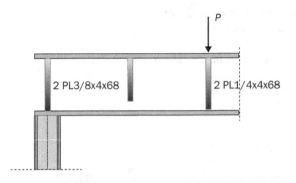

FIGURE 9.40   Final design.

Example 9.4

Steel plate shear walls have become increasingly popular in tall steel buildings. In fact, it can be treated as a cantilever beam built up with plate girders, as shown in Fig. 9.41. A992 grade 50 (F_y= 50 ksi) steel is used. Assume pure shear controls the design. Determine the maximum lateral load, V_u, of the plate girder design.

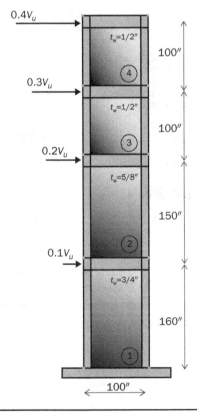

FIGURE 9.41   A steel plate girder shear wall subject to lateral loads.

Solution:

Section 1:

$a = 160''$, $h = 100''$, $t_w = \frac{3}{4}''$

$$\frac{a}{h} = 1.60, \quad \frac{h}{t_w} = 133$$

$$\left(\frac{260}{h/t_w}\right)^2 = \left(\frac{260}{133}\right)^2 = 3.82 > \frac{a}{h} = 1.60 \,(< 3)$$

Thus, $k_v = 5 + \dfrac{5}{(1.60)^2} = 6.953$

$$\lambda_p = 1.10 \sqrt{\frac{k_v E}{F_y}} = 80.267 < \frac{h}{t_w} = 133$$

$$\lambda_r = 1.37 \sqrt{\frac{k_v E}{F_y}} = 100.441 < \frac{h}{t_w} = 133$$

Thus, Section 1 has an elastic web buckling limit state.

$$C_v = \frac{1.51 E k_v}{F_y \left(\dfrac{h}{t_w}\right)^2} = \frac{1.51(29,000^{\text{ksi}})6.953}{(50^{\text{ksi}})(133)^2} = 0.344$$

Tension field action can be developed in this section due to strong support from the fixed support.

$$C_v^* = \frac{1 - C_v}{1.15\sqrt{1 + \left(a/h\right)^2}} = \frac{1 - 0.344}{1.15\sqrt{1 + (1.60)^2}} = 0.302$$

Shear strength:

$$V_{n1} = 0.6 F_y A_w (C_v + C_v^*)$$

$$= 0.6(50^{\text{ksi}})(100'' \times \tfrac{3}{4}'')(0.344 + 0.302)$$

$$= 1.454 \text{ kips}$$

The first-story shear force, $V_{u1} = 0.1 V_u + 0.2 V_u + 0.3 V_u + 0.4 V_u = V_u$

Let $\qquad V_{u1} \le \phi V_{n1}$, or $\quad V_u \le \phi V_{n1} = 0.9 \times 1,454^{\text{kips}} = 1,309 \text{ kips}$

Thus, $V_u \le 1,309$ kips, based on Section 1.

Section 2:

$a = 150''$, $h = 100''$, $t_w = 5/8''$

$$\frac{a}{h} = 1.50, \quad \frac{h}{t_w} = 160$$

$$\left(\frac{260}{h/t_w}\right)^2 = \left(\frac{260}{160}\right)^2 = 2.64 > \frac{a}{h} = 1.50 \, (< 3)$$

Thus, $k_v = 5 + \dfrac{5}{(1.50)^2} = 7.22$

$$\lambda_p = 1.10\sqrt{\frac{k_v E}{F_y}} = 71 < \frac{h}{t_w} = 160$$

$$\lambda_r = 1.37\sqrt{\frac{k_v E}{F_y}} = 89 < \frac{h}{t_w} = 160$$

Thus, Section 2 has an elastic web buckling limit state.

$$C_v = \frac{1.51 E k_v}{F_y\left(\frac{h}{t_w}\right)^2} = \frac{1.51(29,000^{ksi})7.22}{(50^{ksi})(160)^2} = 0.247$$

$$C_v^* = \frac{1 - C_v}{1.15\sqrt{1 + \left(\frac{a}{h}\right)^2}} = \frac{1 - 0.247}{1.15\sqrt{1 + (1.50)^2}} = 0.363$$

Shear strength:

$$V_{n2} = 0.6 F_y A_w (C_v + C_v^*)$$

$$= 0.6(50^{ksi})\left[100'' \times \frac{5}{8}''\right](0.247 + 0.363)$$

$$= 1.144 \text{ kips}$$

The second-story shear force, $V_{u2} = 0.2V_u + 0.3V_u + 0.4V_u = 0.9V_u$

Let $\quad V_{u2} \leq \phi V_{n2}$, or $\quad 0.9V_u \leq \phi V_{n2} = 0.9 \times 1,144^{kips} = 1,030$ kips

Thus, $V_u \leq 1,144$ kips, based on Section 2.

Section 3:

$a = 100''$, $h = 100''$, $t_w = \frac{1}{2}''$

$$\frac{a}{h} = 1.0, \quad \frac{h}{t_w} = 200$$

$$\left(\frac{260}{h/t_w}\right)^2 = \left(\frac{260}{200}\right)^2 = 1.69 > \frac{a}{h} = 1.0 \, (< 3)$$

Thus, $k_v = 5 + \dfrac{5}{(1.0)^2} = 10.0$

$\lambda_p = 1.10\sqrt{\dfrac{k_v E}{F_y}} = 84 < \dfrac{h}{t_w} = 200$

$\lambda_r = 1.37\sqrt{\dfrac{k_v E}{F_y}} = 104 < \dfrac{h}{t_w} = 200$

Thus, Section 3 has an elastic web buckling limit state.

$C_v = \dfrac{1.51 E k_v}{F_y\left(\dfrac{h}{t_w}\right)^2} = \dfrac{1.51(29{,}000^{ksi})10.0}{(50^{ksi})(200)^2} = 0.219$

$C_v^* = \dfrac{1 - C_v}{1.15\sqrt{1 + \left(\dfrac{a}{h}\right)^2}} = \dfrac{1 - 0.219}{1.15\sqrt{1 + (1.0)^2}} = 0.480$

Shear strength:

$V_{n3} = 0.6 F_y A_w (C_v + C_v^*)$

$\qquad = 0.6(50^{ksi})\left(100'' \times \dfrac{1}{2}''\right)(0.219 + 0.480)$

$\qquad = 1{,}049$ kips

The third-story shear force, $V_{u3} = 0.3V_u + 0.4V_u = 0.7V_u$
Let $\qquad V_{u3} \le \phi V_{n3}$, or $\quad 0.7V_u \le \phi V_{n3} = 0.9 \times 1{,}049^{kips} = 944$ kips
Thus, $V_u \le 1{,}349$ kips, based on Section 3.

Section 4 has the same plate girder design as Section 3, in terms of a, h, and t_w as well as F_y.
Thus, $\quad V_{n4} = V_{n3} = 1{,}049$ kips
The fourth-story shear force, $V_{u4} = 0.4V_u$
Let $\qquad V_{u4} \le \phi V_{n4}$, or $\quad 0.4V_u \le \phi V_{n4} = 0.9 \times 1{,}049 = 944$ kips
Thus, $\quad V_u \le 2{,}360$ kips based on Section 4.
Final answer: Section 2 results in the smallest $V_u = 1{,}144$ kips, which is the maximum value that this structure can take. Section 2 (the second floor) will show the web elastic buckling at this load level.

9.9   Problems

9.1   Figure 9.42 shows details of the plate girder in a four-girder structure simply supported and subjected to three concentrated loads. $F_y = 50$ ksi steel is used. Lateral braces are provided at the loading points and supports. Calculate the maximum value of P (30% dead and 70% live load). Assume all stiffeners are properly designed.

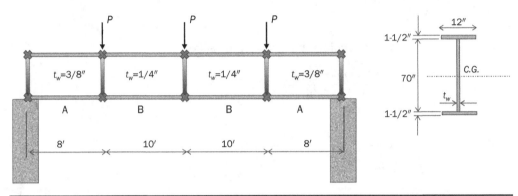

FIGURE 9.42   A four-girder simply supported plate girder.

9.2   The plate girder is needed to support two W12 columns within a 66-ft span, as shown in Fig. 9.43. $F_y = 50$ ksi. Thickness of the plate girder at Sections A and B are $1/2''$ and $3/8''$, respectively. Determine the maximum P (60% dead load and 40% live load). Assume that (1) ITS and bearing stiffeners are properly designed; (2) the proper bracings are provided at ends and at loading positions.

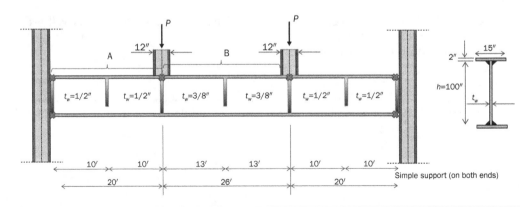

FIGURE 9.43   A plate girder supporting two W12 columns.

9.3   A steel plate girder shear wall becomes increasingly popular in tall steel buildings. In fact, it can be treated as a cantilever beam built up with plate girders as shown in Fig. 9.44. A36 steel (F_y = 36 ksi) is used. Assume pure shear controls the design, determine the maximum lateral load, V_u of the plate girder design. For the same load, V_u, decided in (1), try to reduce thickness by adding more ITS. Lateral and torsional bracings are adequately provided at each loading section.

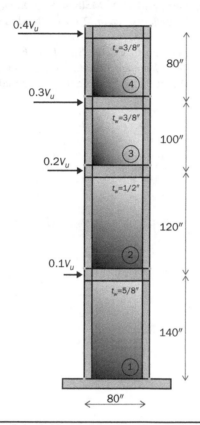

Figure 9.44   A steel plate shear girder subject to lateral earthquake loads.

9.4   Write a python code that computes the shear strength of a plate girder.

Bibliography

Aghayere, A. O., and J. Vigil, *Structural Steel Design: A Practice Oriented Approach,* Pearson Education, Inc., New Jersey, 2015.

AISC, *Seismic Design Manual,* American Institute of Steel Construction, Chicago, IL, 2018.

AISC 341, *Seismic Provisions for Structural Steel Buildings,* ANSI/AISC Standard 341-16, American Institute of Steel Construction, Chicago, IL, 2016.

AISC 358, *Prequalified Connections for Special and Intermediate Steel Moment Frames for Seismic Applications,* American Institute of Steel Construction, Chicago, IL, 2016.

AISC 360, *Specification for Structural Steel Buildings*, ANSI/AISC Standard 360-16, American Institute of Steel Construction, Chicago, IL, 2016.

AISC Manual, *Steel Construction Manual*, 15th ed., American Institute of Steel Construction, Chicago, IL, 2016.

ASCE 7, *Minimum Design Loads for Buildings and Other Structures*, ASCE/SEI 7-16, American Society of Civil Engineers, Reston, VA, 2016.

Salmon, C. G., J. E. Johnson, and F. A. Malhas, *Steel Structures: Design and Behavior—Emphasizing Load and Resistance Factor Design*, 5th ed., Pearson Education, Inc., New Jersey, 2009.

Shen, J., B. Akbas, O. Seker, and C. Carter, *Structural Engineering Handbook—Chapter 8: Design of Structural Steel Members*, 5th ed., McGraw-Hill, New York, 2020.

Timoshenko, S., and S. Woinowsky-Krieger, *Theory of Plates and Shells*, McGraw-Hill, New York, 1959.

Additional Seismic Design Considerations

A.1 Building Configuration and Irregularity

A.1.1 Diaphragm System

Diaphragm system in a building, including its roof, floor, or other membrane or bracing system, is responsible for transferring the lateral forces to the intended vertical seismic force–resisting (SFR) elements, such as moment frame, braced frame, shear walls, etc. The flexibility of the diaphragm directly affects the ways in which the seismic forces are distributed to the SRF elements.

a. *Rigid Diaphragm Condition*

Diaphragms of concrete slabs or concrete-filled metal deck with span-to-depth ratios of 3 or less in structures that have no horizontal irregularities are idealized as "rigid."

b. *Flexible Diaphragm Condition*

Diaphragms constructed of untopped steel decking or wood structural panels are considered as "flexible" in structures, in which the vertical elements are steel or composite steel and concrete braced frames, or concrete, masonry, steel, or composite shear walls. Diaphragms of wood structural panels or untopped steel decks in one- and two-family residential buildings of light-frame construction shall also be considered as "flexible."

c. *Calculated Flexible Diaphragm Condition*

Diaphragms not satisfying the conditions of either "rigid" or "flexible" defined in (a) and (b) A.1 and A.2 might be evaluated by computing in-plane deformation of the diaphragm subjected to a uniformly distributed static lateral load, as shown in Fig. A.1. Practically, a uniformly distributed lateral load in the plane of the diaphragm would be applied between two adjoining SFR elements, and story drifts of SFR elements 1, 2, and the maximum in-plane deflection of the diaphragm, δ_1, δ_2, and δ_M, respectively. Then, we would consider the diaphragm as "flexible" if

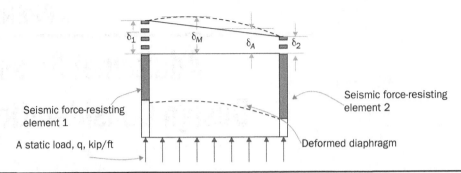

FIGURE A.1   Flexible floor diaphragm example.

$$\delta_M > 2\delta_A \tag{A.1}$$

where $\delta_A = (\delta_1 + \delta_2)/2$. Note that it is left for the designer to make his/her judgment whether the diaphragm can be considered as "rigid" when $\delta_M > 2\delta_A$. However, in the design provisions, we often only need to know whether or not the diaphragm is "flexible."

A.1.2   Irregularity

Any building will be classified as regular or irregular based on the plan and vertical configurations since the seismic performance of irregular structures is much poorer than that of regular structures, particularly under a strong earthquake ground motion. Therefore, the selection of analysis procedures is also based on the regularity of the structure. A building or structure is classified as "regular" if none of the irregularities can be found in the building. These irregularities are listed as follows.

a. *Horizontal irregularity*

Buildings having one or more of the features listed in the following table (Reference: ASCE 7, Table 12.3-1) shall be designated as having horizontal structural irregularity and shall comply with the requirements in design. As a rule of thumb, to avoid torsional irregularity as defined in Table A.1, a structural engineer should keep the eccentricity between the center of mass (C.M.) and center of rigidity (C.R.) in each orthogonal direction to less than 25% of the building dimension normal to the direction of the force. Providing a minimum of four braced frames or moment frames with two in each direction and placing the shear walls, braced frames, moment frames as far away from the C.R. as possible would reduce torsion's adverse impacts (Figs. A.2 and A.3).

b. *Vertical irregularity*

Buildings having one or more of the features listed in Table A.2 (Reference: ASCE 7 Table 12.3-2) shall be designated as having vertical irregularity and shall comply with the requirements in design.

Type	Description
Torsional Irregularity	Maximum story drift, computed including accidental torsion, at one end of the structure transverse to an axis is more than 1.2 times the average of the story drifts at the two ends of the structure (Fig. A.2).
Extreme Torsional Irregularity	Maximum story drift, computed including accidental torsion, at one end of the structure transverse to an axis is more than 1.4 times the average of the story drifts at the two ends of the structure (Fig. A.2).
Reentrant Corner Irregularity	Both plan projections of the structure beyond a reentrant corner are greater than 15% of the plan dimensions of the structure in the given direction (Fig. A.4).
Diaphragm Discontinuous Irregularity	When there are diaphragms with abrupt discontinuities or variations in stiffness, including those having cutout or open areas greater than 50% of the gross enclosed diaphragm area, or changes in effective diaphragm stiffness of more than 50% from one story to the next.
Out-of-Plane Offsets Irregularity	When there are discontinuities in a lateral force–resistance path, such as out-of-plane offsets of the vertical elements.
Nonparallel Systems Irregularity	When the vertical lateral force–resisting elements are not parallel to or symmetric about the major orthogonal axes of the seismic force–resisting systems.

TABLE A.1   Horizontal Structural Irregularities (ASCE 7, Table 12.3-1)

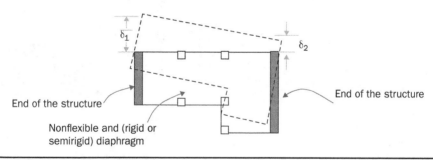

FIGURE A.2   Torsional irregularity exists when $\delta_{max} > 1.2 \, (\delta_1 + \delta_2)/2$ and extreme torsional irregularity exists when $\delta_{max} > 1.2 \, (\delta_1 + \delta_2)/2$.

A.2   Selection of Analysis Procedures in Determining Seismic Response

A structural analysis shall be conducted for all structures using one of the procedures permitted by ASCE 7 on the basis of assigned seismic design category (SDC), structural system, dynamic properties (in terms of the fundamental period of vibration, T), and regularity. These procedures include:

a. Equivalent lateral force analysis (static procedure)
b. Modal response spectrum analysis
c. Linear seismic response history procedure
d. Nonlinear seismic response history procedure

Increasing rigorousness for higher SDC and highly irregular structures

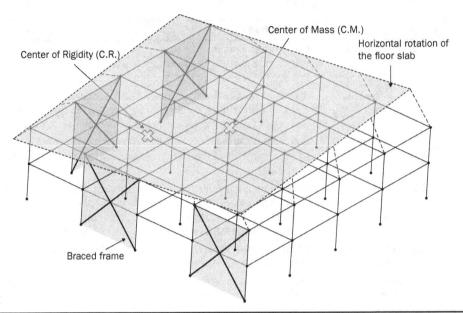

FIGURE A.3   The worst braced frame configuration.

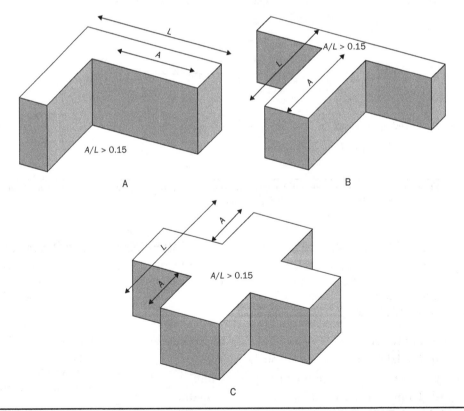

FIGURE A.4   Reentrant corner irregularities for $A/L > 0.15$: (a) L-shaped plan, (b) T-shaped plan, (c) H-shaped plan.

Type	Description
Stiffness-Soft Story Irregularity	There is a story in which the lateral stiffness is less than 70% of that in the story above or less than 80% of the average stiffness of the three stories above (Fig. A.5).
Stiffness-Extreme Soft Story Irregularity	There is a story in which the lateral stiffness is less than 60% of that in the story above or less than 70% of the average stiffness of the three stories above (Fig. A.5).
Weight (Mass) Irregularity	The effective mass of any story is more than 150% of the effective mass of an adjacent story. A roof that is lighter than the floor below need not be considered.
Vertical Geometric Irregularity	Horizontal dimension of the seismic force–resisting system in any story is more than 130% of that in an adjacent story.
In-Plane Discontinuity in Vertical Lateral Force–Resisting Element Irregularity	An in-plane offset of the lateral force–resisting elements is greater than the length of those elements or there exists a reduction in stiffness of the resisting element in the story below.
Discontinuity in Lateral Strength-Weak Story Irregularity	The story lateral strength is less than 80% of that in the story above. The story lateral strength is the total lateral strength of all seismic-resisting elements sharing the story shear for the direction under consideration.
Discontinuity in Lateral Strength-Extreme Weak Story Irregularity	The story lateral strength is less than 65% of that in the story above. The story strength is the total strength of all seismic-resisting elements sharing the story shear for the direction under consideration.

TABLE A.2   Vertical Structural Irregularities (ASCE 7, Table 12.3-2)

The procedures above are in the order of increasing rigorousness in terms of analysis effort, with the Static Procedure being the simplest and Nonlinear Seismic Response History Procedure being the most rigorous. Equivalent Lateral Force Analysis procedure is often referred to as Static Procedure since this procedure does not directly include any dynamic analysis. The other three procedures are all called dynamic procedure since some sort of dynamic analysis is involved in them. ASCE 7, Table 12.6-1 lists the conditions under which types of analysis procedures might be permitted. In design practice, the Static Procedure will be preferred to any dynamic procedure due to its simplicity. Note that the Static Procedure is an approximate procedure for regular buildings, in which the first-mode dynamic vibration is predominant.

A.3   Equivalent Lateral Force (ELF) Procedure

Dynamic analysis has shown that elastic seismic response of a regular structure with low-to-moderate height can be approximately evaluated based on its first-mode response. ELF procedure is directly derived from the elastic response of the first mode in regular multidegree-of-freedom building structures. An overview of the ELF procedure is as follows.

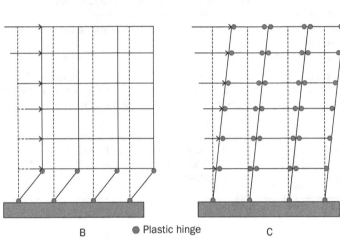

B   ● Plastic hinge   C

Figure A.5 Soft story vertical irregularities: (a) ground floor collapse of a five-story building (1999 Kocaeli earthquake, Turkey), (b) soft story mechanism (generally occurs at ground floor), (c) beam-sway mechanism (strong column–weak beam mechanism) (ductile) (preferred mechanism in earthquake engineering).

A.3.1 Seismic Base Shear

The seismic base shear, V (kips), in a given direction shall be determined in accordance with the following equation:

$$V = C_s W \tag{A.2}$$

where
    C_s = seismic response coefficient (≥ 0.01 in any case)
    W = the effective seismic weight of the structure, kips

Effective seismic weight includes the total dead load and other loads:

    1. In areas used for storage, a minimum of 25% of the reduced floor live load (floor live load in public garages and open parking structures need not be included).

2. Where an allowance for partition load is included in the floor load design, the actual partition weight, or a minimum weight of 10 psf of floor area, whichever is greater.

3. Total operating weight of permanent equipment.

4. Twenty percent of flat roof snow load where flat roof snow load exceeds 30 pounds per square foot.

A.3.2 Vertical Distribution of Seismic Forces

Vertical distribution of seismic forces is mainly based on an inverted triangle with additional lateral load at the roof level.

$$F_x = C_{vx} V \qquad (A.3)$$

$$C_{vx} = \frac{w_x h_x^{\,k}}{\displaystyle\sum_{i=1}^{n} w_i h_i^{\,k}} \qquad (A.4)$$

where

F_x = lateral force induced at floor level x, kips
C_{vx} = vertical distribution factor

$$k = \begin{cases} 1.0, \text{ for } T \le 0.5\text{s} \\ 2.0, \text{ for } T \ge 2.5\text{s} \\ 0.5T + 0.75, \text{ for } 0.5 < T < 2.5\text{s} \end{cases}$$

h_i and h_x = height from the base level "i" or "x" (ft) (Fig. A.6)
n = total number of stories above the ground level
w_i and w_x = effective seismic weight at the floor level "i" or "x," kips
T = fundamental period of the structure, s

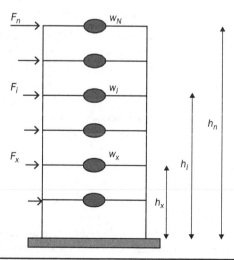

Figure A.6 Lateral load distribution of the base shear.

Figure A.6 illustrates the distribution of the base shear along with the building height. Note that this type of distribution is based on the first mode of vibration. For a building with severe horizontal or vertical irregularity in stiffness, strength, or mass distribution, the distribution of lateral forces is highly irregular and based on individual structures, and the ELF procedure is not permitted in design. For a taller but regular building, higher mode of participation might affect adversely the lateral force distribution, such that more portion of the base shear might be distributed to the upper floors. Factor k is intended to take into higher modes effect approximately up to a certain height.

A.3.3 Relationship between Government and Professionals in Earthquake Hazard Reduction in the United States

The relationship between the U.S. government and structural engineering professionals is typical of the government and any professionals under the governing system in the United States: independent and complementary.

- Government and professionals have independent and complementary roles in earthquake engineering.
- Government funds and coordinates the research, code development, and code enforcement (local level).
- Professional engineering community conducts independent studies of all subjects related to earthquake engineering and formulates codes for all levels of government to use.

Three branches of the federal government: Congress, administration, and courts are all involved in EQ Hazard Reduction missions. Congress has enacted various laws and established programs with regard to reducing earthquake hazard, including:

- After 1971, Congress passed a very important law on a national mission in earthquake hazard reduction, declaring that all states are at the risk of earthquakes, 38 of which are in high seismic risk. The law allows all branches of the federal government to allocate funds for the mission.
- In 1977, Congress established the National Earthquake Hazard Reduction Program (NEHRP) in the effort to coordinate nationwide planning and funding, based on recommendations of professional communities (engineers and researchers).
- The National Science Foundation (NSF), funded by Congress, is mainly responsible for managing research funding through a competition based on merit and impact of the proposals from researchers in universities (annual budget is about $6 billion).
  - Administration (White House) gets involved in the earthquake hazard reduction mission in the form of hazard management mainly through the Federal Emergency Management Agency (FEMA) in the Department of Homeland Security.
- FEMA has funded post-earthquake rescue, recovery, and improvement/implementation of codes after an earthquake (such as 1994 Northridge earthquake, in the form of partnership with the professional engineering community, rather than distributing the funds based on open competitions).

- Local administrative offices (such as City Hall) gets involved in the earthquake hazard reduction mission in the form of enforcing the local governing building code, mainly when deciding building permits. Courts at all levels get involved in the earthquake hazard reduction mission in the form of explaining the governing code at the locality hazard management mainly through Federal Emergency Management Agency (FEMA) in the Department of Homeland Security.

- Practicing engineers and researchers in universities can participate in earthquake hazard reduction either in groups (of professional organizations) or individually with supports from federal funding agencies (NSF, FEMA, NIST, USGS, etc.), or private entities. Professional organizations collect all efforts by professional engineers and university researchers in the United States and have played a critical role in earthquake engineering.

- The major organizations in the United States include: American Society of Civil Engineers (ASCE), Earthquake Engineering Research Institute (EERI), American Institute of Steel Construction (AISC), and American Concrete Institute (ACI).

For the sake of discussion, the main characteristics in earthquake hazard reduction are listed below:

- Because of the complexity of earthquakes and their impact on our society, it is critical to have all parties, government, and professionals to work side by side but absolutely independent of each other, e.g., 1933 Long Beach earthquake, the school building issue; 1994 Northridge earthquake, the steel moment frame issue.

- An effective system in earthquake hazard reduction is always affected by three components: (1) technology, (2) economy, and (3) governing system.

- It is very rare for a society to have effective earthquake engineering based on advanced technology only. For example, building codes in the United States are available for everybody in the world to use, but the hazard is not uniform around the globe.

- Earthquakes and their impact are complex and uncertain in nature, and our understanding is improving as we learn more and more through real earthquakes. Our tuition is to make honest mistakes.

- Transparency is one of the most important factors in improving our understanding of earthquake hazard and reducing its impact on our society. Such transparency is needed in a post-earthquake investigation, in building codes, and the construction process in code-enforcing practice.

- No matter how advanced the code and standard used in the time of construction, they are never going to be sufficient to eliminate the earthquake hazard, only the solid workmanship during construction is the best defense.

- Examples: (1) 1906 San Francisco, 1925 Santa Barbara, and 1933 Long Beach earthquakes: many buildings without codes survived and many collapsed. (2) Wen-Chuan earthquake: many buildings remained functional next to collapsed ones.

A.3.4 Past Earthquakes and Their Impacts on Today's Earthquake Hazard Reduction Practice

The history of earthquake research, development, and implementation is directly connected to earthquakes the United States. Over last about 100 years, there are a few milestones, dividing the development of earthquake engineering into a few periods:

- From 1906 to 1971: From the beginning to middle levels.
- From 1971 to 1994: Developed into an advanced earthquake engineering system, based on traditional criteria and definitions in earthquake engineering.
- From 1994 to 2011: Thinking about the future generations of earthquake engineering systems for the next 100 years.
- From 2011 to the future: Any future systems are the challenge as well as the opportunity due to fast development in technology, economy, as well as population.

Selected Earthquakes (EQ)

- 1906 San Francisco EQ ($M_s = 8.3$)
- 1925 Santa Barbara EQ ($M_s = 6.3$)
- 1933 Long Beach EQ ($M_s = 6.4$)
- 1940 El Centro EQ ($M_w = 6.9$)
- 1971 San Fernando EQ ($M_s = 6.7$)
- 1989 Loma Prieta EQ ($M_s = 7.1$)
- 1994 Northridge EQ ($M_s = 6.7$)

Index

Page numbers followed by "t" refer to tables.
Page numbers followed by "f" refer to figures.
Page numbers followed by "m" refer to movies.